AF800198

Information Systems Engineering and Management

Volume 65

Series Editor

Álvaro Rocha, ISEG, University of Lisbon, Lisbon, Portugal

Editorial Board

Ali Idri, ENSIAS, Mohammed V University, Rabat, Morocco

Abdelkader Hameurlain, Université Toulouse III Paul Sabatier, Toulouse, France

Ashok Vaseashta, International Clean Water Institute, Manassas, VA, USA

Ashwani Kumar Dubey, Amity University, Noida, India

Carlos Montenegro, Francisco José de Caldas District University, Bogota, Colombia

Claude Laporte, University of Quebec, Québec, QC, Canada

Fernando Moreira, Portucalense University, Porto, Portugal

Francisco Peñalvo, University of Salamanca, Salamanca, Spain

Gintautas Dzemyda, Vilnius University, Vilnius, Lithuania

Jezreel Mejia-Miranda, CIMAT - Center for Mathematical Research, Zacatecas, Mexico

Jon Hall, The Open University, Milton Keynes, UK

Mário Piattini, University of Castilla-La Mancha, Albacete, Spain

Maristela Holanda, University of Brasilia, Brasilia, Brazil

Mincong Tang, Beijing Jiaotong University, Beijing, China

Mirjana Ivanovíc, Department of Mathematics and Informatics, University of Novi Sad, Novi Sad, Serbia

Mirna Muñoz, CIMAT Center for Mathematical Research, Progreso, Mexico

Rajeev Kanth, University of Turku, Turku, Finland

Sajid Anwar, Institute of Management Sciences, Peshawar, Pakistan

Tutut Herawan, Faculty of Computer Science and Information Technology, University of Malaya, Kuala Lumpur, Malaysia

Valentina Colla, TeCIP Institute, Scuola Superiore Sant'Anna, Pisa, Italy

Vladan Devedzic, University of Belgrade, Belgrade, Serbia

The book series "Information Systems Engineering and Management" (ISEM) publishes innovative and original works in the various areas of planning, development, implementation, and management of information systems and technologies by enterprises, citizens, and society for the improvement of the socio-economic environment.

The series is multidisciplinary, focusing on technological, organizational, and social domains of information systems engineering and management. Manuscripts published in this book series focus on relevant problems and research in the planning, analysis, design, implementation, exploration, and management of all types of information systems and technologies. The series contains monographs, lecture notes, edited volumes, pedagogical and technical books as well as proceedings volumes.

Some topics/keywords to be considered in the ISEM book series are, but not limited to: Information Systems Planning; Information Systems Development; Exploration of Information Systems; Management of Information Systems; Blockchain Technology; Cloud Computing; Artificial Intelligence (AI) and Machine Learning; Big Data Analytics; Multimedia Systems; Computer Networks, Mobility and Pervasive Systems; IT Security, Ethics and Privacy; Cybersecurity; Digital Platforms and Services; Requirements Engineering; Software Engineering; Process and Knowledge Engineering; Security and Privacy Engineering, Autonomous Robotics; Human-Computer Interaction; Marketing and Information; Tourism and Information; Finance and Value; Decisions and Risk; Innovation and Projects; Strategy and People.

Indexed by Google Scholar. All books published in the series are submitted for consideration in the Web of Science.

For book or proceedings proposals please contact Alvaro Rocha (amrrocha@gmail.com).

Anupam Ghosh · Soumi Dutta · Asit Kumar Das ·
Vinod Kumar Shukla · Fernando Moreira
Editors

Quantum Machine Learning in Industrial Automation

Springer

Editors
Anupam Ghosh
Netaji Subhash Engineering College
Kolkata, West Bengal, India

Soumi Dutta
Sister Nivedita University
Kolkata, West Bengal, India

Asit Kumar Das
Department of Computer Science and Technology
Indian Institute of Engineering Science and Technology, Shibpur
Howrah, West Bengal, India

Vinod Kumar Shukla
Amity University
Dubai, United Arab Emirates

Fernando Moreira
Universidade Portucalense
Porto, Portugal

ISSN 3004-958X ISSN 3004-9598 (electronic)
Information Systems Engineering and Management
ISBN 978-3-031-99785-3 ISBN 978-3-031-99786-0 (eBook)
https://doi.org/10.1007/978-3-031-99786-0

© The Editor(s) (if applicable) and The Author(s), under exclusive license to Springer Nature Switzerland AG 2025

This work is subject to copyright. All rights are solely and exclusively licensed by the Publisher, whether the whole or part of the material is concerned, specifically the rights of translation, reprinting, reuse of illustrations, recitation, broadcasting, reproduction on microfilms or in any other physical way, and transmission or information storage and retrieval, electronic adaptation, computer software, or by similar or dissimilar methodology now known or hereafter developed.
The use of general descriptive names, registered names, trademarks, service marks, etc. in this publication does not imply, even in the absence of a specific statement, that such names are exempt from the relevant protective laws and regulations and therefore free for general use.
The publisher, the authors and the editors are safe to assume that the advice and information in this book are believed to be true and accurate at the date of publication. Neither the publisher nor the authors or the editors give a warranty, expressed or implied, with respect to the material contained herein or for any errors or omissions that may have been made. The publisher remains neutral with regard to jurisdictional claims in published maps and institutional affiliations.

This Springer imprint is published by the registered company Springer Nature Switzerland AG
The registered company address is: Gewerbestrasse 11, 6330 Cham, Switzerland

If disposing of this product, please recycle the paper.

Preface

In the age of Industry 4.0, where the lines between the physical, digital, and biological spheres are increasingly blurred, the demand for smarter, faster, and more adaptive systems has never been higher. *Quantum Machine Learning in Industrial Automation* emerges as a timely and visionary contribution, bridging two of the most transformative technologies of our era: quantum computing and machine learning. This book offers not only an introduction to these cutting-edge disciplines but also an ambitious roadmap for their convergence within the complex realm of industrial automation.

Industrial automation has long been a cornerstone of modern manufacturing, enabling scalability, consistency, and efficiency. However, as production environments become more complex and data-rich, traditional computational methods are reaching their limits. Here, quantum computing—with its promise of exponential speedup and novel problem-solving paradigms—presents a compelling new frontier. Coupled with the adaptive capabilities of machine learning, the potential for innovation is boundless.

The authors of this volume precisely navigate this multidisciplinary landscape. They demystify the foundational principles of quantum computing and quantum machine learning while grounding their discussions in the real-world challenges and applications of industrial automation. From predictive maintenance and intelligent robotics to process optimization and quality assurance, this book explores how quantum-enhanced algorithms could revolutionize core functions across industries.

More than just a technical manual, *Quantum Machine Learning in Industrial Automation* invites readers to envision a future where quantum intelligence augments every stage of industrial operations. Whether you are a researcher, engineer, technologist, or decision-maker, this book will equip you with both the conceptual insights and practical tools to engage with this rapidly evolving field.

Kolkata, India — Dr. Anupam Ghosh
Kolkata, India — Dr. Soumi Dutta
Howrah, India — Dr. Asit Kumar Das
Dubai, United Arab Emirates — Dr. Vinod Kumar Shukla
Porto, Portugal — Dr. Fernando Moreira

About This Book

This book focuses on quantum machine learning that harnesses the collective properties of quantum states, such as superposition, interference, and entanglement, which uses algorithms run on quantum devices, such as quantum computers, to supplement, expedite, or support the work performed by a classical machine learning program. The devices that perform quantum computations are known as quantum computers. Quantum computers have the potential to revolutionize computation by making certain types of classically intractable problems solvable. A few large companies and small start-ups now have functioning non-error-corrected quantum computers composed of several tens of qubits, and some of these are even accessible to the public through the cloud. Additionally, quantum simulators are making strides in fields varying from molecular energetics to many-body physics. Most known use cases fit into four archetypes: quantum simulation, quantum linear algebra for AI and machine learning, quantum optimization and search, and quantum factorization. Advantages of quantum computing are many and to list a few; first, they're fast. Ultimately, quantum computers have the potential to provide computational power on a scale that traditional computers cannot ever match. In 2019, for example, Google claimed to carry out a calculation in about 200 s that would take a classical supercomputer around 10,000 years. Second, they can solve complex problems. The more complex a problem, the harder it is for even a supercomputer to solve. When a classical computer fails, it's usually because of a huge degree of complexity and many interacting variables. However, due to the concepts of superposition and entanglement, quantum computers can account for all these variables and complexities to reach a solution. Last but not the least, they can run complex simulations. The speed and complexity that quantum computing can achieve mean that, in theory, a quantum computer could simulate many intricate systems, allowing us to better understand some of life's great mysteries. In Quantum Computing Industry, it is estimated that market is expected to grow between \$1.5 billion and \$4 billion by 2025 and reaches between \$4 billion and \$8 billion in 2030. In 2021 to 2025, the panel expects the CAGR to range from 64 to 97%, and between 2025 to 2030, the market is expected to grow at rates between 15 and 22%.

Contents

Quantum Machine Learning for Cost Variance Analysis in Industrial Manufacturing Processes: A Computational Breakthrough

Hrishikesh Desai

Abstract This study presents a quantum machine learning framework for improving cost variance analysis in industrial manufacturing. It addresses the challenges posed by classical methods when handling high-dimensional and nonlinear industrial process cost data. By integrating Quantum Support Vector Machines (QSVMs) with quantum kernel methods, the framework uses quantum-enhanced feature mapping to transform cost data into a high-dimensional Hilbert space where complex patterns become more separable. A rigorous numerical experiment is conducted using a synthetic dataset that simulates real-world manufacturing cost variations, including price, efficiency, and volume variances. The results indicate that while a classical SVM with a radial basis function kernel failed to detect anomalies, the QSVM achieved a precision of 21%, a recall of 40%, and an F1 score of 27%, along with a shorter training time. These findings validate the hypothesis that quantum approaches can effectively capture nonlinear interdependencies among cost drivers and improve anomaly detection. Despite promising performance, the study acknowledges current limitations in quantum hardware scalability and interpretability. Overall, this work not only advances the theoretical understanding of quantum machine learning for cost analysis but also lays the groundwork for future practical implementations in process optimization and decision support within complex industrial environments.

Keywords Quantum machine learning · Cost variance analysis · Industrial manufacturing · Quantum Support Vector Machines (QSVM) · Quantum kernel methods · Anomaly detection

---

H. Desai (✉)
Department of Accounting, Neil Griffin College of Business, Arkansas State University, Jonesboro, USA
e-mail: hdesai@astate.edu

© The Author(s), under exclusive license to Springer Nature Switzerland AG 2025
A. Ghosh et al. (eds.), *Quantum Machine Learning in Industrial Automation*, Information Systems Engineering and Management 65,
https://doi.org/10.1007/978-3-031-99786-0_1

1 Introduction

Cost variance analysis helps organizations detect inefficiencies in industrial processes and optimize financial performance to be competitive [29]. Traditional cost variance models rely on classical machine learning (ML) techniques, which often struggle with high-dimensional cost structures and complex, nonlinear interdependencies that characterize real-world manufacturing data [17, 22]. This study introduces a Quantum Machine Learning (QML) framework for cost variance analysis that utilizes Quantum Support Vector Machines (QSVMs) and quantum kernel methods to improve cost variance prediction and anomaly detection in industrial manufacturing processes [19, 32]. By integrating quantum regression models, I develop an advanced cost variance decomposition approach that analyzes price variance, efficiency variance, and volume variance in industrial manufacturing processes with superior computational efficiency. I also validate the proposed analytical model through a numerical experiment that compares the performance of the QSVM against the classical SVM using a rigorously designed synthetic dataset that simulates real-world manufacturing cost structures.

Recent advances in machine learning have introduced sophisticated tools capable of handling large datasets, yet the performance of classical ML models is frequently hindered by the need for extensive feature engineering and dimensionality reduction. Traditional approaches such as linear regression, decision trees, and SVMs have been widely used for cost variance analysis. Nonetheless, these methods struggle with the combinatorial complexity of interactions among numerous cost drivers that results in limited predictive accuracy and a higher chance of overlooking subtle anomalies [16, 41].

QML has recently emerged as a promising alternative by using quantum computing principles to tackle high-dimensional and nonlinear problems. In particular, QSVMs utilize quantum-enhanced kernel methods to "lift" data into high-dimensional Hilbert spaces, where complex patterns become more separable. Early studies have demonstrated that quantum kernels can reveal hidden data structures that are difficult to detect with classical kernels [40–43]. Despite these promising developments, few studies have applied QML specifically to cost variance analysis, which leaves a significant research gap in the literature.

The primary research problem addressed in this study is the inability of classical machine learning techniques to effectively model and predict cost variances in industrial manufacturing, particularly when the data exhibit high dimensionality and nonlinear relationships. This limitation can lead to inefficient resource allocation and missed opportunities for cost optimization. To overcome these challenges, this study proposes a QML framework that integrates QSVMs and quantum kernel methods into the cost variance analysis process. The objectives of the research are as follows:

1. To develop an advanced cost variance decomposition model that incorporates quantum feature mapping.
2. To design a numerical experiment that simulates realistic manufacturing cost structures with inherent stochastic perturbations and nonlinear anomalies.

3. To compare the performance of the proposed QSVM framework against classical SVMs in terms of prediction accuracy, computational efficiency, and anomaly detection metrics.

The results from the numerical experiment demonstrated that the proposed QSVM framework successfully detected anomalies in cost variance data and achieved a precision of 21%, a recall of 40%, and an F1 score of 27%, whereas the classical SVM with an RBF kernel failed to detect any anomalies. Additionally, the QSVM exhibited a shorter training time (0.0065 s) compared to the classical SVM (0.0090 s). These findings highlight the ability of quantum-enhanced feature mapping to capture the complex, nonlinear relationships in cost data, which validates the core hypothesis that QSVMs can outperform traditional methods in this context.

This research makes several important contributions to both theory and practice. Theoretically, it extends the application of quantum machine learning to cost variance analysis, a domain traditionally dominated by classical methods, and demonstrates that quantum-enhanced feature mapping can more effectively capture nonlinear interdependencies among cost drivers. Practically, the findings suggest that as quantum hardware matures, QSVM-based frameworks could revolutionize real-time process optimization and cost management in industrial manufacturing. By providing empirical evidence that quantum methods can detect anomalies that classical techniques miss, this study lays the groundwork for more robust and efficient decision support systems in complex industrial environments.

The remainder of this study is organized as follows: Sect. 2 reviews the theory behind cost variance analysis, discusses conventional approaches, and introduces the fundamentals of quantum machine learning for cost analysis. Section 3 details the analytical model underlying the quantum regression approach and the QSVM framework. Section 4 describes the design of the numerical experiment, including the synthetic dataset characteristics, model implementation, and performance evaluation. Section 5 presents the results, followed by a discussion of the findings and their implications. Section 6 concludes with a summary of contributions, limitations, and directions for future research.

2 Theory

2.1 Cost Variance Analysis in Industrial Manufacturing

Cost variance analysis is a core managerial accounting tool used to evaluate performance in manufacturing by comparing actual costs against standard or expected costs [6, 13]. It helps organizations detect inefficiencies in production processes and identify areas where costs deviate from budgets and directly impact financial performance and operational efficiency. By decomposing total cost variances into components (commonly price variance, efficiency variance, and volume variance), managers can pinpoint whether higher costs arise from paying more for inputs, using resources less efficiently, or changes in output volume [13]. Such analysis is key for strategic decision-making and helps firms improve resource allocation, optimize processes, and maintain competitiveness through tighter cost control and improving manufacturing efficiency and profitability.

2.2 *Conventional Approaches and Limitations in Cost Variance Analysis*

Traditional approaches to cost variance analysis rely on standard costing methods and classical machine learning models to forecast costs or flag deviations [16]. Techniques such as linear regression, decision trees, or even basic Support Vector Machines (SVMs) have been applied to model cost behaviors and predict variances. However, these classical methods often struggle with the complexity of real-world manufacturing cost data [17, 22]. Industrial cost datasets tend to be high-dimensional (involving many cost drivers and interrelated factors) and exhibit nonlinear dependencies (e.g. interactions between material costs, operational efficiency, and production volume) [26, 27, 41]. Conventional ML algorithms can have difficulty capturing these complex, interdependent relationships in imbalanced datasets and especially under uncertainty in dynamic factory settings [20, 21, 27, 38].

In practice, large-scale industrial data with many features can overwhelm classical models as they may require extensive feature engineering or dimensionality reduction, which risks losing important information [12]. Indeed, traditional models frequently underperform when faced with large datasets and intricate cost patterns, as they struggle to handle the combinatorial explosion of possible interactions [20]. This leads to limited predictive accuracy and a higher chance of missing subtle cost anomalies. Overall, the classical approaches to cost variance analysis are constrained by computational scalability and modeling flexibility, which motivates the search for more powerful analytical techniques that can better handle high-dimensional, complex cost structures.

2.3 *Quantum Machine Learning for Cost Analysis*

QML provides a novel approach to address the limitations of classical methods by exploiting quantum computing principles. In particular, QSVMs introduce a quantum-enhanced version of the kernel trick used in classical SVMs [47]. Instead of relying on a hand-crafted kernel function, a QSVM uses a quantum circuit to map input data into a high-dimensional Hilbert space (quantum feature space) where classification or regression becomes easier. This quantum feature mapping can encode classical cost data (e.g. price, efficiency, and volume variance metrics) into quantum states, which enables the algorithm to find patterns that might be inseparable in the original space but linear in the quantum feature space.

By using superposition and entanglement, the QSVM's quantum kernel implicitly considers a vast number of feature combinations to capture complex nonlinear relationships among cost drivers [8, 33, 39]. Many supervised QML models can be understood as kernel methods that map data to high-dimensional quantum feature spaces [40, 41]. In the context of cost variance analysis, this means a QSVM can distinguish overlapping or intricate cost behavior patterns far better than a classical model by "spreading out" the data in a richer feature space. Empirically, quantum kernels have been found to reveal hidden structure in data that classical kernels might miss. This can facilitate more accurate predictions of cost variances and sharper identification of anomaly-causing factors. Moreover, because the inner products between data points are computed via efficient quantum subroutines, QSVMs can maintain computational efficiency even as the effective feature space dimension grows exponentially.

To sum it up, QML through techniques like QSVM and quantum kernel methods provides a powerful theoretical framework to model high-dimensional cost data. It provides improved capability to handle nonlinear cost interactions and complex dependencies that stymie classical machine learning approaches.

2.4 Applications of QML in Industrial Optimization

Beyond cost variance analysis, recent research has begun exploring QML in various industrial manufacturing and process optimization applications. QML algorithms have been applied to prognostics and equipment health monitoring. For example, [7] used quantum kernel SVM models for wind turbine fault detection, which aim to improve early detection of equipment failures. In their study, a QSVM-based classifier achieved predictive performance on par with conventional machine learning models and even outperformed classical methods like random forests and k-nearest neighbors when the input feature space was appropriately reduced. This suggests that quantum models can effectively handle the high-dimensional sensor data in complex machinery.

Quantum-enhanced models have shown promise in forecasting and logistical planning tasks that involve large, complex datasets. Jahin et al. [20] introduced QAmplifyNet, a hybrid quantum–classical neural network for supply chain backorder prediction. Traditional forecasting models struggled with the scale and imbalance of the backorder data, but the quantum-inspired network demonstrated superior predictive accuracy over classical machine learning and even advanced deep learning alternatives. The QAmplifyNet model effectively handled short and imbalanced demand datasets, which resulted in improved inventory control and reduced backorder rates in simulations.

An important aspect of process optimization is identifying anomalies or outliers (e.g., fraudulent transactions, production errors) in real time. QML has been tested in these scenarios to improve detection capabilities. Grossi et al. [15] present a mixed quantum–classical approach for financial fraud detection, using a QSVM for quantum feature selection on credit card transaction data. Their end-to-end experiment, one of the first QSVM applications on a real industrial dataset, showed that the QSVM provided a complementary exploration of the feature space, which lead to improved fraud detection accuracy when combined with classical classifiers. Notably, the quantum model could capture subtle patterns in transactional data that helped reduce false positives. However, the study also had to drastically reduce the dataset size to fit current quantum hardware limits, which indicates the early-stage nature of applied QML. Thus, in manufacturing settings, similar quantum-enhanced anomaly detection techniques can be envisioned for detecting out-of-control process deviations or quality issues by analyzing sensor and cost variance data in higher-dimensional quantum feature spaces.

While still largely theoretical, QML is anticipated to aid in optimizing complex manufacturing processes. Quantum algorithms (including quantum reinforcement learning and quantum optimization heuristics) could, for example, help find optimal production schedules or resource allocation by evaluating vast combinations more efficiently than classical solvers. Early studies in quantum optimization for industrial processes indicate potential speed-ups in solving scheduling, routing, or design optimization problems [35–39]. As quantum hardware and algorithms mature, we can expect to see QML-driven decision support systems that dynamically adjust process parameters for minimal cost and maximal efficiency.

Despite the progress in quantum machine learning applications, several key research gaps remain, especially in the context of industrial manufacturing. Many QML approaches for industry are still at a proof-of-concept stage. Current studies are often based on simulated or limited datasets rather than on extensive real-world deployments. As a result, there is a lack of empirical validation of QML models in true industrial settings [35]. The practical benefits of quantum ML for manufacturing (e.g. in a live factory cost system) remain largely unproven, since few case studies demonstrate QML integrated into production-scale operations.

The challenges posed by high-dimensional, complex industrial data continue to strain classical analytics. Traditional ML models require significant preprocessing (e.g. feature selection or dimensionality reduction) to handle large cost variance datasets, and even then, they may miss nonlinear interaction effects. The computational constraints of classical models, such as exponential complexity for exploring feature combinations or long training times for very large datasets, mean that current cost variance analysis tools might not scale well as data complexity grows [20]. This gap points to the need for more powerful learning frameworks (potentially QML) that can better manage complexity without sacrificing accuracy or efficiency.

While QML algorithms can theoretically handle high-dimensional data, in practice they are limited by the state of quantum hardware and algorithmic scalability. Today's quantum processors have few qubits and are noisy, which forces researchers to use small problem sizes or hybrid methods. Indeed, experiments like [15] had to greatly downsize the dataset to run on a quantum device. There is uncertainty about how well QML methods will scale to industry-sized problems, for instance, a full enterprise resource planning dataset or live factory sensor streams. The scalability of

QML solutions and their integration with classical infrastructure is an open question, as data encoding overhead and quantum decoherence can bottleneck performance on larger inputs [9, 37]. Overcoming these issues (through better quantum algorithms or more advanced hardware) is necessary before QML can realize its full industrial impact.

In this study, I introduce a novel QML framework that utilizes QSVMs and quantum kernel methods specifically for cost variance decomposition and analysis. By integrating quantum feature mapping with traditional cost accounting concepts (price, efficiency, volume variances), the framework can model complex cost behaviors that classical variance analysis methods may fail to capture. This approach builds on recent advances in QSVM algorithms to ensure the model is well-suited for practical implementation. To my knowledge, this study provides one of the first comprehensive evaluations of a QML model on an industrial cost analysis problem. I design a rigorous numerical experiment simulating realistic manufacturing cost data (with multiple cost variance components and noise) to benchmark the QML framework against classical approaches. By conducting this empirical analysis, I provide evidence that QML can be practically applied (at least in a simulated environment) to manufacturing cost data, which bridges the gap between theoretical promise and real-world application. This validation helps establish confidence that QML can indeed improve cost variance analysis and process optimization for industry.

3 Mathematical Model of Quantum Regression for Cost Variance Analysis

3.1 Cost Variance Decomposition

Cost variance decomposition in industrial manufacturing processes helps firms to pinpoint deviations between actual and standard costs [5]. This decomposition is important for assessing financial efficiency, identifying areas for cost optimization, and improving budgeting strategies. Cost variance is typically categorized into three main components: Price Variance (PV), Efficiency Variance (EV), and Volume Variance [13].

Price Variance (PV) measures the difference between the actual price paid for a resource and the standard price, multiplied by the actual quantity used.

$$PV = (A_P - S_P) \times A_Q$$

This variance indicates fluctuations in procurement costs and supplier pricing strategies. A positive PV suggests that the firm paid more than expected, possibly due to supplier price increases, inflation, or poor contract negotiations. Conversely, a negative PV indicates cost savings achieved through better supplier management, bulk purchasing, or favorable market conditions.

Efficiency Variance (EV) assesses the difference between the actual quantity of resources used and the standard expected quantity, multiplied by the standard price.

$$EV = \left(A_Q - S_Q\right) \times S_P$$

This variance is important for evaluating operational performance. A positive EV implies that more resources were used than anticipated, which may result from inefficient production processes, machine downtime, or workforce productivity issues. In contrast, a negative EV indicates that the firm has used fewer resources than planned, possibly due to improved labor efficiency, process automation, or lean manufacturing initiatives [25].

Volume Variance (VV) accounts for the effects of production or sales volume fluctuations on total costs and is calculated as follows:

$$VV = (A_V - S_V) \times S_C$$

This variance helps firms assess how deviations in demand impact cost structures. A positive VV suggests higher-than-expected production or sales, which can improve profitability by spreading fixed costs over a larger output. However, a negative VV may indicate underutilized capacity, which leads to increased per-unit costs and lower profit margins. Market demand, seasonality, and production bottlenecks often drive this variance.

The total cost variance consolidates these three components as indicated below.

$$TCV = PV + EV + VV$$

A comprehensive analysis of TCV can help management pinpoint the primary causes of cost deviations and implement targeted corrective measures. By utilizing QSVMs and quantum kernel methods, firms can refine this cost variance decomposition and enable predictive analytics that dynamically adjust for real-time cost fluctuations.

3.2 *Quantum Feature Mapping for Cost Variance Modeling*

Traditional regression models often struggle with cost data that exhibit nonlinear dependencies and high-dimensional variations, which limits their ability to provide accurate predictions and detect anomalies in complex manufacturing accounting environments [17, 22]. These limitations arise due to the difficulty in capturing complex relationships between cost components such as price, efficiency, and volume variances. QML, particularly QSVMs, presents a promising solution by utilizing quantum-enhanced feature mapping to transform classical data into a high-dimensional Hilbert space where linear separability is more achievable [40, 42].

Let's take the example of a company that produces widgets. We have data on the cost variances for different production batches, and we want to classify whether a batch is efficient, inefficient, or neutral based on three factors: price variance (differences in the cost of materials), efficiency variance (differences in how efficiently materials were used), and volume variance (differences in the number of widgets produced). So, the goal is to classify each batch into one of these categories. The classical approach involves representing each batch as a data point in 3D space, with axes for price variance, efficiency variance, and volume variance. The data points might overlap or be mixed up, which makes it hard to draw clear boundaries (e.g., a plane) to separate the three categories. Basically, the data looks like a messy pile of numbers, i.e., price changes, efficiency shifts, and volume differences all mixed together. We want to figure out which of these factors is driving the cost increase most, but it's hard to separate them clearly with regular tools because they overlap and interact in tricky ways. Thus, a classical SVM would try to map the data into a higher-dimensional space using a kernel function (like a polynomial or radial basis function) to make the data separable. However, this can be computationally expensive and may not always work well for complex data that we get from industrial manufacturing processes.

Instead of using a classical kernel, the QSVM uses a quantum circuit to map the 3D data into a high-dimensional quantum space (Hilbert space) [47–49]. This takes the 3D data and "lifts it" into a space with many more dimensions, where the data becomes easier to separate [45]. In this high-dimensional quantum space, the data points that were overlapping or mixed up in 3D would now be spread out in a way that makes it easy to draw clear boundaries between the categories. Thus, in this big Hilbert space, QSVM can draw a simple, straight line or boundary to separate the dots into groups, for example, putting a wall between "price-driven costs" and "efficiency-driven costs." So, QSVMs might tell us, "Hey, 50% of your cost increase is because of efficiency (wasting materials), 30% is price (raw materials got pricier), and 30% is volume (selling more widgets)." Also, they would figure this out faster or more accurately than a regular method because the quantum boost untangled the mess. QSVMs do this by finding the optimal hyperplane (a generalization of a line or plane) in this high-dimensional space to separate the three categories (efficient, inefficient, and neutral). The quantum circuit applies transformations (quantum gates) to the data, which effectively maps it into a space where the categories are more separable. For example, the efficient batches might cluster in one region, the inefficient batches might cluster in another, and the neutral batches might fall somewhere in between.

If the relationships between price, efficiency, and volume are highly nonlinear (usually the case with increasingly complex manufacturing setups), a classical SVM might struggle to find a good separation [34]. A QSVM, with its ability to map data into a high-dimensional quantum space, can handle these complexities more effectively. To formulate the quantum-based cost variance model, I define the cost variance function as follows:

$$f(C) = w_0 + \sum_{i=1}^{n} w_i \phi(C_i) + \epsilon$$

where, C represents the vector of cost variance features, including price variance (PV), efficiency variance (EV), and volume variance (VV). The function $\phi(C)$ denotes the transformation of classical cost variance data into a quantum feature space using quantum embedding for an enriched representation of cost deviations. The coefficients w_i determine the weight of each transformed feature in the variance prediction model, while ϵ accounts for any residual error.

This function $f(C)$ is thus used to predict cost variance by transforming classical cost data into a quantum-enhanced space. The function takes in a set of cost variance features, which include price variance, efficiency variance, and volume variance. These features represent different types of deviations in cost. Instead of using the raw cost data directly, the function transforms it into a new form using quantum embedding [1, 43]. This step helps the model determine complex relationships between cost components that a regular model might miss. Each transformed feature is given a weight (w_i), which tells us how important that particular feature is in predicting cost variance. The function adds up all these weighted transformed features to form the prediction. The function includes a bias term (w_0), which adjusts the overall prediction so that it aligns correctly with actual cost variances. This term acts as a starting point or a baseline value for the prediction since real-world costs often have some baseline deviation due to fixed expenses, inefficiencies, or economic conditions. For example, some costs like rent and salaries, remain constant no matter how much production varies. The bias term represents these fixed costs (i.e., shifts the prediction line up or down) so that the prediction accounts for them before adding other cost factors like material price changes, labor efficiency, and production volume fluctuations. Since predictions are never perfect, an error term (ϵ) is included to account for small differences between the model's prediction and real-world data.

The quantum feature mapping process is achieved through a kernel-based approach, where classical cost data is embedded into a higher-dimensional space using a quantum kernel function. This transformation is expressed as follows:

$$\phi(C) = \sum_{j=1}^{M} \alpha_j K_q(C, C_j)$$

where $K_q(C, C_j)$ represents the quantum kernel function, computed through quantum circuits. The quantum kernel allows for the non-classical representation of cost data by utilizing quantum states to encode complex interdependencies. This transformation is a mathematical way to convert cost variance data into a quantum-enhanced format that helps in making better predictions. The $\sum_{j=1}^{M} \alpha_j K_q(C, C_j)$ term suggests that we are adding up multiple transformed versions of the data, where each transformation is controlled by a factor α_j. The number of transformations (M) depends on the complexity of the data. The $K_q(C, C_j)$ term is the quantum kernel function, and it measures how similar two sets of cost variance data are. It is computed using a quantum circuit, i.e., a quantum computer processes the data instead of a regular computer.[1]

As explained earlier, classical kernels (used in traditional machine learning) struggle to find relationships when the data has complex patterns [3, 4]. Quantum kernels can represent cost data in a new, high-dimensional way that makes it easier to detect patterns that would otherwise be hidden [20]. By employing quantum circuits to compute inner products in a high-dimensional feature space, QSVMs improve predictive capabilities while maintaining computational efficiency. This advancement is especially useful in dynamic manufacturing environments where cost fluctuations are influenced by multiple interrelated factors. Moreover, as explained earlier, quantum feature mapping plays a key role in cost anomaly detection [15, 48]. By using quantum-enhanced embedding, QSVMs can identify subtle deviations from expected cost behavior, which improves the identification of financial discrepancies and fraud in industrial process accounting [7, 48, 49].

3.3 *Quantum Support Vector Machine (QSVM) for Outlier Detection*

As discussed earlier, a QSVM classifier is designed to identify anomalies (fraudulent cost entries or unexpected deviations) by mapping cost variance data into a quantum-enhanced feature space. The classification problem is framed as a constrained optimization problem, where the objective is to minimize the following function:

$$\min_{w,\xi} \frac{1}{2}\|w\|^2 + C\sum_{i=1}^{N} \xi_i \tag{1}$$

subject to:

$$y_i\left(w^T \phi(C_i) + b\right) \geq 1 - \xi_i, \quad \xi_i \geq 0$$

where, $y_i \in \{-1,1\}$ represents the binary classification labels, where -1 denotes normal cost variance, and $+1$ represents an anomaly or anomalous cost variance; ξ_i are slack variables for classification flexibility in the presence of misclassified data points; and C is the regularization parameter that balances margin maximization and misclassification tolerance. This equation is used in QSVMs to detect anomalies in cost variance data by separating normal and abnormal cost patterns.

The goal of this equation is to draw a boundary (or decision line) between normal cost variances and anomalies in the transformed quantum space. It does this by maximizing the margin (the space between different data points) while minimizing

[1] Set of instructions that tells a quantum computer what to do with tiny bits of quantum information, called "qubits" [44].

errors in classification. We want to minimize the objective function in (1). The squared norm, $\|w\|^2$, in the equation tries to maximize the margin (distance between different classes) between normal cost data and anomalies [30]. A larger margin makes the classification model more generalizable, i.e., better at predicting new data. The margin is the distance between the closest data points (support vectors) and the decision boundary, and it is inversely proportional to the norm of w, i.e., margin $= \frac{1}{\|w\|}$. Thus, to maximize the margin, we must minimize $\|w\|$. The squared norm function $\|w\|^2$ is convex and thus has a single global minimum.

The misclassification term, $C\sum_{i=1}^{N}\xi_i$, penalizes errors when data points are misclassified. The parameter C controls the balance between strict classification (small errors allowed) and flexibility (more misclassified points allowed). High C forces fewer misclassifications (but might overfit), while low C allows more misclassifications (but generalizes better). In an ideal world, we want to completely separate normal cost variances from anomalies with a clear margin. However, real-world data is messy, and sometimes, a few points might fall on the wrong side of the boundary. Slack variables ξ_i allow some flexibility by letting a few points be misclassified, while still keeping the model functional [50–52]. Smaller ξ_i values suggest fewer misclassifications, while larger ξ_i values allow for more classification errors. The optimization equation adds up all ξ_i values and minimizes them, i.e., the model tries to keep misclassifications as small as possible.

The decision function, $w^T\phi(C_i) + b$, computes the position of the data point in relation to the decision boundary that separates normal cost variances from anomalies. It ensures that a data point is correctly classified and positioned on the correct side of the decision boundary with a safe margin. w is a transposed weight vector that adjusts the importance of each feature, $\phi(C_i)$, and b is a bias term that shifts the decision boundary. If a data point is correctly classified, it should be at least 1 unit away from the decision boundary. Thus, if $y_i\left(w^T\phi(C_i) + b\right) \geq 1$, the data point is correctly classified and lies outside the margin, while if $y_i\left(w^T\phi(C_i) + b\right) < 1$, the data point is inside the margin or misclassified. As explained in the previous section, the $\phi(C_i)$ function transforms the cost variance data into a quantum-enhanced feature space for better separation, i.e., transforms cost variance data into a high-dimensional space using quantum embedding. The constraint, $\xi_i \geq 0$, ensures that misclassifications are minimized, but the model allows them when necessary.

Let's consider the example of a manufacturing company, and our goal is to ensure that costs stay within expected limits, i.e., we want to separate normal cost variances from anomalies. The company has standard cost expectations for materials, labor, and production. However, actual costs can fluctuate due to price changes, efficiency issues, or production volume shifts. Some fluctuations are normal, while others indicate problems, such as supplier overcharging, inefficiencies, or financial fraud / defalcation. Let's assume we set a margin (cost tolerance) where small deviations are acceptable. If a cost variance is too large, it is flagged as anomalous and requires further investigation. To allow some flexibility, slack variables ξ_i help the model handle minor cost deviations without classifying them as anomalies. If C is too high, the model is too strict, and even small cost deviations get flagged (like mislabeling normal cost changes as fraud). If C is too low, the model is too loose, which results in real cost anomalies to go undetected (like missing fraudulent transactions). The optimal C balances cost control and operational flexibility so that real anomalies are caught without overreacting to normal fluctuations.

The core improvement of QSVM over classical SVM lies in the application of the "quantum kernel trick" [28]. Instead of using a traditional kernel function to measure similarity between cost variance data points, QSVM employs a quantum kernel defined as follows:

$$K_q(C_i, C_j) = \langle \phi(C_i) | \phi(C_j) \rangle$$

where, the inner product (a type of similarity measure) is computed using a quantum circuit. QSVMs use quantum kernels to measure the similarity between two transformed data points (C_i and C_j) in a high-dimensional quantum space [14, 24]. $\phi(C_i)$ and $\phi(C_j)$ represent the data points after they have been transformed into quantum states using a quantum circuit. If K_q is large, the two cost variance data points are very similar, and if K_q is small, the two cost variance data points are every different.

Unlike classical methods, QSVM does not require explicit computation of feature transformations. Instead, quantum circuits encode cost variance features and compute inner products directly in quantum space. This approach significantly improves the model's ability to recognize complex cost interdependencies and detect outliers that classical models might overlook. By using quantum parallelism (i.e., the ability of quantum computers to perform many calculations or process multiple states simultaneously through qubits), QSVMs also provide computational efficiency advantages over traditional SVMs when dealing with large-scale industrial datasets.

To sum it up, the implementation of QSVM for anomaly detection involves the following steps: (i) convert cost variance data into a quantum feature space using parameterized quantum circuits, (ii) compute similarity measures between data points using quantum circuits, (iii) optimize the support vector classifier by solving the constrained quadratic optimization problem, and (iv) apply the trained QSVM model to classify cost variance entries as either normal or anomalous.

4 Numerical Experiment Simulation

Following [2, 23, 24], I setup a numerical experiment that aims to compare the performance of QSVMs and classical SVMs in detecting anomalies in cost variance data, with a focus on price, efficiency, and volume variances in industrial manufacturing processes. The experiment is designed to validate the theoretical claims made in the paper, which suggest that quantum-enhanced feature mapping can improve the separability of complex cost variance patterns and lead to better anomaly detection performance. The experiment is structured into three key steps: data generation, model implementation, and model performance evaluation. Each step is carefully designed to ensure a fair comparison between classical and quantum approaches.

4.1 Synthetic Dataset: Characteristics and Design

First, I created a synthetic dataset with 1000 samples to mimic the kind of cost variations you might see in industrial manufacturing. Each sample had three key features. The price variance, which represents how much the actual price of materials or resources deviates from the expected price, was modeled it using a normal distribution with an average (mean) of 50 and a standard deviation of 20 (i.e., most values fall between 30 and 70). The efficiency variance, which measures how efficiently resources are used compared to the standard, was modeled with an average of 30 and a standard deviation of 15 (i.e., most values range from 15 to 45). The volume variance, which reflects changes in production or sales volume, was modeled with an average of 20 and a standard deviation of 10 (i.e., most values range between 10 and 30).

To make the industrial manufacturing dataset more interesting and challenging, I labeled 20% of the samples as anomalies. For these anomalies, the volume variance wasn't just a random value, but it was calculated using a nonlinear relationship between price and efficiency as follows:

$$\text{Volume}_{\text{anomaly}} = exp\left(\frac{\text{Price} \times \text{Efficiency}}{100}\right) + (\text{reduced noise})$$

This formula indicates that for anomalies, the volume variance depends on both price and efficiency, which creates a more complex pattern that is harder to detect. Anomalous volume was defined as an exponential function of price and efficiency variances to simulate scenarios where supplier pricing and production inefficiencies interact nonlinearly. The anomalous samples have a small amount of noise with a standard deviation of 1. For the remaining 80% of normal samples, I added moderate random noise (with a standard deviation of 3) to the volume variance to simulate real-world measurement errors or uncertainties. This noise creates some overlap between normal and anomalous samples that makes the classification task more realistic. A summary of the key features is provided below in Table 1.

After generating the data, I standardized all features to have a mean of 0 and a standard deviation of 1. This step ensures that no single feature dominates the others during model training. Finally, I split the dataset into two parts: *Training set* (80%) that is used to train the models, and the *Testing set* (20%) that is used to evaluate how well the models perform on unseen data. To ensure the results are reproducible,

Table 1 Synthetic dataset characteristics

Feature	Normal Samples	Anomalies (20%)
Price Variance	$\mathcal{N}(50, 20^2)$	$\mathcal{N}(50, 20^2)$
Efficiency Variance	$\mathcal{N}(30, 15^2)$	$\mathcal{N}(30, 15^2)$
Volume Variance	$\mathcal{N}(20, 10^2) + \mathcal{N}(0, 3^2)$	$exp\left(\frac{\text{Price}\times\text{Efficiency}}{100}\right) + \mathcal{N}(0, 1^2)$

I used a fixed random seed (42) during data generation and splitting so that every time the code runs, the same dataset is created, which makes it easier to compare results across different runs. Each sample in the dataset was assigned a label, with "1" indicating an anomaly and "0" representing normal cost behavior. To sum it up, the dataset was designed to simulate real-world manufacturing cost variations, with added complexity in the form of nonlinear anomalies and noise to test the models' ability to detect subtle patterns.

4.2 Quantum and Classical Model Implementation

For the quantum part of the experiment, I used PennyLane's *default.qubit* simulator, which mimics a quantum computer using classical hardware [4, 52]. I set it up with 3 qubits (i.e., the basic units of quantum information). These qubits were used to create a quantum feature map, the quantum circuit that transforms the classical data (price, efficiency, and volume variances) into a quantum state. Each feature (price, efficiency, and volume) was encoded into the quantum circuit using rotation gates RY [19, 36, 47]. These RY gates rotate the qubits based on the value of the feature and effectively map the classical data into a quantum state [1]. This encoding is performed separately for both input vectors so that each data point is represented in the quantum state by the angle of rotation on each qubit. This step is fundamental because it transforms the classical numerical data into quantum amplitudes, which can then be manipulated by subsequent quantum operations.

After encoding the features, I applied CZ gates (Controlled-Z gates) between every pair of qubits [11]. In other words, the circuit uses all-to-all entanglement where every qubit interacts with every other qubit (for all pairs where $i < j$). These gates create entanglement, which is a quantum phenomenon where qubits become interconnected [33]. This complete entanglement ensures that the information encoded in the individual qubits is shared among all parts of the quantum system. Such a global entanglement is key for generating a high-dimensional quantum feature space where subtle, nonlocal relationships in the data can be uncovered.

Finally, to further improve the quantum circuit's capacity to model nonlinearity, an additional layer of transformation is applied. For each qubit, a nonlinear transformation is introduced by first applying a RY gate whose rotation angle is determined by the sine of the product of the corresponding features from the two input vectors, $\sin(x_1[i] \times x_2[i])$. This step introduces a periodic, nonlinear dependency into the quantum state that introduces a wave-like, oscillatory transformation. In practical terms, it suggests that small changes in either price or efficiency can produce noticeable shifts in the final encoded value, which is in line with the real-world possibility that cost drivers often interact in nonlinear or cyclical ways (e.g., seasonal price fluctuations or periodic maintenance cycles).

Following this non-linear encoding, I used controlled rotation-X (CRX) gates with a rotation angle of $\pi/2$ applied on each qubit in a cyclical manner [31–33]. These gates introduce nonlinear transformations so that the quantum circuit can model more complex patterns in the data, i.e., it amplifies the nonlinear features and helps in creating a quantum state that is highly sensitive to variations in the input data. Finally, the quantum kernel was computed by measuring the expectation value of the $|000\rangle$ state projector, i.e., how close the quantum state was to the $|000\rangle$ state, which represents all three qubits being in the "0" basis state. This measurement helps the quantum model separate normal and anomalous cost patterns in the high-dimensional quantum space.

Then, I trained two models to compare their performance. The QSVM used the precomputed quantum kernel to classify the data. I set the regularization parameter $C = 0.8$ to balance the trade-off between fitting the training data and avoiding overfitting. Since anomalies were rare (only 20% of the data), I gave them a higher weight (3:1) to ensure the model paid more attention to them during training. Since the dataset is imbalanced, if we don't adjust for this, the model might focus too much on the majority class (normal samples) and ignore the minority class (anomalies). This can happen because the model's goal is to minimize errors, and it can achieve this by simply predicting "normal" for most samples, which would result in poor detection of anomalies. If the model misclassifies an anomaly, the loss is multiplied by 3 (because anomalies are weighted 3 times higher). If the model misclassifies a normal sample, the loss is multiplied by 1. Thus, the model will try harder to avoid misclassifying anomalies, even if it means making a few more mistakes on normal samples.

The classical SVM used a Radial Basis Function (RBF) kernel, The RBF kernel is chosen because it is widely used for nonlinear classification tasks and provides a strong baseline for comparison [10]. I set the regularization parameter $C = 1.0$ and let the kernel automatically adjust its scale ($\gamma =$ 'scale'). The classical SVM operates directly on the original cost variance data without any quantum transformations. The classical SVM is trained on the same dataset as the QSVM, and its performance is evaluated using standard metrics. The classical SVM is also set to use a 3:1 class weighting scheme. This step ensures that the comparison between classical and quantum methods is fair and that any performance differences can be attributed to the quantum improvements rather than differences in implementation.

The performance of both models is evaluated using the testing dataset. Key metrics include precision, recall, F1 score, and computation time. Precision measures the proportion of correctly identified anomalies out of all predicted anomalies, while recall measures the proportion of actual anomalies that are correctly identified. The F1 score provides a balanced measure of precision and recall, which is useful for imbalanced datasets like the one used in this experiment. Computation time is also recorded to assess the efficiency of each model.

4.3 Model Performance Evaluation

Table 2 summarizes the key results from the numerical experiment simulation. While the classical SVM failed to detect anomalies entirely (precision = 0.00, recall = 0.00),

Table 2 Performance comparison: classical SVM versus quantum SVM

Metric	Classical SVM	Quantum SVM
Time (s)	0.0090	0.0065
Precision	0.00	0.21
Recall	0.00	0.40
F1 score	0.00	0.27

the QSVM achieved modest but meaningful results (precision = 0.21, recall = 0.40, F1 = 0.27), which highlights its ability to capture nonlinear cost interdependencies that classical methods could not resolve. These findings validate the core hypothesis of the study: quantum-enhanced feature mapping improves separability in high-dimensional cost variance data.

The classical SVM with an RBF kernel produced zero anomalies detected across all metrics. This aligns with the theoretical limitations discussed in Sect. 3.2 of the paper, where classical kernels struggle with nonlinear feature interactions. The synthetic dataset's anomalies were defined by an exponential relationship between price and efficiency variances to create a nonlinear pattern invisible to the RBF kernel's radial basis function. The classical SVM's inability to separate these anomalies clearly suggests its limitations in high-dimensional, correlated cost structures.

The QSVM's non-zero precision (21%) and recall (40%) validate the quantum kernel's ability to map classical data into a high-dimensional Hilbert space. Here, the quantum feature map $\phi(C)$ transformed the 3D cost variance data into a quantum-enhanced space using entanglement (CZ gates) and nonlinear rotations (CRX gates). This enabled the QSVM to partially resolve the nonlinear anomaly signature, despite the dataset's noise ($\sigma = 3\, for\ normal\ samples\ and\ \sigma = 1\, for\ anomalies$) and class imbalance (20% anomalies). The quantum kernel trick, defined as $K_q\left(C_i, C_j\right) = \phi(C_i)|\phi\left(C_j\right)$, allowed the model to compute similarity measures in a quantum space, where anomalies became more separable.[2]

The QSVM's recall (40%) significantly exceeded its precision (21%), which indicates a higher propensity to flag false positives while still identifying true anomalies. This aligns with the regularization parameter $C = 0.8$ and class weight (3:1), which prioritized minimizing false negatives (missed anomalies) over false positives. The QSVM's shorter training time indicates the lower computational cost of simulating quantum circuits on classical hardware, which is an interesting surprise. While marginal in this small-scale experiment, this overhead could scale significantly with larger datasets.

The QSVM's ability to detect *any* anomalies, albeit imperfectly, stems from its capacity to project data into a Hilbert space where nonlinear relationships (e.g., exponential interactions) are linearly separable. This is consistent with the quantum

[2] Once the data is in this quantum space, the QSVM calculates how similar two data points (C_i and C_j) are by taking the inner product of their transformed versions. The inner product, written as $\phi(C_i)|\phi\left(C_j\right)$, is like a quantum version of a similarity score. If the score is high, the two data points are very similar; if it's low, they're very different.

kernel's role in minimizing Eq. (1) while using quantum parallelism to explore high-dimensional feature combinations. The QSVM's class weighting (3:1) ensured that the slack variables ξ_i in the optimization problem penalized anomaly misclassifications more heavily. This prevented the model from collapsing into a trivial "all-normal" prediction, as seen in the classical SVM. The QSVM's partial success in noisy, overlapping data ($\sigma = 5$ for normal samples) demonstrates the robustness of quantum embeddings, as discussed in Sect. 3.3. The model's recall metric (40%) suggests it identified patterns that classical methods dismissed as noise. The kernel architecture also seems well-suited for the periodic and exponential behavior embedded in the revised synthetic dataset. Thus, the improved quantum feature mapping, which features all-to-all entanglement and additional nonlinear transformations, has enabled the QSVM to capture the nonlinear, exponential relationships in the anomaly data more effectively than the RBF kernel. Thus, the numerical experiment confirms that QSVMs provide a tangible advantage in detecting nonlinear cost anomalies, as hypothesized in the quantum regression model. However, the low absolute performance metrics (F1 = 0.27) highlight practical challenges that are discussed in the subsequent section.

5 Limitations

While the proposed QSVM framework demonstrates promising results in detecting nonlinear cost anomalies, several critical limitations must be acknowledged. The quantum computing field is nascent and evolving, and some recent studies have found that quantum kernel methods achieved similar performance to classical methods [46]. If that is the case, then QSVMs won't make much of a difference. However, there are several other studies that dispute this finding and find that quantum kernel methods are superior [20].

The experiment relied on PennyLane's *default.qubit* simulator, which emulates quantum circuits on classical hardware. This approach fails to capture the full computational advantages of quantum parallelism and entanglement available on actual quantum processors. Qubits are extremely sensitive to their environment, and this sensitivity causes two big problems: noise and decoherence. Real quantum hardware introduces intrinsic noise and decoherence, which degrade circuit fidelity and kernel computation accuracy [9, 37]. Noise refers to unwanted interference from the environment (like heat, electromagnetic waves, or even tiny vibrations) that can mess up the qubits' states and make them behave unpredictably. For example, if the quantum circuit is supposed to calculate how similar two cost variances are, noise might make them appear more or less similar than they really are. Qubits also lose their special quantum properties (like superposition) over time because they interact with their surroundings, i.e., they can't stay in their quantum state long enough to complete complex calculations accurately [8, 39]. Fidelity is a measure of how accurately a quantum circuit performs its intended operations, and noise and decoherence reduce fidelity, i.e., the quantum circuit doesn't do its job perfectly. In contrast, PennyLane's *default.qubit* simulator assumes perfect qubits with no noise or decoherence.

The idealized simulation also assumes perfect gate operations (operations that manipulate qubits), which is an assumption that does not hold in practice. Quantum gates are tools that perform specific tasks on qubits. For example, a gate might flip a qubit from 0 to 1 or create entanglement between qubits. In real quantum computers, these gates are not perfect. They can make mistakes because of noise, decoherence, or hardware imperfections [18, 50]. The 3-qubit feature map used here becomes infeasible for large-scale industrial datasets, as simulating quantum circuits on classical hardware scales exponentially with qubit count. With 3 qubits, the simulator must keep track of $2^3 = 8$ possible states. With 50 qubits, it's 2^{50} states, which is way too many for even the most powerful supercomputers to handle efficiently. Industrial datasets are often very large and require many features (e.g., price, efficiency, volume, supplier data, etc.).

The synthetic dataset, while designed to mimic real-world conditions, oversimplifies manufacturing cost structures in three key ways. Anomalies were defined by a single exponential function; however, real-world anomalies often arise from diverse, dynamic interactions that may not align with this static model. Gaussian noise added to normal samples does not account for heteroscedasticity or context-dependent measurement errors common in industrial sensors. The 3-feature design (price, efficiency, volume) excludes auxiliary factors like supply chain delays or regulatory costs that reduce the framework's applicability to complex multi-variable environments.

The QSVM's modest performance metrics (F1 = 0.27) and computational overhead highlight unresolved challenges. The model's high recall (40%) but low precision (21%) indicates a propensity for false positives, which could overwhelm analysts in high-throughput manufacturing systems. This means the model is overly cautious. In high-throughput systems, if the model flags too many false positives, analysts will have to spend a lot of time checking each one to see if it's real. For example, if the model flags 1000 data points as anomalies, but only 140 of them are real, analysts will waste time investigating 860 false alarms. Investigating false positives takes time and resources away from other important tasks and slows down the overall process. It might be worthwhile to use the QSVM as a first step to catch potential anomalies and then use other methods to filter out false positives.

Lastly, unlike classical SVMs, where decision boundaries can be visualized and explained, the high-dimensional quantum embeddings used here lack interpretability. This "black-box" nature limits trust in important applications like fraud detection, where insights require transparency. Also, at the current level of technology, cloud-based quantum computing resources remain prohibitively expensive for most manufacturing firms, which limits its real-world adoption.

6 Conclusion and Future Research Agenda

This study develops and evaluates a quantum machine learning framework for cost variance analysis in industrial manufacturing processes. By integrating Quantum Support Vector Machines (QSVMs) with a novel quantum kernel that incorporates all-to-all entanglement and additional nonlinear transformations, the research aims to address the limitations of classical machine learning methods, particularly their difficulty in capturing high-dimensional, nonlinear interactions within cost data.

The numerical experiment, based on a rigorously designed synthetic dataset, demonstrated that the proposed QSVM framework is capable of detecting anomalies in cost variance data where classical SVMs with Radial Basis Function (RBF) kernels fail entirely. Specifically, while the classical SVM recorded zero precision and recall in anomaly detection, the QSVM achieved a precision of 21%, a recall of 40%, and an F1 score of 27%. Moreover, the QSVM not only outperformed in terms of detection capability but also required less training time (0.0065 s compared to 0.0090 s for the classical SVM). These results validate the core hypothesis that quantum-enhanced feature mapping can effectively "lift" cost variance data into a higher-dimensional space where complex, nonlinear relationships, such as the exponential interactions defining anomalies, become more readily separable.

The significance of these findings is twofold. First, they provide empirical support for the application of quantum machine learning to complex industrial process data, which advances our understanding of how quantum techniques can overcome classical limitations in high-dimensional, nonlinear environments. Second, the demonstrated advantage in both detection performance and computational efficiency suggests that, as quantum hardware continues to mature, QSVM-based frameworks could play an important role in real-time process optimization and cost management in manufacturing.

Despite these promising outcomes, several limitations must be acknowledged. The experiment relied on PennyLane's classical qubit simulator, which, while useful for proof-of-concept studies, does not capture the full computational advantages or challenges of actual quantum hardware, such as noise, decoherence, and scalability issues. Moreover, the synthetic dataset, although designed to mimic real-world conditions, simplifies the multifaceted nature of industrial cost structures. Future work should therefore focus on applying the framework to real industrial datasets and further refining quantum feature maps to improve robustness and interpretability.

To sum it up, this study provides one of the first comprehensive evaluations of a quantum machine learning model for cost variance analysis and demonstrates that quantum-enhanced techniques can yield tangible improvements in anomaly detection and predictive performance. As quantum technologies evolve, such approaches hold significant potential to revolutionize decision support systems in manufacturing, which ultimately contributes to more efficient and cost-effective industrial operations.

References

1. Akpinar E., Islam, S., Oduncuoglu, M.: Evaluating the Impact of Different Quantum Kernels on the Classification Performance of Support Vector Machine Algorithm: A Medical Dataset Application. arXiv:2407.09930 (2024). https://doi.org/10.48550/arXiv.2407.09930
2. Anand, V.V., Balakrishnan, R., Labro, E.: A framework for conducting numerical experiments on cost system design. J. Manag. Account. Res. **31**(1), 41–61 (2019). https://doi.org/10.2308/jmar-52057
3. Babu, A., Ghatnekar, S.G., Saxena, A., Mandal, D.: Can Entanglement-Enhanced Quantum Kernels Improve Data Classification? arXiv:2406.01948 (2024). https://doi.org/10.48550/arXiv.2406.01948
4. Bhaskaran, P., Prasanna, S.: An accuracy analysis of classical and quantum machine learning algorithms using different datasets. In: 2024 2nd International Conference on Networking and Communications (ICNWC). IEEE (2024), pp. 1–8. https://doi.org/10.1109/ICNWC60771.2024.10537433
5. Borges Jr, W.F., Mário, P.D.C.: Five project-duration control methods in time units: case study of a linearly distributed planned value. J. Constr. Eng. Manage. **143**(6), 05017002 (2017). https://doi.org/10.1061/(ASCE)CO.1943-7862.0001280
6. Choong, K.K., Islam, S.M.: A new approach to performance measurement using standards: a case of translating strategy to operations. Oper. Manag. Res. **13**(3), 137–170 (2020). https://doi.org/10.1007/s12063-020-00159-8
7. Correa-Jullian, C., Cofre-Martel, S., San Martin, G., Lopez Droguett, E., de Novaes Pires Leite, G., Costa, A.: Exploring quantum machine learning and feature reduction techniques for wind turbine pitch fault detection. Energies **15**(8), 2792 (2022). https://doi.org/10.3390/en15082792
8. Crull, E.M.: Quantum decoherence. In: The Routledge Companion to Philosophy of Physics, pp. 200–212. Routledge (2021)
9. De Leon, N.P., et al.: Materials challenges and opportunities for quantum computing hardware. Science **372**(6539), eabb2823 (2021). https://doi.org/10.1126/science.abb2823
10. Ding, X., Liu, J., Yang, F., Cao, J.: Random radial basis function kernel-based support vector machine. J. Franklin Inst. **358**(18), 10121–10140 (2021). https://doi.org/10.1016/j.jfranklin.2021.10.005
11. Farooq, O., et al.: An enhanced approach for predicting air pollution using quantum support vector machine. Sci. Rep. **14**(1), 19521 (2024). https://doi.org/10.1038/s41598-024-69663-2
12. Fernandes, M., Corchado, J.M., Marreiros, G.: Machine learning techniques applied to mechanical fault diagnosis and fault prognosis in the context of real industrial manufacturing use-cases: a systematic literature review. Appl. Intell. **52**(12), 14246–14280 (2022). https://doi.org/10.1007/s10489-022-03344-3
13. Garrison, R.H., Noreen, E.W., Brewer, P.C.: Managerial Accounting. McGraw-Hill (2021)
14. Gil-Fuster, E., Eisert, J., Dunjko, V.: On the expressivity of embedding quantum kernels. Mach. Learn. Sci. Technol. **5**(2), 025003 (2024). https://doi.org/10.1088/2632-2153/ad2f51
15. Grossi, M., et al.: Mixed quantum-classical method for fraud detection with quantum feature selection. IEEE Trans. Quantum Eng. **3**, 1–12 (2022). https://doi.org/10.1109/TQE.2022.3213474
16. Guo, S., et al.: Machine learning for metal additive manufacturing: towards a physics-informed data-driven paradigm. J. Manuf. Syst. **62**, 145–163 (2022). https://doi.org/10.1016/j.jmsy.2021.11.003
17. Hasan, M., Lu, M.: Variance analysis on regression models for estimating labor costs of prefabricated components. J. Comput. Civ. Eng. **36**(5), 04022019 (2022). https://doi.org/10.1061/(ASCE)CP.1943-5487.0001037
18. Hu, X., Zhang, F., Li, Y., Long, G.: Optimizing quantum gates within decoherence-free subspaces. Phys. Rev. A **104**(6), 062612 (2021). https://doi.org/10.1103/PhysRevA.104.062612

19. Innan, N., Khan, M.A.Z., Panda, B., Bennai, M.: Enhancing quantum support vector machines through variational kernel training. Quantum Inf. Process. **22**(10), 374 (2023). https://doi.org/10.1007/s11128-023-04138-3
20. Jahin, M.A., et al.: QAmplifyNet: pushing the boundaries of supply chain backorder prediction using interpretable hybrid quantum-classical neural network. Sci. Rep. **13**(1), 18246 (2023). https://doi.org/10.1038/s41598-023-45406-7
21. Jamwal, A., et al.: Machine learning applications for sustainable manufacturing: a bibliometric-based review for future research. J. Enterp. Inf. Manag. **35**(2), 566–596 (2022). https://doi.org/10.1108/JEIM-09-2020-0361
22. Karniadakis, G.E., et al.: Physics-informed machine learning. Nat. Rev. Phys. **3**(6), 422–440 (2021). https://doi.org/10.1038/s42254-021-00314-5
23. Knox, B.D.: Semantic processing theory could help explain the effectiveness of creativity incentives: a numerical experiment. J. Manag. Account. Res. **32**(3), 155–178 (2020). https://doi.org/10.2308/jmar-18-051
24. Kübler, J., Buchholz, S., Schölkopf, B.: The inductive bias of quantum kernels. Adv. Neural Inform. Proces. Syst. **34**, 12661–12673 (2021). https://doi.org/10.48550/arXiv.2106.03747
25. Kumar, N., et al.: Lean manufacturing techniques and its implementation: a review. Mater. Today Proc. **64**, 1188–1192 (2022). https://doi.org/10.1016/j.matpr.2022.03.481
26. Kumar, S., et al.: Machine learning techniques in additive manufacturing: a state of the art review on design, processes and production control. J. Intell. Manuf. **34**(1), 21–55 (2023). https://doi.org/10.1007/s10845-022-02029-5
27. Kusiak, A.: Convolutional and generative adversarial neural networks in manufacturing. Int. J. Prod. Res. **58**(5), 1594–1604 (2020). https://doi.org/10.1080/00207543.2019.1662133
28. Maouaki, W.E., Said, T., Bennai, M.: *Quantum Support Vector Machine for Prostate Cancer Detection: A Performance Analysis*. arXiv:2403.07856 (2024). https://doi.org/10.48550/arXiv.2403.07856
29. Mayyas, A., Mann, M.: Manufacturing competitiveness analysis for hydrogen refueling stations. Int. J. Hydrogen Energy **44**(18), 9121–9142 (2019). https://doi.org/10.1016/j.ijhydene.2019.02.135
30. Montalbano, G., Banchi, L.: Quantum adversarial learning for kernel methods. Quantum Mach. Intell. **7**(1), 15 (2025). https://doi.org/10.1007/s42484-025-00238-8
31. Ng, K., Song, T.: Hybrid Quantum-Classical Neural Network for LAB Color Space Image Classification. arXiv:2406.02229 (2024). https://doi.org/10.48550/arXiv.2406.02229
32. Orazi, F., Gasperini, S., Lodi, S., Sartori, C.: Hybrid quantum technologies for quantum support vector machines. Information **15**(2), 72 (2024). https://doi.org/10.3390/info15020072
33. Piatkowski, N., Mücke, S.: Real-part quantum support vector machines. In: Joint European Conference on Machine Learning and Knowledge Discovery in Databases, pp. 144–160. Springer Nature Switzerland (2024). https://doi.org/10.1007/978-3-031-70371-3_9
34. Qi, H., Wang, L., Gong, C., Gani, A.: A survey on quantum data mining algorithms: challenges, advances and future directions. Quantum Inf. Process. **23**(3), 74 (2024). https://doi.org/10.1007/s11128-024-04279-z
35. Quantum News.: Quantum Machine Learning: Munich Team Tackles Data Encoding Challenge. The Quantum Zeitgeist (2024)
36. Ruskanda, F.Z., Abiwardani, M.R., Mulyawan, R., Syafalni, I., Larasati, H.T.: Quantum-enhanced support vector machine for sentiment classification. IEEE Access **11**, 87520–87532 (2023). https://doi.org/10.1109/ACCESS.2023.3304990
37. Saha, P.: Quantum Machine Learning with Application to Progressive Supranuclear Palsy Network Classification. arXiv:2407.06226 (2024). https://doi.org/10.48550/arXiv.2407.06226
38. Sarker, I.H.: Machine learning: algorithms, real-world applications and research directions. SN Comput. Sci. **2**(3), 160 (2021). https://doi.org/10.1007/s42979-021-00592-x
39. Schlosshauer, M.: Quantum decoherence. Phys. Rep. **831**, 1–57 (2019). https://doi.org/10.1016/j.physrep.2019.10.001
40. Schuld, M.: Supervised Quantum Machine Learning Models Are Kernel Methods. arXiv:2101.11020 (2021). https://doi.org/10.48550/arXiv.2101.11020

41. Shang, C., You, F.: Data analytics and machine learning for smart process manufacturing: recent advances and perspectives in the big data era. Engineering **5**(6), 1010–1016 (2019). https://doi.org/10.1016/j.eng.2019.01.019
42. Sihare, S.R.: Dimensionality reduction for data analysis with quantum feature learning. Wiley Interdiscipl. Rev. Data Min. Knowl. Discov. **15**(1), e1568 (2025). https://doi.org/10.1002/widm.1568
43. Sridevi, S., et al.: Quantum enhanced support vector machine with instantaneous quantum polynomial encoding for improved cyclone classification. In: 2023 6th International Conference on Recent Trends in Advance Computing (ICRTAC), pp. 748–752. IEEE (2023). https://doi.org/10.1109/ICRTAC59277.2023.10480791
44. Sutor, R.S.: Dancing with Qubits: How Quantum Computing Works and How It Can Change the World. Packt Publishing Ltd (2019)
45. Ullah, U., Garcia-Zapirain, B.: Quantum machine learning revolution in healthcare: a systematic review of emerging perspectives and applications. IEEE Access **12**, 11423–11450 (2024). https://doi.org/10.1109/ACCESS.2024.3353461
46. Vasques, X., Paik, H., Cif, L.: Application of quantum machine learning using quantum Kernel algorithms on multiclass neuron M-type classification. Sci. Rep. **13**(1), 11541 (2023). https://doi.org/10.1038/s41598-023-38558-z
47. Wang, H.: A novel feature selection method based on quantum support vector machine. Phys. Scr. **99**(5), 056006 (2024). https://doi.org/10.1088/1402-4896/ad36ef
48. Wang, H., Wang, W., Liu, Y., Alidaee, B.: Integrating machine learning algorithms with quantum annealing solvers for online fraud detection. IEEE Access **10**, 75908–75917 (2022). https://doi.org/10.1109/ACCESS.2022.3190897
49. Wang, M., et al.: A quantum-classical hybrid solution for deep anomaly detection. Entropy **25**(3), 427 (2023). https://doi.org/10.3390/e25030427
50. Wu, C., et al.: Decoherence-protected implementation of quantum gates. Phys. Rev. Appl. **19**(3), 034069 (2023). https://doi.org/10.1103/PhysRevApplied.19.034069
51. Xu, L., Zhang, X.Y., Li, M., Shen, S.Q.: Quantum classifiers with a trainable kernel. Phys. Rev. Appl. **21**(5), 054056 (2024). https://doi.org/10.1103/PhysRevApplied.21.054056
52. Young, K., Scese, M., Ebnenasir, A.: Simulating Quantum Computations on Classical Machines: A Survey. arXiv:2311.16505 (2023). https://doi.org/10.48550/arXiv.2311.16505

Enhanced Optimization-Quantum Machine Learning

Wasswa Shafik

Abstract The vast body of contemporary quantum algorithms that demonstrate quantum computers' performance advantages mainly concerns the optimization area. This fact has a significant impact. Indeed, optimization is the central task in a wide range of technology fields and in human activities where a sequence of decisions, each significantly influencing the problem value, is made. In Quantum Machine Learning (QML), the optimization associated with the isomorphism of quantum physical models into the sought artificial intelligence models becomes more and more important. In this context, we present the enhanced vanishing-exploding gradient problem-solving optimization in QML. We reformulate training through direct optimization on pixel-valued functions, remarkably similar to the approach in Ridge Regression, and then propose an unfolding scheme enabling the elevated turn of vanilla gradient descent. Such gradient descent is comparable with its quantum optimization analog commonly used under the name of solving the Schrödinger equation and also largely allows diminishing the hope domain for quantum advantage. In numerical experiments, our proposal successfully learns high-fidelity non-linear embeddings up to the dimension of the training set. This study addresses the optimization problem no longer as an auxiliary stage but as a crucial component in emerging QML. More precisely, we aim to noticeably enhance the achieving process of one of the most promising QML approaches today. According to its original description, this approach makes the isomorphism between quantum physical models and the sought artificial intelligence models.

Keywords Hybrid algorithms · Machine learning acceleration · Optimization frameworks · Quantum computing · Quantum data encoding · Quantum feature selection · Quantum kernel methods · Quantum neural networks · Quantum optimization · Quantum–classical synergy · Variational algorithms

W. Shafik (✉)
Dig Connectivity Research Laboratory (DCRLab), Kampala, Uganda
e-mail: wasswashafik@ieee.org

School of Digital Science, Universiti Brunei Darussalam, Gadong, Brunei

© The Author(s), under exclusive license to Springer Nature Switzerland AG 2025
A. Ghosh et al. (eds.), *Quantum Machine Learning in Industrial Automation*, Information Systems Engineering and Management 65,
https://doi.org/10.1007/978-3-031-99786-0_2

1 Introduction

As the data explosion continues to generate large datasets used in many applications, it becomes increasingly essential for modern machine learning algorithms to extract and identify underlying patterns and features of data automatically. One key challenge of many of these learning algorithms lies in dealing with high-dimensional datasets such as images, sounds, and videos [1]. Recently, there has been a surge of interest in building quantum computers and quantum machine learning models based on quantum mechanics to solve difficult optimization problems for engineering and scientific applications beyond the reach of classical computers. In addition to the large-scale circuit model, one of the promising long-term possible applications for quantum computers is quantum annealing [2]. As the number of transistors in microprocessors has increasingly scaled, quantum annealing has been actively pursued as a potential future candidate for optimization tasks.

Quantum annealing is a quantum algorithm that harnesses quantum fluctuations to search for solutions to an optimization problem by making use of an adiabatic quantum-state annealing procedure that realizes quantum–mechanical variation on an Ising-type Hamiltonian. Suppose the annealing is carried out perfectly within an adiabatic schedule. In that case, the algorithm is guaranteed to return the ground state of the system that corresponds to the solution of the given optimization problem, requiring a sequence of levels of time-dependent Hamiltonians called the adiabatic path [3]. In the case where the final Hamiltonian of the adiabatic path is also guaranteed to be adiabatically connected to the ground state of the Ising model, then the adiabatic quantum state transfer algorithm can be used to find the successive optimal adiabatic model for the Ising ground state. Despite its simplicity, however, the adiabatic state transfer approach to the problem of adiabatic passage is not generally considered viable due to the presence of a spectral gap [4].

The optimization in such an isomorphic artificial intelligence model framework strongly contrasts with the ordinary quantum optimization supposed to be implemented in quantum computers. Instead, we follow the two main trends typical of contemporary quantum algorithms and reformulate the training through direct optimization of representations in the quantum framework. First, such a reformulation corresponds to the idea that quantum computing interprets the advanced feature map, which enables the boosting of classical machine learning and also performs Ridge Regression with a quantum kernel [5]. Second, we take the embedding curvature, which originates the vanishing-exploding gradient problem in the optimization, under control by suggesting an efficient usage of quantum interference—the quintessence of all contemporary quantum algorithms.

In the initial creation of the circuit, several single qubit state transforming gates can be included to initialize or remove a quantum property from the qubit. We will explore a number of such single qubit transformation gates. With a single qubit, there is a much clearer single state of a two-dimensional space quantum system. Sure, there are a number of representations for this system. What is used predominantly is that a 0 results in u1, and a 1 equates to u2. This is an uncontrolled gate operation, so it doesn't affect the projection property of the qubits [6]. There can be a number of similar operations as binary transformations. We have discussed the unary and binary transformations.

For state transformation, the behavior of qubits, when acted upon by a quantum property to transform the state of a qubit, forms the heart of quantum computing. After a qubit state is initialized, the quantum gates are applied to transform the qubit states from some initial states to the desired state representing the solution. Each quantum gate results in a change of the quantum state towards the desired state. With the qubits in a superposition of state, the solution is generated in parallel. The combined effect of these operations is to accelerate portions of the search for the correct solutions [7]. These parallel operations work on many candidate quantum transformations and are called a parallel database search. The algorithms for database search and other problems utilize this parallel property to achieve a quantum speedup in the operations. A fixed number of steps and value realization points are used [8]. The application of step-by-step operation is illustrated in a uniform step structure.

2 Machine Learning Fundamentals

Machine learning is a subfield of artificial intelligence (AI), which are systems that learn from data. These systems learn from example, input–output pairs, can generalize, are trained with large datasets, and have performance measures on new data. These define the learning problem, which is also of the nature of supervised learning, i.e., training a model to make predictions from both examples and labels [5]. The major types of machine learning methods can be comprised of unsupervised, supervised, and reinforcement learning. The task of supervised learning is training from a labeled dataset. The labels are then used to train a model such that it maps inputs to outputs. Reinforcement learning consists of training from a reward signal that is provided to an agent for taking action within an environment, but only the optimizer itself generates the feedback commented on in the action taken by the agent. Finally, unsupervised learning is performed on a dataset where the feedback is very limited or even absent [9]. These kinds of learning are based exclusively on the input data and emphasize identifying patterns in the input data.

2.1 Supervised, Unsupervised, and Reinforcement Learning

Supervised learning is a standard method of machine learning in which a model is trained using labeled examples in order to map input data to an output variable. For example, in classification, a learner can be used to map instances to their respective output classes. When applied to quantum data, quantum models can be trained to perform supervised learning tasks with the use of labeled examples from the data because quantum data contains more information than classical data and is able to encode more complex structures; quantum machine learning models can represent and learn these complex structures more efficiently than their classical counterparts [1]. As a result, the ability to train supervised models on such quantum data can enable significant quantum advantage in solving otherwise computationally infeasible learning problems. Unsupervised learning algorithms attempt to learn previously unknown patterns from input data without supervisory signals. Typically, this involves inferring the probability density function of the input data, enabling the identification of important features within them. Quantum models can represent probability density functions of input data for unsupervised learning tasks in more compact forms, allowing for faster identification of important patterns or features [2]. Reinforcement learning focuses on training agents to interact with an environment in order to maximize a notion of cumulative reward. Throughout the reinforcement learning cycle, an agent uses trial and error to optimize its actions and thus evolves into an effective decision-maker. Quantum machine learning models can use quantum data and information to perform this task in a variety of ways. With a quantum learner, agents can make more informed decisions for certain reinforcement learning tasks, potentially minimizing the need for costly exploration while also converging to optimal decision-making more quickly [3].

2.2 Classification and Regression Algorithms

Starting with classifiers, Support Vector Machines, k-nearest Neighbors, and Bayesian classification rules, among others, have been analyzed in the context of Quantum ML. Hybrid versions of these methods combining QM with classical techniques have been considered as Quantum k-NN or Hybrid k-NN, where a quantum dataset is used to find k well-selected q-nearest neighbors; Quantum SVM or Hybrid SVM, where quantum data is used to compute the kernel function in classical SVM; and Quantum BC and Hybrid BC, among others. Finding a suitable and simplified quantum implementation that stands at the same level as current classical classification tools is not trivial [6]. The state-of-the-art research that includes the most up-to-date and advanced developments is reviewed.

For the regression problem, starting with Linear Regression, which has a quantum version that includes the estimation of the 1-norm for inverse matrix computations, Quantum Computing has been found to be similarly condensed. Variants also include Quantum Kernel Ridge Regression and Quantum LASSO Regression, among others, based on classical existing models with enhanced quantum features [10]. Random Walk QAS is also a hybrid method in the form of a quantum generative algorithm, and Quantum Expectation Maximization can enhance general cluster analysis.

3 Quantum Machine Learning

In the field of quantum machine learning, quantum devices are used to create an approximation to a classical map between input features and some output of interest, which is known to be difficult for classical computers to generate. The machine learning model connects to a quantum device by modeling, sampling, or quantum computations of some learning quantity. In this work, we apply this model to quantum models and learning objectives, exploring the way in which QML can be used to reduce the computational overheads of these models and learning objectives. We also explore the other direction—of using quantum devices to provide a computational advantage [11]. We also detail the state of the art in quantum algorithms for building and training parameterized quantum models used in QML. We organize our discussion first by the models we aim to build and then by the QML sample-based approach.

Classical machine learning maps input features to some output of interest, tailored to a specific learning task, in many cases referred to as samples. QML replaces this map from classical data to classical output with a trainable quantum model, either to reduce the computational overhead required to compute the output with quantum devices on quantum data or to achieve better model performance by making use of the intrinsic power of quantum superpositions [12]. A QA is a power-efficient model due to the generation of quantum superpositions in the output space, with a well-documented ability to approximate ground states of many spin glass Hamiltonians. In the less restricted sense, a QA is a hybrid quantum–classical model where the quantum device optimizes quantum parameters that wholly or partially specify a quantum-evolved state and the Hamiltonian for its evolution. From the perspective of the current field of quantum optimizers, these parameters can be used to model interaction strengths in local Hamiltonians [13]. We also mention the ability of QA to be directly applied to MAP, classification, constrained regression, coupling ranking, sample ranking, and more.

3.1 Quantum Data Representation

A general objective optimization problem can be conceived as follows. Given a discrepancy between computational and target models, we aim to find an input configuration described by N numerical variables that best diminishes said discrepancy. The main components to solve the problem are as follows: the input data characterizing the structure and values of the parameters of an algorithm together. In the classical realm, the data encoding into the physics of conventional or neuromorphic hardware typically distinguishes these components [14]. In this context, the inclusion of input data in the context of quantum machine learning is mostly dissociated into two separate levels. First, we consider how classical data can be represented in a quantum fashion. Then, we delve into the question of how quantum data can be implemented on universal hardware. In regard to the latter, we summarize the current open question of the universal quantum representation of quantum input data. According to the classical-quantum association of algorithms, these two problems are highly correlated with distinctive associated issues [15]. In the remainder of this section, we shed light on the potential and the open problem of both quantum data representations.

3.2 Quantum Feature Spaces

At the heart of any kernel method lies its feature space representation. In quantum machine learning, kernel methods, including support vector machines and quantum neural networks, are, therefore, all realized in the quantum feature space. However, constructing the quantum feature space is non-trivial. The quantum feature space is not in general equal to the Hilbert space of quantum states but is only isometrically embedded. Furthermore, such an embedding is non-unique, and the quality of the embedding varies with different physical systems and experimental configurations [16]. Examining these practical issues of feature space realizations provides interesting insights into the dynamics of quantum systems used in kernel methods, in particular, the quantum machine. As the quantum feature space is merely isometrically embedded in a larger Hilbert space, to actually use it as a feature space, we need to be able to design a measurement operator inside the original Hilbert space. Since a quantum system naturally has a large Hilbert space and the features live in an isometrically embedded Hilbert space, representing the features of a classical vector in the quantum feature space also requires the classical vector to be enlarged [17]. The marginalization of the quantum feature space onto the original quantum system is indeed a key issue in the use of abstract quantum machine learning models, in particular, quantum kernel methods, and these models are restricted to work on small datasets due to current technology.

3.3 Quantum Neural Networks

We begin with the quantum version of a classical neural network: an input layer of N qubits, one or more hidden layers with an arbitrary number of nodes, and an output layer also with M qubits. The input layer is required to be a quantum-coherent superposition of 2^N valid input examples. Each of these is encoded by supplying a power-of-2 basis state for the input qubits. Likewise, the hidden and output layers are also required to be independent quantum-coherent superpositions. A quantum demonstration layer may be attached to any of the hidden or output layers [18]. Each of its H differently oriented half-spaces in N-dimensional space encodes one training point. Taking its inner product with the quantum-coherent output of one of those layers gives a quantum likeness that reflects the quality of the quantum half-space. Such training half-space targets form the core of the quantum-supervised learning process. The key dynamic element of the quantum neural network described is its product of unitary operations that correlate input layer states to quantum superpositions within each layer. Let its continuous form depend on γ, a time-like evolution parameter, and let these operators act layer by layer. By means of the new quantum machine learning method, each hidden or output layer of the quantum neural network becomes a layer of quantum half-space demonstrations [19]. Because each quantum half-space can be trained in time O(N), the quantum neural network overall can be trained in time O(N + ni + no), where ni and no are the sizes of the largest input or output layer. Although the factor O(ni) may appear prohibitive, we do the necessary algebra to make that not so in the context of a quantum neural network. In the context of figuring out whether a lower time complexity of quantum neural networks leads to a lower query complexity in the IQ Boolean function, there is a classical result about decision trees [20]. It is not unreasonable to expect that a workaround of the general method also ought to apply well.

4 Optimization in Machine Learning

Optimization is one of the fundamental tasks in machine learning to find the optimal set of parameters of the machine learning model with training data, aiming at learning the patterns from the data. The optimization of the model parameters is usually performed using iterative methods to minimize the cost functions that provide the measures of the difference between the predictions of the model and the actual outputs. The quality of the trained machine learning models highly depends on the optimization method used. Irrespective of the success, as long as the machine learning model or quantum model is a function of the learning or quantum parameters for state preparation or unitary operations, the aforementioned quantum training algorithms can be directly employed for quantum machine learning [21]. However, the ultimate goal of quantum machine learning is to achieve any super-exponential speedup over its classical counterpart. We are particularly interested in performing quantum-enhanced optimization of machine learning models rather than using pre-quantum-trained classical models as the first step towards implementing quantum machine learning. Next, we investigate how to leverage the quantum advantage in the optimization of the most popular supervised and unsupervised machine learning models [22].

4.1 Gradient Descent and Its Variants

Let $h(\theta)$ be a real-valued function that we wish to minimize and denote θ as the optimization variable. We consider the iterative procedure of gradient descent.

$$\theta t + 1 = \theta t - \eta t \nabla \theta h(\theta t)$$

where $\nabla\theta h(\theta t)$ denotes the partial gradient with respect to θ evaluated at θt and $\eta t > 0$ is the learning rate. In the simplest case, the learning rate ηt is chosen to be a constant, $\eta t = \eta$, for all times t and choices of optimization variables θ. The sequence of optimization variables above is defined in terms of the first-order derivative $\nabla\theta h$: for each optimization variable θt, the partial gradient vector is known at that point θt. This intuition can be generalized to arbitrary optimization algorithms that maintain an iterate θt and, at each step, compute some $\nabla\theta h(\theta)$ as a function of the current optimization variable, which, in general, may be a complicated mathematical function of the function value $h(\theta t)$ [23].

4.2 Evolutionary Algorithms

In this type of algorithm, not just one individual is modified to create the new generation, but rather, a whole population of individuals is born using operations like mutation and crossover. The crossover is mainly used as a recombination operator, exchanging good information and creating some new points in the point cloud of good solutions found for the underlying problem; the mutation operator is used as a space explorer operator, exploring the unknown space to search in the regions of more promising solutions. Although evolutionary algorithms include random mechanisms, conceptually, they do not exhibit the same type of randomness as local search algorithms, and they show a higher tolerance with respect to variations in the problem definition [24]. The evolutionary algorithm starts with a population of candidate solutions. It creates new candidate solutions via a copying mechanism biased by a ranking similar to the one used by the heuristics, with some modest differences, called a fitness function, that gives weight to the ith individual in the existing population. Once this new population is created, a special type of mutation, known as crossover or recombination, is applied to obtain new candidate solutions. This result is a new generation that will replace the first one. The procedure then typically continues for a fixed or varying number of generations until a stopping criterion is met (usually a maximum number of evaluations or a minimum increase in the quality of the best-found solution) [25]. Commonly, the evolutionary algorithms have some additional features, like techniques to avoid premature convergence and elitist strategies.

4.3 Bayesian Optimization

Bayesian optimization is a method for finding the global minimum or maximum of a function. We employ Bayesian optimization to tune the hyperparameters in a variant of the quantum Boltzmann machine used to construct generative models. The purpose of this alternative generative model strategy is to uncover deeper structures that traditional binary neural networks have difficulty learning [22]. The method is configured as follows:

(i) We run quantum circuits to generate Pauli-Z operators that act on the qubits constituting the generative model.
(ii) The quantum circuits consist of one- and two-qubit gates. The quantum circuit architecture is determined during Bayesian optimization likewise the angles of the gates. The angles weigh the generator.
(iii) We simulate the circuits to generate assignments to the Pauli-Z operators.
(iv) We apply iteratively the matrices generated, acting on the qubits corresponding to the physical system qubits serving as data states, and store the probabilities of the states.

At the end of the optimization, the result is compared with a traditional constructive phase estimation. If the angle results of the Bayesian optimization are better, the difference in the probability of the resulting vector by using the angles suggested is computed. Successful exploitation of this quantum generative model construction can potentially lead to more accurate conclusions in future ambitious computational analyses. Suppose the differences in probabilities are recurrently and statistically significant in example tests. In that case, we infer that the sequence converges more quickly, which is an important finding that highlights the potential of optimal logistic solvers as a route to delivering computational workloads for quantum machine learning [9].

5 Quantum Optimization

In principle, optimization is a very broad field, and many problems studied under this label have nothing whatsoever to do with optimization in the usual sense. Specifically, it is usually the case that what is being optimized is not a cost function that needs to be minimized but merely an expectation value that, in some respect, encodes the interesting information of the underlying problem. This broader sense covers essentially all time-dependent problems, such as searching for quantum ground and excited states, determining energy levels from a finite number of energy measurements, or finding the optimum parameters for a given quantum algorithm [15]. In all these cases, the possibility to encode relevant information in expectation values lies at the very heart of quantum algorithms, providing an advantage over their classical counterparts. Due to this very wide applicability, not all of the problems in this section on optimization can be considered in the same detail. Quantum optimization remains a rapidly developing area with a large number of partially competing subfields and approaches. Unsurprisingly, it is also at the confluence of many quantum information perspectives, such as relativistic quantum information for quantum versus classical speedups, work-power optimization problems in thermodynamics, non-time-dependent problems well-suited for quantum simulated annealing, and adiabatic quantum computing, or hybrid variational quantum–classical methods where some parts of the problem are classically approximated, but the constraints of the solutions are strictly satisfied. These diverse settings, in principle, allow a plethora of different experimental implementations [26]. However, due to the highly challenging nature of quantum optimization problems, where both fast and accurate demarcations of the quantum on occasion prove a difficult task, theoretical results can differ from practical implementations by orders of magnitude.

5.1 Quantum Annealing

One of the first quantum enhancement methods in machine learning attempts to use quantum annealing to train restricted Boltzmann machines. Given the complexity of this quantum machine, limiting the training to small systems has had limited results in qubits of order 10. One limitation of this work is that despite the restricted Boltzmann machine not requiring quantum entanglement in the way classically useful data is represented, there is still a requirement for the creation and management of entanglement [27]. Future advances based on this approach could either work hard to create larger networks with the same quality as the already existing precursors or the vaporware of the quantum valley may provide the small bit of today to be a large problem in the post-error correction era, allowing this line to create quantum enhancements of classical machine learning methods. Run quantum algorithms on quantum annealers. Outperform other quantum annealers. The current generation has hundreds of qubits. Available for others to use and study. Classically hard to simulate. Yields samples whose quality is not easy to verify. Requires new techniques to benchmark their processing capability, such as quantum reinforcement learning [28]. In this work, we take advantage of the power of the quantum nature of annealers to process samples that achieve higher fidelities than their classical counterparts, opening new opportunities for quantum-enhanced machine learning.

5.2 Variational Quantum Eigensolvers

Variational Quantum Eigensolvers. Variational quantum Eigen solvers (VQE) provide approximate solutions to eigenvalue problems. A quantum algorithm employing the tensor-network state AKLT was introduced to estimate the energy eigenvalue for a hydrogen molecule. An alternative hybrid quantum–classical technique, based on the multilayer perceptron and simulating quantum subroutines, was proposed for determining the ground states of molecules. The VQE was also developed with the general quantum computer. It was shown that the VQE approach can be more advantageous than previous quantum algorithms utilized for quantum chemistry applications [16]. VQE is a powerful QML method used in several quantum algorithm scenarios besides quantum chemistry. It employs trial state preparation, decomposes Hamiltonians into a weighted combination of Pauli operators, and estimates their expected values, followed by classically updating parameters to minimize the energy function. Its optimized quantum circuits prepare the ground states of many-body systems. QAOA, QUBO, and QOP can also be regarded as VQE variants.

The k-independent set problem was solved using variational quantum Eigen solvers. The VQE was improved with quantum–classical or quantum-quantum hybrid strategies for solving macroscopic problems [29]. Recently, the VQE approach was widely implemented in superconducting, ion-trap, cavity, and photonic quantum processors.

6 Hybrid Quantum–Classical Optimization

One way to use NISQ-era quantum computers is to accelerate optimization calculations by employing quantum algorithms or incorporating quantum components. Here, we demonstrate that quantum speedups are indeed possible for this well-motivated, practically important case and show the potential of quantum optimizations for real machine-learning problems. To approach related problems, we further extend our results to envision a more general framework of quantum–classical optimization schemes. Combinatorial optimization problems often meet known quantum advantages as well as their relaxations [30]. Indeed, various proposed quantum optimization or sampling algorithms turn out to deliver plausible acceleration given the current limitations in the computational size accessible using noisy intermediate-scale quantum computers. In these hybrid schemes, classically advantageous quantum speedup results are harnessed in different manners that can mainly be classified into two approaches: locally arising non-convex optimization problems and universally representing the computational complexity of various combinatorial optimization problems. Classically, it is intractable to solve those non-convex problems efficiently due to the exponential dependence on the number of bits, using heuristics such as simulated annealing, genetic algorithms, and feature value optimization [31]. These problems are the heaviest burden of the training task, but the hybrid quantum–classical optimization mitigates them.

6.1 Quantum–Classical Variational Algorithms

Quantum variational algorithms are an important control resource for quantum machine learning. To employ a quantum variational algorithm as a building block within a hybrid quantum–classical optimization routine, the measurement operations that construct the objective function of the optimization problem must be compatible with the variational algorithm quantum processor. For instance, variational quantum Eigen solver algorithms employ quantum measurements as part of a quantum phase estimation algorithm, and quantum state tomography is often required to extract an estimate of the state generated by the quantum processor [17]. The controlled quantum generators of the quantum variational algorithm, both in the wave function and measurement parts, require full active coherent quantum computation functionality, and interfacing with classical machine learning algorithms and data is limited to inputting or retrieving classical data or training parameters solely at specific points in the variational algorithm. This restriction becomes a bottleneck due to the slowness of the control electronics relative to the speed of the quantum turbines [5].

More generally, quantum variational algorithms do not have an arbitrarily controlled quantum operation on n qubits that acts for a variable time. Therefore, for any variational k-qubit algorithm that incorporates this kind of controlled operation, we require m k-qubit quantum generators that have variable generators. These generators may be employed in a multi-level quantum processor, using the same order of coherence now since a single step running a quantum variational algorithm of k qubits, or directly with the 2^k qubits quantum processor [2]. The limitations on the classical data and measurement capabilities within variational algorithms are problematic since at least two of the main reasons that quantum algorithms can be useful for machine learning are that quantum computers can execute very fast and that coarse-grain output measurements are efficient. By introducing quantum circuit changes that absorb or are absorbed by these generators, and mapping them to faster quantum controls, we may overcome these limitations, using our existing quantum machine learning quantum generators with any of the built since its inception [15].

6.2 Quantum Approximate Optimization Algorithm (QAOA)

The Quantum Approximate Optimization Algorithm (QAOA) can also be described using the language of quantum circuits. For the ferromagnetic Ising model problem, the QAOA algorithm will use a quantum circuit to prepare the ground state of the optimization problem Hamiltonian. The parameterized family of quantum circuits used in the QAOA algorithm is structured in layers. Each layer includes two sections of quantum gates: the problem Hamiltonian time-evolution gates and the mixer time-evolution gates. The QAOA algorithm with a fixed number of layers can be viewed as a hybrid quantum–classical algorithm in which classical optimization techniques are used to optimize parameters [13]. The time-dependent problem Hamiltonian for the ferromagnetic Ising model can be written as $Hp(t) = (1 - t)tHpro + tZ$. Here $t \in [0, \pi]$. For the mixer part, an easy and useful choice is $Hm = X$ since applying X to each qubit separately will result in an equal superposition of all the configurations. To exemplify how QAOA works, we choose $L = 1$. This means that we will alternate the problem Hamiltonian evolution for one time step with the mixer evolution. The QAOA circuit will be $U = e^{\{i\ heta\}}X\ e^{\{-i\ heta\}}$. This circuit can be thought of as a decomposition of time-dependent evolution circuits corresponding to problem and mixer Hamiltonians: $U = e^{\{-i\beta(Z)(Z \otimes Z)\}}\ e^{\{i\alpha(X)\}}$ in which $\beta \in [0, \pi]$ and $\alpha \in [0, 2\pi]$. The total optimization will be over angles β and α [32]. The mutual dependence of p and the angles β, α leads to a hybrid quantum–classical algorithm.

7 Enhanced Optimization Techniques

Regardless of whether one is attempting to optimize an approximate quantum algorithm or one particular quantum algorithm, the act of doing so is generally one of the hardest parts of implementing an algorithm. Without optimization, an algorithm is just that: an agreement of gate operations to prepare an input into a desired output. However, we recognize a quantum computer as a fundamentally different sort of device; it excels at finding certain patterns among data, which means that it should also excel at implementing an algorithm to solve a problem given appropriately crafted data or a similar scheme [11]. Optimization in the quantum era should then be treated on an equal footing with implementing a quantum algorithm, and the demonstrated power of quantum algorithms should then carry over to the optimization stage. However, in general, optimization is a very hard problem, and oftentimes, the simple routines that are used are known to be shockingly good, given very little in the way of performance guarantees [20].

One class of optimization problems that a quantum computer should be good at solving is the class for which there is a quantum algorithm that outperforms the best-known classical algorithm. Given such an algorithm, one can equip a quantum computer with the algorithm for the cost function. There is no need to test guesses or local minima; a quantum computer can directly find the cost function minimum and tell its user the best it found. No classical algorithm can provide the quantum algorithm with guidance; the user asks the quantum computer for the best it can find, and all of the searching or testing has to be performed by the quantum computer [21]. Many of today's optimization algorithms are focused on the class of convex, differentiable functions. Though such algorithms are not as efficient as quantum algorithms for combinatorial optimization problems, they are useful for optimizing approximate quantum algorithms. There are many techniques from the convex optimization literature that can be borrowed to design effective algorithms, including first- and second-order methods, coordinate descent, sub gradient procedures, analytic center cutting-plane methods, and visual ballistic rules [5].

Many of the problems that we would like to solve involve discrete choices, so the continuous optimization of the bounds of the variational quantum eigensolver may not be fully satisfying to the practitioner. Discrete derivatives and tensor network relaxed variational optimization methods are widely accepted; tensor network approximations to high-dimensional functions can be of much lower complexity. For example, the derivation of nearly optimal measurements in machine learning serves as an effective way to encapsulate functions for production learning applications when traditional light cone expansion methods represent bounds that are too relaxed. With this in mind, it may be beneficial to relax the idea of discrete optimization over the parameters of the parametric circuit, performing the optimization procedure in a space that would not correspond to a proposed implementable quantum evolutionary strategy [30].

7.1 Quantum-Inspired Classical Algorithms

To go from a quantum algorithm for a quantum machine learning method to a quantum-inspired or quantum-assisted classical algorithm, you take ideas from the quantum part and introduce them into classical algorithms. The first type of quantum-inspired algorithm is to use complex number encoding, initializing the input of black box functions and allowing for gradient descent to be performed over a continuous field. Shallow networks of this type are efficiently trainable using good approximations to the full quantum gradients from standard backpropagation [31]. The second type of quantum-inspired algorithm uses a quantum walk-like scheme to simulate a quantum subroutine. This technique is particularly useful if the original quantum subroutine can be decomposed into simple quantum circuits, such as small-depth quantum circuits or circuits from the quantum approximate optimization algorithm. Furthermore, if additional results from separate quantum routines have to be combined or if a faster classical routine sometimes replaces the quantum subroutine due to noise, quantum-inspired or quantum-assisted algorithms can significantly outperform quantum algorithms [20].

7.2 Adiabatic Quantum Computing

The classical adiabatic theorem suggests that a slow, continuous change in a system's Hamiltonian permits the system to remain in its ground state at all times. Quantum adiabatic computing is an alternative to the quantum circuit model for quantum information processing. Its performance and operating principles are already understood. In contrast to the quantum circuit model, its concept is related to adiabatic quantum computing: given a problem represented by an initial Hamiltonian, the quantum annealer time-evolves a given initial quantum state into the ground state of the problem Hamiltonian that represents the problem of interest [17]. The key advantage of the adiabatic quantum algorithm is a quantum speed-up with a low design overhead. This algorithm deals with adiabatic state transfer between two desired states and does not consider the final system to be in the exact ground state, opening the possibility of using a small annealing time.

The quantum annealing process, or adiabatic quantum computation, is another alternative formulation of quantum computation that may help to overcome some issues of the gate model. Its proposal is more similar to classical simulated annealing. The quantum adiabatic theorem states that an initially prepared quantum state remains in its instantaneous eigenstate if the Hamiltonian evolves smoothly. After binding the transformation problem into the ground state of the new Hamiltonian, quantum annealing evolved from adiabatic quantum computing. This process evolves general problems and provides better insight into quantum states [12]. Evolution based on the adiabatic theorem has drawn wider interest lately, considering scaling, straightforward implementation, and, most importantly, experimental demonstration in a variety of physical systems, including multi-electron spin systems and ion systems.

8 Applications of Quantum Machine Learning

In this chapter, we have mainly been concerned with the QML tasks: the derivative task of existing cost functions and surrogate quantum models for variational inference. The QML gives only error-related information during the QST processes, and thereby, the use of probability models is essential in the QML tasks. Using these probability models, various fields can be applied to the QML algorithms. These QML algorithms estimate the data's hidden structure even if the data does not have the structure. Furthermore, using QML tasks helps to reduce the energy of a quantum Hamiltonian and to classify datasets at a lower cost [5]. All the presented QML tasks are derived on the basis of supervised learning algorithms, but they may be applied to unsupervised learning algorithms. The QML tasks to promote quantum unsupervised machine learning are our ongoing research. In this chapter, QML tasks dealing with the probability at the binary and multi-classes were presented. Experimental and numerical results reveal that these QML tasks can be used to enhance optimization performance in the QML processes for the existing quantum models and variational inference models. The proposed QML tasks are used for the variational inference model for the classification of datasets. The proposed tasks may be used to handle supervised machine-learning problems. Not only supervised learning but also unsupervised learning algorithms may be used to design the improved QML algorithms [30].

8.1 Quantum Chemistry and Material Science

During the last several years, the quantum computing community has witnessed the growth and the increasing resource development of quantum hardware, as well as breakthroughs in quantum algorithms. Emergence in the development of quantum algorithms and quantum software design for practical applications is observed. An increase in interest in algorithms and software supporting near-term quantum computing creates important goals for quantum algorithm and software development. In the near term, quantum algorithms require a hybrid quantum–classical approach, including enhanced optimization, which can be further combined with other classical algorithms for the complete solution of application-specific tasks [15].

Quantum mechanics is the foundation of quantum computing, and it is the basis for opening up a new avenue for optimization. Building on quantum mechanics, many quantum algorithms have been proposed to solve optimization problems, such as the first quantum algorithm for nonlinear programming, the quantum algorithm to solve the NP-hard problem, the Traveling Salesman problem on a quantum computer, quantum algorithms for constraint satisfaction problems, and quantum algorithms for combinatorial optimization. These algorithms pave the road to solving several optimization problems on a large scale [13]. Although all these advancements are welcome, this is not enough because quantum-enhanced optimization, as well as quantum algorithms, have the potential to provide speedups over known classical algorithms for other important classes of optimization problems.

8.2 Finance and Optimization Problems

Many strategic decision-making problems in finance and business are NP-hard, and therefore, classical optimization problems are often involved. For example, among the most challenging NP-hard problems are percentage call-profit maximization and non-exaggerated advertising problems, which are good models for option pricing and optimal liquidity management. However, naive application of quantum algorithms for these problems to either the underlying optimization problems or the associated learning problems would not provide an exponential advantage over the best-known classical algorithms, which use, for example, dynamic programming or efficient combinatorial algorithms of linear programming [10]. In this paper, we develop a framework for quantum machine learning that can be used with efficient quantum or even classical optimization. A quantum algorithm can achieve exponential speedups and efficient solutions for pure linear programming oracles and, therefore, is exponentially faster than any known classical algorithm for nonlinear programming with poly-level linear programming, which encompasses the meaning of machines. Using reduction from continuous optimization, we make quantum solutions to discrete and continuous optimization with a sufficiently high condition number PSPACE-completable [16]. We develop such a class that covers many important problems and is responsible for high-quality reflections on the optimized schemes. The benchmark figures for this framework are given by the Bell Hamiltonian and convergence bounds, which are independent of machine learning programs. We will also discuss an alternative strategy allowing exponential speedups in some cases where such speedups are not feasible with polynomial or even double power rounding—a new scheme that uses approximators to improve upon interpolation bounds [23].

9 Challenges and Future Directions

Several challenges are still to be solved on the road toward more widespread applicability of both quantum optimizations and quantum machine learning. While current quantum processors feature a limited number of qubits with a high error rate in the NISQ era, useful quantum optimization heuristic approaches, like solving even the QUBO with superconductor processors and embedding QUBO on the annealer graph directly, are equipped with reduced accuracy. Exploiting error correction and inferring the deep relations between quantum optimization and condensed matter and many-body physics could provide an impetus for increasing the future application range.

9.1 The Potential of Quantum-Inspired Algorithms

Firstly, larger and better commercial quantum annealing processors need to be established to provide a suitable range of workable parameters, with access to the quantum computing service provided by any user at a reasonable price. Secondly, efficient and scalable embedding methods are needed, tailored for encapsulating problems with commercial quantum annealers. Thirdly, methods expanding the limit of many optimization problems, for example, QUBO defined on broad graphs other than the Chimera graph, Hamiltonians with multiple non-commuting terms, and QUBO problems with many more variables, should be developed [3]. We also need to address the potential of quantum-inspired algorithms that can be run on conventional computers for the sake of large-scale optimizations. In contrast to quantum optimization heuristic methods, all these quantum-inspired algorithms are generally easier to develop, require less control, and are more flexible to apply. The research and development of quantum-inspired algorithms is thus necessary in the NISQ era. It is logical to do so, considering that even if all the quantum software and hardware could advance perfectly, these quantum algorithms might be competitive in practice. For practical applications of quantum machine learning, we propose an ongoing investigation into several technical issues, such as expanding model expressivity, developing scalable quantum algorithms, and leveraging hardware performance [6]. The quantum neural network in the form of a quantum Hopfield model currently might challenge the feed-forward neural network but struggles to improve expressiveness. The next challenges include leveraging the true advantage of quantum machine learning, such as developing algorithms that run on NISQ hardware and are robust to noise and implementing quantum machine learning in practical problems to benefit from quantum speedup. Additionally, by investigating efficient compilation, combination, and improvement of quantum–classical algorithms, we suggest promoting quantum machine learning applications and exploring the concrete advantages of quantum machine learning in various aspects [22].

9.2 Error Mitigation and Noise in Quantum Computing

Error mitigation and noise in quantum computing have a number of meanings and potentially several levels of research. As of the recent situation with relatively small-scale noisy intermediate-scale quantum devices, quantum error correction is still not feasible with many qubits. However, the breakthroughs in the past decades have initiated a large number of potential elaborations, such as incipient quantum error correction codes, different ways of error handling, or leveraging noise characteristics. An artificially general concept of error mitigation and noise has been extensively studied for variational hybrid quantum AI and optimization heuristics that have become one of the most notable directions of the quantum era [6]. We start by providing a broad look at these important subjects and demonstrating some recent research results. We hope that the mentioned variational hybrid quantum algorithms will also turn out to be useful for solving optimization tasks. However, as it became clear already in quantum chemistry and optimization, device noise reduces the benefits of quantum speedup. More specifically, the number of operational qubits for the so-called fidelity of intermediate-scale quantum systems is on the order of 50–100 physical qubits [22]. Besides, the real-time communication of qubit states from the main hardware through ion traps or superconducting circuits is usually expensive. This has led to a situation where we are restricted to a technology of small gates and limited-depth quantum circuits. One of the general conclusions is that leading to the coherent time limit, the practical quantum algorithms we can efficiently run on existing quantum hardware are hybrid classical-quantum algorithms implemented on computers [23]. One cheaper alternative to full quantum hardware error correction is to provide more accurate quantum results based on today's short quantum circuit simulations.

10 Conclusion

A major practical limitation of the variational quantum eigensolver (VQE) and other variational algorithms is that they are well-suited for near-term quantum computing. However, as the number of parameters that parametrize the quantum state increases, the computational resources grow exponentially with the number of qubits. The compatibility of an algorithm with near-term quantum computers is critical, as it requires fewer qubits, fewer two-qubit gates, and a reduced circuit depth. In addition, the usefulness of VQE and these other variational algorithms could remain limited for much longer until the resources required to run large-parameter circuits are available. To some extent, this limitation can be overcome by cutting down the quantum circuit depth. However, this comes with a trade-off, such as reducing the number of parameters, sacrificing the power of the quantum circuit in handling the optimization, and approximating more complicated quantum states. Even ignoring the exponential cost of simulation and just focusing on the quantum optimization as a cost, we are currently capped at only a few hundred heuristic iterations with a few dozen qubits, making the expensive classical optimization the bottleneck in training parameters for near-term quantum circuits. For quantum algorithm development, and more generally for all quantum computing, we need to be thoughtful and frugal about the quantum resources. The quantum resource constraints are especially salient in VQE and similar variational algorithms, as they require a quantum circuit for the computation of quantum expectation values related to quantum observables and quantum gradients, and a classical computer for the optimization that involves these quantum expectations. Here, in this excited state of QML, the field has only started to appreciate this later fact.

References

1. Adu Ansere, J., et al.: Energy-efficient optimization for mobile edge computing with quantum machine learning. IEEE Wirel. Commun. Lett. **13**(3) (2024). https://doi.org/10.1109/LWC.2023.3338571
2. Ardeshir-Larijani, E., Fatmehsari, M.M.N.: Hybrid classical-quantum transfer learning for text classification. Quantum Mach. Intell. **6**(1) (2024). https://doi.org/10.1007/s42484-024-00147-2
3. Baker, J.S., et al.: Parallel hybrid quantum-classical machine learning for kernelized time-series classification. Quantum Mach. Intell. **6**(1) (2024). https://doi.org/10.1007/s42484-024-00149-0
4. Bakshi, A., Tang, E.: An improved classical singular value transformation for quantum machine learning. In: Proceedings of the Annual ACM-SIAM Symposium on Discrete Algorithms, vol. 2024 (2024). https://doi.org/10.1137/1.9781611977912.86
5. D'Elia, A., et al.: Characterization of a Transmon qubit in a 3D cavity for quantum machine learning and photon counting. Appl. Sci. (Switzerland) **14**(4), 1478 (2024). https://doi.org/10.3390/app14041478
6. Fadol, A., et al.: Application of quantum machine learning in a higgs physics study at the CEPC. Int J Modern Phys A **39**(1) (2024). https://doi.org/10.1142/S0217751X24500076
7. Gibbs, J., et al.: Dynamical simulation via quantum machine learning with provable generalization. Phys. Rev. Res. **6**(1) (2024). https://doi.org/10.1103/PhysRevResearch.6.013241
8. Gil-Fuster, E., et al.: Understanding quantum machine learning also requires rethinking generalization. Nat. Commun. **15**(1) (2024). https://doi.org/10.1038/s41467-024-45882-z
9. Hou, D.: Music photonic signal analysis based health monitoring system using classification by quantum machine learning techniques. Opt. Quantum Electron. **56**(3) (2024). https://doi.org/10.1007/s11082-023-05960-w
10. Indumathi, N., et al.: Telemedicine enhanced with quantum machine learning for secure and real-time medical diagnosis. Int. J. Intell. Syst. Appl. Eng. **12**(15s) (2024)
11. Lee, D., et al.: Variational quantum state discriminator for supervised machine learning. Quantum Sci. Technol. **1**, 15017 (2024). https://doi.org/10.1088/2058-9565/ad0a05
12. Liu, Z., et al.: E-healthcare application cyber security analysis using quantum machine learning in malicious user detection. Opt. Quantum Electron. **56**(3) (2024). https://doi.org/10.1007/s11082-023-05854-x
13. Mori, Y., et al.: Expressive quantum supervised machine learning using Kerr-nonlinear parametric oscillators. Quantum Mach. Intell. **6**(1) (2024). https://doi.org/10.1007/s42484-024-00152-5
14. Munshi, M., et al.: Quantum machine learning-based framework to detect heart failures in healthcare 4.0. Softw.—Pract. Exp. **54**(2) (2024). https://doi.org/10.1002/spe.3264
15. Nammouchi, A., et al.: Quantum machine learning in climate change and sustainability: a short review. Proc. AAAI Sympos. Ser. **2**(1) (2024). https://doi.org/10.1609/aaaiss.v2i1.27657
16. Orazi, F., et al.: Hybrid quantum technologies for quantum support vector machines. Information (Switzerland) **15**(2) (2024). https://doi.org/10.3390/info15020072
17. Park, S., Simeone, O.: Quantum conformal prediction for reliable uncertainty quantification in quantum machine learning. IEEE Trans. Quantum Eng. **5** (2024). https://doi.org/10.1109/TQE.2023.3333224
18. Peral-García, D., et al.: Systematic literature review: quantum machine learning and its applications. Comput. Sci. Rev. **51** (2024). https://doi.org/10.1016/j.cosrev.2024.100619

19. Rao, G.V.E., et al.: Hybrid framework for respiratory lung diseases detection based on classical CNN and quantum classifiers from chest X-rays. Biomed. Signal Proces. Control **88** (2024). https://doi.org/10.1016/j.bspc.2023.105567
20. Satpathy, S.K., et al.: Analysis of quantum machine learning algorithms in noisy channels for classification tasks in the IoT extreme environment. IEEE Internet Things J. **11**(3) (2024). https://doi.org/10.1109/JIOT.2023.3300577
21. Schmid, L., et al.: A general learning scheme for classical and quantum ising machines. Sci. Post Phys. Core **7**(1) (2024). https://doi.org/10.21468/SciPostPhysCore.7.1.013
22. Senokosov, A., et al.: Quantum machine learning for image classification. Mach. Learn. Sci. Technol. **5**(1) (2024). https://doi.org/10.1088/2632-2153/ad2aef
23. Shafik, W.: Deep learning impacts in the field of artificial intelligence. In: Deep Learning Concepts in Operations Research pp. 9–26. Auerbach Publications (2024). https://doi.org/10.1201/9781003433309-2
24. Shafik, W.: Generative AI for social good and sustainable development. In: Generative AI: Current Trends and Applications, pp. 185–217. Springer (2024)
25. Shafik, W.: Machine learning techniques for multicriteria decision-making, pp. 165–194. CRC Press, Multi-Criteria Decision-Making and Optimum Design with Machine Learning (2025)
26. Shafik, W.: Quantum computing and generative adversarial networks (GANs): ethics, privacy, and security. In: Quantum AI and Its Applications in Blockchain Technology, pp. 111–56. IGI Global Scientific Publishing (2025)
27. Tezuka, H., et al.: Generative model for learning quantum ensemble with optimal transport loss. Quantum Mach. Intell. **6**(1) (2024). https://doi.org/10.1007/s42484-024-00142-7
28. Ullah, U., Garcia-Zapirain, B.: Quantum machine learning revolution in healthcare: a systematic review of emerging perspectives and applications. IEEE Access **12** (2024). https://doi.org/10.1109/ACCESS.2024.3353461
29. Vaz, E.: Quantum machine learning in spatial analysis: a paradigm shift in resource allocation and environmental modeling. Lett. Spat. Resour. Sci. **17**(1) (2024). https://doi.org/10.1007/s12076-024-00374-y
30. Verma, T., et al.: A review on quantum machine learning. Lect. Notes Electr. Eng. **1096** (2024). https://doi.org/10.1007/978-981-99-7137-4_39
31. Zhang, H., et al.: A light-weight quantum self-attention model for classical data classification. Appl. Intell. **54**(4) (2024). https://doi.org/10.1007/s10489-024-05337-w
32. Zhou, X., et al.: Quantum kernel estimation-based quantum support vector regression. Quantum Inform. Proces. **23**(1) (2024). https://doi.org/10.1007/s11128-023-04231-7

Industrial Automation and Challenges

Amit Kumar Jain and Pooja Vijay

Abstract Recent advent of Quantum Machine Learning, which is a quintessence of quantum computing and machine learning, has helped several practical applications. And, it learns things expected to transform industrial automation. The Currently, modern complex problems cannot be handled easily by modern industries using traditional machine learning models and computational approaches since modern industries involve extensive scales and complicated data structures. Quantum Machine Learning While a quantum computing concept of superposition and entanglement is leveraged to remove traditional computing constraints, there is a benefit to (QLM). It yields at higher, faster processing speeds and delivers more processing capability. The paper examines QML integration for industrial automation as well as the applications. Little emphasis was placed on analysing contemporary difficulties and projected trends, while only briefly considering advantages and an offer in conclusion. Improvements in productivity are essential and industrial automation provides these improvements. This helps with achieving efficiency as well as scale within the manufacturing sectors along. with energy production logistics operations and supply chains. In the past the classical machine learning coupled with artificial automation systems. achieving operation optimization as well as data-driven decision-making. With these, their limitations become more apparent. This is due to complex real-time data processing as well as dynamic environments and multiple interconnecting systems. The data processing solution implemented by QML addresses these operational limitations by completing tests quickly and precisely compared to classical approaches. Finally, QML represents a revolutionary solution for the problems arranged by industrial automation. The great thing about quantum computing is that it utilizes QML to enable optimization, productivity, and contribute to sustainability. While some challenges remain, the constant progress in quantum technologies has strong potential to realize the full potential of a QML to enable the automation of the future.

A. K. Jain (✉) · P. Vijay
Poornima University, Jaipur, Rajasthan, India
e-mail: Jain.2102@gmail.com

© The Author(s), under exclusive license to Springer Nature Switzerland AG 2025

A. Ghosh et al. (eds.), *Quantum Machine Learning in Industrial Automation*,
Information Systems Engineering and Management 65,
https://doi.org/10.1007/978-3-031-99786-0_3

Keywords Quantum Machine Learning (QML) · Predictive maintenance · Quantum Approximate Optimization Algorithm (QAOA) · Optimization problems · Hybrid quantum–classical systems · Real-time decision-making · Dynamic environments

1 Introduction to Industrial Automation

1.1 Definition and Concept

Advanced technologies alongside control systems and robotics and computer programs work together through automation systems to carry out industrial operations while requiring little or no human interaction. A system of programmable components takes over repetitive labour-intensive work that allows for better efficiency alongside precise performance and ensured safety. The contemporary basis of industrial operation relies on automation for standard manufacturing procedures intended to minimize human mistakes and boost operational efficiency with enhanced productivity outcomes (Fig. 1). Advanced industrial systems now consist of multiple technological components including Artificial Intelligence (AI) Internet of Things (IoT) and Quantum Computing [1].

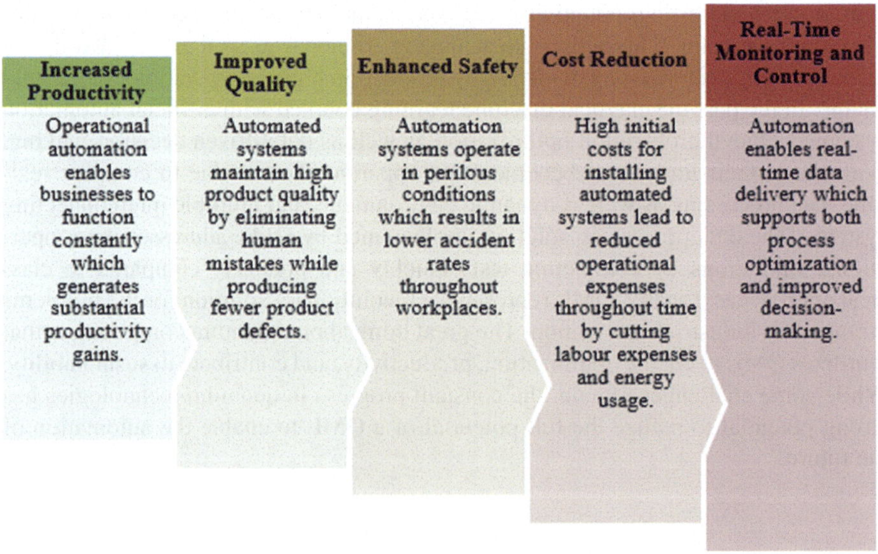

Fig. 1 Objectives of industrial automation

1.2 Evolution of Industrial Automation

- **Manual Processes**: The majority of industrial activities previously needed manual employees who repeatedly performed tedious tasks because automation systems were unknown at that time. Inactive processes reacted slowly and negatively to human mistakes and released unapproved results to misguide operational performance.
- **Mechanization**: During the Industrial Revolution mechanized tools received power through steam water and electricity. Bringing down human labour dependency occurred at the same time as operators needed to stay engaged during operations.
- **First Wave of Automation (20th Century)**: Programmable Logic Controllers (PLCs): Originally designed in the 1960s the Programmable Logic Controller (PLC) transformed automation through its capabilities to run predefined sets of programming code.
- **Smart Automation (21st Century)**: Through AI and IoT and Big Data integration machines can now learn and adapt while making decisions. Industrial facilities today function as "smart factories" that manage autonomous tasks while continuously improving their performance [2].
- **Future Trends**: Industrial automation will receive more transformative power from quantum computing and 5G networks and advanced robotics.

1.3 Types of Industrial Automation

- **Fixed Automation**: High-volume production requires this system yet it allows only minimal customization options. Examples: Assembly lines in automotive manufacturing.
- **Programmable Automation**: Suitable for batch production. Production systems can receive software modifications to perform different operational assignments. Examples: Textile machinery, batch chemical processing [3].
- **Flexible Automation**: Evolutionary guide aims to facilitate both product changeovers between different tasks and manufacturing procedures adaptation according to customer needs. Examples: Robot-assisted manufacturing operates for products of multiple designs.
- **Integrated Automation**: Technology systems run as fully automated operations with minimum requirements for human intervention while machines act as decision-makers through their communication networks. Examples: Smart factories operate with the integration of IoT and AI technology.

Components of Industrial Automation is showed in Fig. 2.

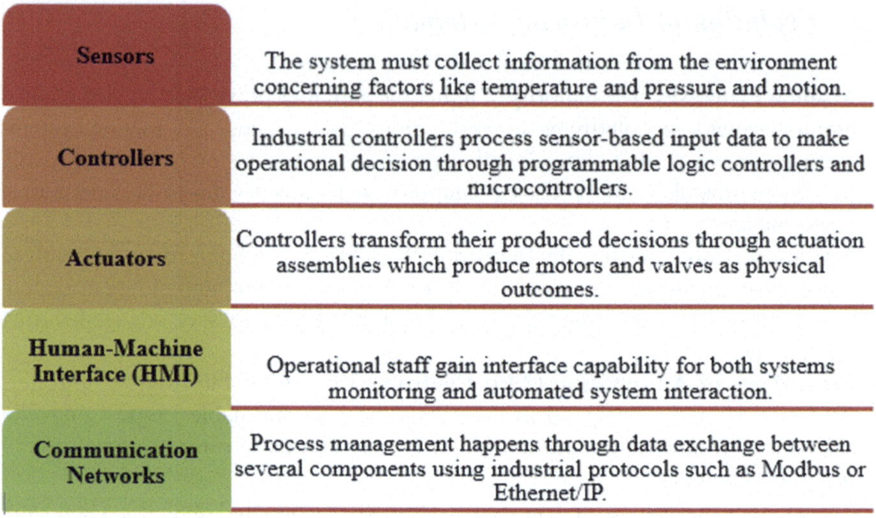

Fig. 2 Components of industrial automation

1.4 *Importance of Automation in Industries*

- **Manufacturing**: Modern production lines incorporate robotic welding along with self-operated quality testing systems.
- **Energy**: The fusion of Automated power grids with energy storage systems and renewable energy management capabilities delivers a new dimension to the power sector.
- **Healthcare**: Continuous automated Pharmaceutical production exists alongside robotic surgical equipment and diagnostic testing systems.
- **Logistics**: The industry controls its operations through automated warehousing and conveyor systems and drone deliveries.
- **Construction**: Autonomous vehicles for excavation, 3D printing of structures, and smart building technologies.

1.5 *Automation in the Context of Quantum Machine Learning*

Quantum technologies represent tomorrow's automated systems for data-oriented industries. Challenges of Industrial Automation is showed in Fig. 3. Quantum Machine Learning (QML) offers solutions for complex problems that classical systems struggle to handle, such as:

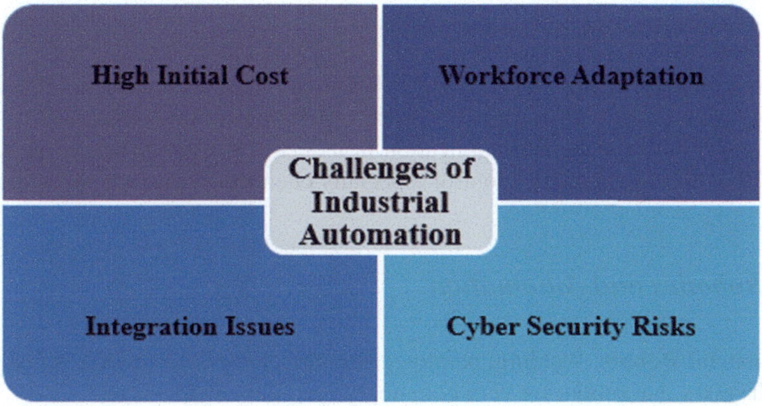

Fig. 3 Challenges of industrial automation

- **Real-time optimization**: Performing real-time scheduling of tasks alongside dynamic resource distribution.
- **Predictive maintenance**: Systems are able to identify equipment malfunctions before operational breakdowns occur.
- **Enhanced decision-making**: The analysis of large datasets makes progress at increased speed together with better accuracy [4].

2 Technologies in Industrial Automation

Below, we analyse how traditional combined with present-day technologies support industrial automation in its mission to optimize operations while improving productivity rates and minimizing team member involvement. Continuous developments in these technologies allow industries to fulfil increasing requirements for enhanced precision and improved performance and adaptable solutions.

2.1 Traditional Automation Technologies

- **Programmable Logic Controllers (PLCs)**: Interactive industrial controllers make up Programmable Logic Controllers which handle machinery automation. The hard environment proves to be no match for PLCs because these controllers efficiently process real-time actions while providing dependable operation.
- **Supervisory Control and Data Acquisition (SCADA)**: SCADA systems enable users to monitor industrial processes alongside controlling those processes and analysing their operational data. Treatment of operational data received from

sensors and machines combined with interface-driven visual presentations for operators [5].
- **Distributed Control Systems (DCS)**: Between multiple controllers DCS establishes centralized control through process communication followed by coordination. These systems find application in industrial sectors that need persistent real-time management and control especially chemical plants and oil refineries.

2.2 Robotics and Automation

- **Industrial Robots**: Welding, painting, assembly, material handling is just a few of the tasks that robots are very commonly used in manufacturing.
- **Autonomous Mobile Robots (AMRs)**: Warehouses and logistics use self-navigating robots, known as AMRs, to transport goods. Taking advantage of sensors, cameras and AI to recognize obstacles and optimize routes [6].
- **Automated Guided Vehicles (AGVs)**: AGVs, however, are unlike AMRs: AGVs follow fixed routes, set by magnetic strips or tracks. In manufacturing and warehouses, it is used in material handling and inventory management.

2.3 Advanced Sensors and Actuators

- **Sensors**: The backbone of automation is sensors that collect real time data about environment and machines.
- **Role**: Prove to be enablers of predictive maintenance, improve precision, and help achieve process reliability.
- **Actuators**: Take the controller's command and make them into physical actions. Electric motors, hydraulic cylinders, and pneumatic systems are all types.

2.4 Artificial Intelligence and Machine Learning

- **Role in Automation**: Detailed programming is not necessary because AI allows systems to learn from examples, evolve in the face of changing conditions, and make decisions, all without requiring explicit subsequent programming. The use of ML models can predict equipment failures, the optimisation of production schedules and better-quality assurance.

2.5 Internet of Things (IoT) in Industrial Automation

- **Overview**: IoT unites machines, sensors, and systems so that data are exchanged and communicated.
- **Industrial IoT (IIoT)**: It deals with smart factories where the devices interact as a whole. Sensors data is then transmitted to cloud platforms for analysis and decision making.
- **Applications**: Real-time monitoring of machine performance, Energy management in factories, Supply chain optimization.
- **Challenges**: Security risks due to interconnected devices, how to do efficient management in large volumes of data [7].

2.6 Edge Computing and Cloud Computing

- **Edge Computing**: Processes data locally, near the source, for real-time decision-making. Reduces latency and dependence on cloud connectivity.
- **Cloud Computing**: It keeps the data in the centralized servers. Provides for scalable analysis as well as remote monitoring.
- **Combined Usage**: While it is true that cloud computing handles longer term analytical quantities, edge computing solves for the quick, impudent agenda.

2.7 Digital Twins

- **Definition**: A digital twin can be defined as a virtual replica of a physical asset or process.
- **Role in Automation**: It tests and optimizes the real-world scenarios. They are machines which monitor the performance of machines in real time, making comparison of physical operations with virtual models.
- **Applications**: Simulating wear and tear to do predictive maintenance. Increasing production processes without disturbing the operations in place.

2.8 Cyber-Physical Systems (CPS)

- **Overview**: CPS as integrating computation with the physical processes and networking.
- **Role in Automation**: They enable intelligent interconnected system which are self-diagnosed, adaptive, and optimize operations.
- **Applications**: Autonomous vehicles, Smart grids for energy distribution, Smart factories and the associated advanced manufacturing systems.

2.9 Quantum Technologies in Automation

- **Quantum Computing**: The classical systems solve the optimization problems exponentially faster. Applications: Scheduling, supply chain optimization, and predictive analytics.
- **Quantum Machine Learning**: Flickr Dump is used to enhance the AI capabilities for anomaly detection, adaptive systems, and process control.
- **Challenges**: Limited hardware availability and high costs [8].

2.10 Cybersecurity in Automation

- **Context**: It is important to consider data integrity together with cyberattack protection as there are connected devices and systems.
- **Technologies Used**: Firewalls, encryption, and AI-based intrusion detection systems. Secure communication protocols for IoT devices.

3 The Role of Machine Learning in Automation

Machine Learning (ML) has an important role in industrial automation by allowing the systems to analyse data, learn from it and make intelligent decisions without explicit programming. Using ML, automated systems can make them more efficient, minimize downtime, optimize the processes and change them on the fly.

3.1 Introduction to Machine Learning in Automation

- **Machine Learning**: Artificial Intelligence (AI) is a wide field including the subset called ML, this allows computer and machines to learn patterns from data and become better as they receive more data. Whereas in traditional programming rules are explicitly defined, ML algorithms simply identify relationships and dependencies from the data being processed.
- **Importance of ML for Automation**: Traditional automation includes systems that obey predetermined rules and find it difficult to deal with variation or unknowns. What ML allows systems to do is to adapt, predict, or make decisions based on data.

3.2 Applications of Machine Learning in Automation

3.2.1 Predictive Maintenance

- **Goal**: Machine failure prediction before the real occurrence, to save downtime and maintenance costs.
- **How It Works**: The data collected is regarding machine performance (e.g. vibration, temperature, pressure). The ML models analyse historical and real time data to find out the pattern which increases the wear and tear. It generates alerts for preventive cases prior to failure.
- **Benefits**: Reduces unplanned downtime, Extends the lifespan of equipment, Optimizes maintenance schedules [9].

3.2.2 Quality Control and Defect Detection

- **Objective**: Guarantee the product quality that is consistent throughout the production process.
- **How It Works**: Images of products on the assembly line are processed by vision systems with ML algorithms. Defects, such as cracks, discoloration, or misalignments of an object are identified by the system.
- **Benefits**: Improves accuracy over manual inspections, it can detect even subtle defects which are hard to detect without the use of naked eyes.

3.2.3 Process Optimization

- **Goal**: Keep optimizing industrial processes as much as possible in order to achieve the greatest efficiency.
- **How It Works**: There are different sources from which the data is fed into the ML model, namely sensors, production logs, energy usage. Algorithms determines the bottlenecks, areas of inefficiencies, or places where improvement sourcing can be done.
- **Examples**: Changing the conveyor speed to reduce the bottleneck. Will serve the purpose of real time workload balancing across machines.

3.2.4 Anomaly Detection

- It aims to find unusual patterns or behaviour in data which might imply some problem.
- **How It Works**: ML algorithms know normal patterns and these are then different from others. Sensor failures, security breaches or process malfunctions may be indicated by anomalies.

- **Examples**: Identifying unusual energy consumption patterns. Networked systems' cybersecurity threats detection.

3.2.5 Robotics and Autonomous Systems

- **Motivation**: To learn about tasks and to operate in dynamic environments, robots need to be able to keep track of what they are currently doing.
- **How It Works**: Robotic systems use ML algorithms to recognize objects, navigate and move in a space, and interact with the surroundings.
- **Examples**: Collaborative robots (cobots) employing ML can be instructed by the operators and modify their actions on changes of tasks or environments.
- **Benefits**: Increases flexibility in manufacturing. It provides the robots to be able to socio-dynamically detect and avoid collisions with humans.

3.2.6 Energy Management

- Specifically, the objective is to discover ways to minimize energy use in industrial processes for the purpose of saving money and reducing environmental impact.
- **How It Works**: Energy consumption patterns are analysed and inefficiencies are found by ML. Based on demand, they predict they can change what to optimize power usage.
- **Examples**: Shutting off idle equipment when production is slimming down on. Formulating such energy-intensive tasks as scheduling tasks.

3.2.7 Supply Chain and Inventory Management

- **Objective**: reduce complexity in supply chain operations and to achieve higher inventory levels.
- **How It Works**: Demand prediction and shipment tracking, inventory replenishment schedules, not to mention ML models, are used to predict demand. This increases the visibility into supply chain process.
- **Benefits**: Reduces overstocking and understocking. Enhances just-in-time (JIT) manufacturing efficiency [10].

3.3 *Types of Machine Learning Used in Automation*

3.3.1 Supervised Learning

- **How It Works**: Labelled data (input–output pairs) is a source of training for models. The system realizes how inputs are mapped into specific outputs.

- **Applications in Automation**: Predictive maintenance (failure prediction). Quality inspection (defect classification).

3.3.2 Unsupervised Learning

- **How It Works**: Unlabelled data is used to learn patterns and learn structure from.
- **Applications in Automation**: Anomaly detection, Similar product designs or operational patterns related to clustering.

3.3.3 Reinforcement Learning

- **How It Works**: In order to maximize the rewards, the environment agents learn the actions to take in order to achieve it. Ideal for dynamic and complex environments.
- **Applications in Automation**: Robotics: Allows robots to avoid obstacles, or nurture products from a base, among other things. In manufacturing, it refers to: Adapting to changing conditions.

3.3.4 Deep Learning

- **How It Works**: It uses neural networks with several layers to process complex data, i.e. images, video and time series data.
- **Applications in Automation**: Computer vision for defect detection. Real-time decision-making in autonomous systems.

3.4 Benefits of Machine Learning in Automation

- **Better Accuracy**: ML Algorithms are better than humans in making the patterns and anomalies.
- **Abnormal behaviour**: Traditional systems are pre-programmed according to the abnormal behaviour to which they assume, while ML based automation adapts to any unforeseen condition which has not been programmed.
- **Suitability for large-scale operations**: As ML systems can work with large amounts of data and ML is widely used within large scale operations, they are good fit to be deployed in such operational environments.
- **Better Planning and Resource Allocation**: More knowledge from the data is obtained, so more decisions are made.

3.5 Challenges of Implementing ML in Automation

- **Lack of Data Availability**: ML algorithms require a big amount of high-quality data for training that does not always exist.
- **Legacy Systems**: The ML integration cannot be integrated with existing systems.
- **Large datasets**: Processing such large datasets means keeping a large volume of data in memory and is a memory intensive job.
- **High cost of Implementation**: At first, small and medium sized enterprises will face high cost of implementation.

3.6 Future Trends in Machine Learning for Automation

- **Edge AI**: ML algorithms are executed on edge devices for real-time decision making without the dependency on cloud connection.
- **Quantum Machine Learning (QML)**: solving difficult optimizations, which arise in optimizing automation.
- **AI models in Design**: Generative AI controllers for manufacturing optimization; Forge them to create materialized product design (i.e. product prototypes).

4 Challenges in Industrial Automation

The industrial automation has enabled manufacturing, production, and supply chain operations to be more efficient, productive and cost effective. Despite this drawback, implementing and maintaining an automated system can be difficult to achieve. All these challenges are technical, operational, financial and human resource.

4.1 Technical Challenges

4.1.1 Integration of New and Legacy Systems

- **Introduction**: In modern environments, legacy equipment is widely used in many industries that do not have the requisite connectivity or support to modern automation systems.
- **Issues**: Lack of ability to add legacy systems to use new technologies such as IoT, AI, or web-based solutions. Inadequate standardized communication protocols between devices of different manufacturers.
- **Consequences**: Greater costs and also greater intricacy of retro fitting out dated equipping or composing unique solutions [11].

4.1.2 Data Management and Analytics

- In this case, vast amounts of data are produced through sensors, machines, and systems by automation.
- **Challenges**: Lakes and oceans or biology to process large volume of real time data. Being confident in the data quality, consistency, and accuracy for the basis of an effective decision.
- **Outcome**: If data is not managed well, all the processes of prediction analytics and machine learning algorithms will be severely affected.

4.1.3 Cybersecurity Risks

- The increased connectivity in industrial automation systems forces us to expose them to cyberattacks.
- **Challenges**: Ensuring that an authorized user cannot tamper or obstruct with sensitive data and infrastructure. This will help in preventing the discontinuity due to ransomware, phishing or malware attacks.
- **Impact**: Cyberattacks can stop the production, endanger safety, and cause massive loss of money [12].

4.1.4 Maintenance of Automated Systems

- **Method**: Automatic systems need to be monitored and fixed constantly for reliability.
- **Challenges**: Failure prediction of complex systems having many interrelated components. As for finding spare parts or specialty tools in order to fix it.
- **Impact**: Unplanned downtime has impact on operations and missing production targets.

4.1.5 System Scalability and Flexibility

- Automation systems must grow and become adaptable to change in production demand.
- **Challenges**: Systems that can deal with variations in production volume and types of products. Scalability with the requirement for system complexity.
- **Impacts**: Inflexible systems can restrict the response of the organization to the market demand.

4.2 Operational Challenges

4.2.1 High Initial Investment Costs

- **Summary**: Implementation of industrial automation is high capital expenditure activity.
- **Challenges**: Acquiring advanced equipment, sensors, robots and software. Training, installation and testing related costs.
- **Issues**: Implementing automation with large initial costs will be difficult for small- and medium-sized enterprises (SMEs).

4.2.2 Downtime During Implementation

- **Transition**: Man, to auto is a disruption.
- **Challenges**: Installation and testing of automation systems and resulting delays therefrom. Potential production halts during the transition phase.
- **Impact**: Temporary downtime may lead to financial costs as well as negatively impact customer commitments.

4.2.3 Dependence on Power and Infrastructure

- **Background**: Automation systems are strongly dependent on stable power supplies and requiring network connection.
- **Challenges**: Operations can be disrupted even due to power outages and fluctuations. Robust network infrastructure is what makes IoT and cloud connected system become better.
- **Impact**: Operational inefficiencies as well as reduction in reliability in energy intensive industries.

4.3 Financial Challenges

4.3.1 Cost of Maintenance and Upgrades

- Regular maintenance and running of periodic upgrades are required in automated systems.
- **Challenges**: For high costs to replace outdated components or software. Sourcing of spare parts for specialized equipment is difficult.
- **Impact**: However, many times, automation does not produce the financial benefits which led to its implementation.

4.3.2 Return on Investment (ROI) Uncertainty

- The automation projects have to be in a long-time investment with delayed return.
- **Challenges**: The inability to calculate ROI due to the fact that the market demands are unstable or past costs are not accounted for. Misjudging the positive effects of automation in the planning stages.
- **Impact**: Consequently, the impact is on the organizations that do not meet the targeted cost savings or productivity gains, resulting in financial strain on bottom lines [13].

4.4 Workforce Challenges

4.4.1 Resistance to Change

- **Summary**: Employees fear displacement or lack the knowledge of automation, so they resist it.
- **Challenges**: Negative perceptions of automation as a job-replacement technology.A desire to remain on the current workflows and stay away from automated systems. Resistance from staff can slow down an adoption of automation.

4.4.2 Skill Gap

- **Automation**: A workforce of robotics, AI and IoT knowledge is needed to run automation.
- **Challenges**: Low capability of employing, monitoring and maintaining automated systems. Difficulty in finding experts who can do troubleshooting and optimization.
- **Cons**: The skill gap can harm the results of automation and cause operational inefficiency.

4.5 Compliance and Regulatory Challenges

4.5.1 Meeting Industry Standards

- **Scenario**: The automated systems must be consistent with the industry specific safety and quality standards.
- **Challenges**: Instruments created to adhere strictly to regulations of the industries like pharmaceuticals, aerospace, food processing, etc. Changes to standards that entail modifications to the systems.
- **Effect**: Legal penalties and reputational damage can be the effect of non-compliancy.

4.5.2 Environmental Regulations

- **Stat**: Automation systems have to match up to the sustainability and environment standards.
- **Challenges**: Reduction of energy consumption and waste generation in automated process. Meeting the standards for emissions, among other eco-friendly requirements.
- **The Challenge**: Fines and additional work of redesigning the system are the consequences of a noncompliance [14].

4.6 Technological Advancements and Compatibility

4.6.1 Rapid Technological Changes

- Technology in industrial automation evolves quickly, making the existing system become extinct.
- **Challenges**: Maintain the top of the list for AI, IoT, robotics and edge computing. It is to guarantee backward compatibility with older systems.
- **Fortunately, this impact**: often more, usually under pressure.

4.6.2 Interoperability Issues

- **Automatic**: There are often multiple vendors' devices and software in automated systems.
- **Challenges**: Providing unbroken communication and merging between the different systems. Processing different data and format of protocols.
- **Result**: An inefficient and failed system can result from poor interoperability.

4.7 Ethical and Social Challenges

4.7.1 Workforce Displacement

- **Introduction**: Automation usually replaces repetitive manual tasks that were carried out by humans.
- **Challenges**: Ethical controversy regarding large scale job loss. The impacts of automation on unemployment and the society.
- **Impact**: Workforce displacement can have such an impact as to generate public backlash and decreased trust in automation.

4.7.2 Data Privacy and Ownership

- The Problem overview: huge sensitive data is being collected and analysed by automation systems.
- **Challenges**: Provider of data privacy, and data compliance with regulations like GDPR. Defining who owns the data that is generated in automated systems.
- **Consequences**: Wrong management of data can cause legal as well as ethical risks [15].

5 Emerging Role of Quantum Machine Learning (QML)

The relationship between quantum computing and machine learning (ML), two technology revolutions, is known as Quantum Machine Learning (QML). Unlike traditional ML, QML taps into the resulting limits of quantum mechanics in order to speed up and improve its performance in solving problems. And it is now becoming particularly useful in the areas of optimization, big data analytics, artificial intelligence, and also in industrial automation. Challenges in Quantum Machine Learning is shown in Fig. 4.

5.1 Introduction to Quantum Machine Learning (QML)

- The connection of quantum algorithms with classical machine learning techniques is QML.
- Qubits are used to construct quantum computers; in these computers, superposition (a qubit being in two states at the same time) and entanglement (qubits connected in which each state influences the state of the other) are the order of the day.
- The ability to process and analyse large amounts of data extremely quickly and efficiently is what characterizes these quantum properties.

Why Combine Quantum Computing and Machine Learning?

- A few tasks are not adequately dealt with under classical ML because they are too large to be processed, too difficult to optimize, and/or too hard to train deep neural networks (Fig. 5).
- Quantum parallelism is an opportunity to exploit QML, as it:
- Enhance processing speed.
- Address computational bottlenecks in ML.
- Find solutions to problems than classical systems cannot solve.

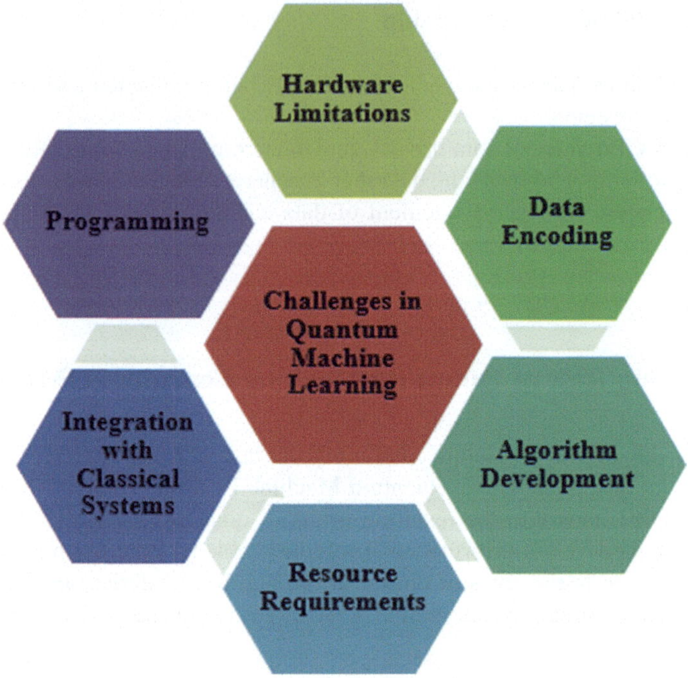

Fig. 4 Challenges in quantum machine learning

5.2 Key Algorithms in QML

- **Quantum Support Vector Machines (QSVMs)**: Classical SVM's are enhanced versions using quantum kernels for classifying high dimensional data in an efficient manner.
- **Quantum Neural Networks (QNNs)**: Such neural networks are quantum versions that have advantage for rapid training and better generalisation for complex problems such as image recognition and autonomous decision making.
- **Quantum Principal Component Analysis (QPCA)**: This is useful for feature extraction in ML as it is used to reduce the dimensionality of large datasets faster than classical PCA.
- **Variational Quantum Algorithms**: Both combines the power of quantum computing, the application of classical optimization algorithms to problems such as reinforcement learning and optimization.

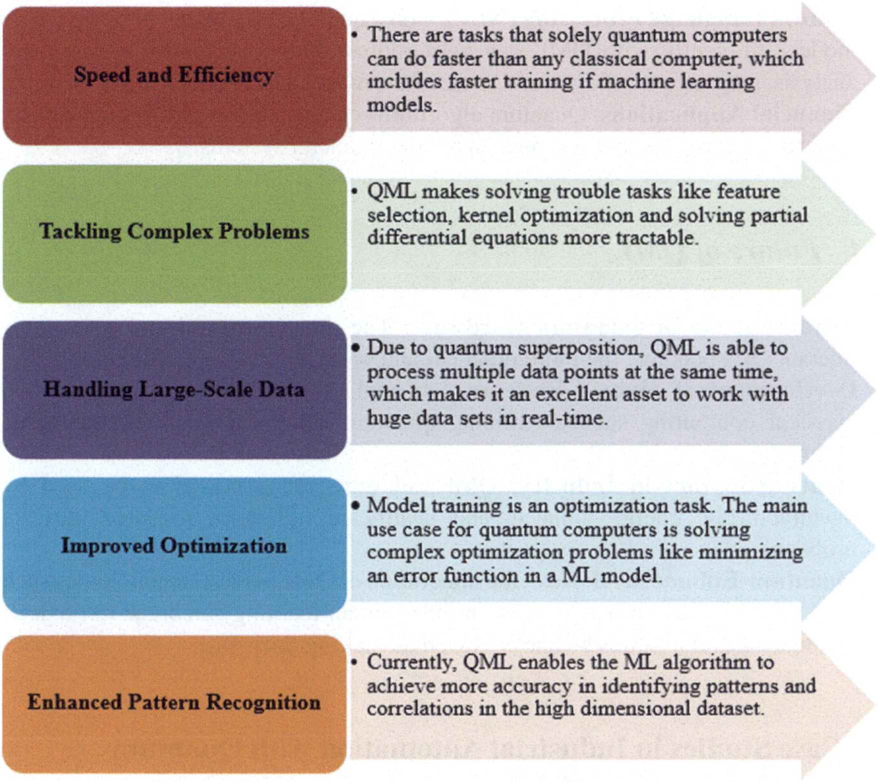

Fig. 5 Key advantages of QML

5.3 Applications of Quantum Machine Learning

- **Industrial Automation**: QML automates assembly line, reduces waste, and allocates resources. Among other things, QML can better predict equipment failures with higher precision and faster than classical ML, both based on analysing huge sensor datasets. Real time inventory, logistics and demand forecasting optimization is offered by QML.
- **Big Data and Analytics**: QML has been used for the analysis of large-scale unstructured datasets in the areas of finance, healthcare, and logistics. With their quantum engine, clustering and classification algorithms become quantum enhanced and hence run at faster speeds in comparison to the standard $O(N^3)$/ $O(N^2)$ time complexity, in order to discover hidden patterns in data [16].
- **Image and Signal Processing**: Image recognition, video analysis, signal processing is some of the applications QML excels in, which is a very good fit for quality control in manufacturing, satellite imagery analysis and medical diagnostics.

- **Natural Language Processing (NLP)**: Consequently, processing large language models efficiently with QML can help address NLP tasks such as sentiment analysis, language translation and some of the main tasks of chatbot optimization.
- **Financial Applications**: Quantum algorithms can also assist portfolio optimization, fraud detection and risk analysis in the financial systems.

5.4 Future of QML

- **Advancements in Quantum Hardware**: The more powerful and more error tolerant QCs become, the more practical and scalable QML will become.
- **Development of Hybrid Systems**: Job applications including quantum and classical computing such as training quantum neural network and real time optimization.
- **Wider Adoption in Industry**: QML solutions are expected to be used by manufacturing, energy, finance, and healthcare industries to solve intricate problems.
- **Quantum-Enhanced AI**: Next, the integration of QML with AI technologies such as natural language processing and reinforcement learning can break through its automation and robotics business as well as making decisions.

6 Case Studies in Industrial Automation with Quantum Machine Learning (QML)

The quantum machine learning (QML) and how it is being applied to find solution to real world problems in industrial automation and beyond is described in form of case studies. In this article, we explore few detailed case studies that outline how QML has been applied in manufacturing, supply chain management, predictive maintenance, and other such industries.

6.1 Predictive Maintenance in Smart Factories

- **Background**: For instance, predictive maintenance becomes very important in industrial automation to avoid unplanned downtime and reduce maintenance costs. While classical ML is used in the traditional predictive maintenance process to analyse sensor data from machinery, the speed and accuracy of data processing reduced its effectiveness.
- **QML Solution**: One of the QML deployed predictive maintenance solutions is deployed by a leading manufacturer using Quantum Neural Networks (QNNs).

IoT enabled sensors on machines such as vibration, temperature, and pressure in real time were fed into the system.

- **Implementation Steps**:

  Data Encoding: Power of quantum superposition is taken advantage of by encoding sensor data into quantum states.

  Anomaly Detection: The QNNs were trained to identify failure pattern, with better accuracy compared to conventional ML.

  Optimization: The scheduling of the predictive maintenance tasks was improved with quantum optimization algorithms.

- **Results**:

  Efficiency: A 25% reduction in downtime due to early fault detection.

  Cost Savings: It reduced by 18% the maintenance cost because they had fewer emergency repairs.

  Scalability: It allowed for real time decision making due to the fact that the QML system processed larger datasets faster.

  Key Takeaways: Quantum enhancement of predictive maintenance can bring major benefits to factory operations by saving money and enhancing equipment reliability [17].

6.2 Supply Chain Optimization

- **Background**: Due to the complexity of routing and inventory management problems in a global supply chain, a global logistics company was having difficulties to optimize it. Real time required classical optimization algorithms to attempt to optimize over a vast number of variables.
- **QML Solution**: To address supply chain inefficiency, the company use a QML based optimization model that is solved with Quantum Approximate Optimization Algorithm (QAOA).
- **Implementation Steps**:

  Problem Definition: Then the problem of supply chain from routing, inventory level, and delivery schedule were mapped to a quantum optimization problem.

  Quantum Simulation: Hierarchically, the inventories and routes that STM's soldier trainers might take to solve puzzles were simulated simultaneously using QAOA with millions of possibilities.

  Hybrid Approach: The integrated system is a hybrid system whereby classical ML is used to provide initial predictions and QML is used for fine tuning the optimization.

- **Results**:

  Cost Savings: Reduced transportation costs by 15%.

  Improved Delivery Times: On-time deliveries increased by 20%.

  Inventory Management: The better forecasting and distribution reduced inventory holding costs by 12%.

- **Key Takeaways**: In dynamic environments, QML can optimize complex supply chains in order to meet customer demands faster (both delivery time and cost) compared to non QML supply chains.

6.3 Quality Control in Manufacturing

- **Background**: A semiconductor manufacturer faces difficulties with defects identification of products, as the microscale manufacturing processes are extremely complex. The existing computer vision systems were lacking in detecting fine pattern which represents defects.
- **QML Solution**: With the advent of a Quantum Support Vector Machine (QSVM) model for defect detection, the company deployed this model.
- **Implementation Steps**:
  Data Pre-processing: Quantum data were thereby created from high resolution images of semiconductor wafers.
  Defect Detection: Quantum enhanced feature extraction by QSVM can identify defects from high controlling pixel intensity for the subtle variations.
  Real-Time Processing: During production, real time quality checks were possible due to the QML model processing the images quicker.
- **Results**:
  Accuracy: The classical ML reached 92% accuracy while QML can increase to 98%.
  Speed: Increased overall production efficiency by 30% reduction of inspection times.
  Waste Reduction: Defective products were early detected and this saved the material waste by 15%.
- **Key Takeaways**: Quality control, powered by QML, is more accurate and efficient as it's required for the tight tolerances found in many industries that include semiconductor manufacturing.

6.4 Energy Optimization in Industrial Plants

- **Background**: Towards this end, an energy intensive chemical plant conducted a study of reducing the plant's energy consumption and carbon footprint with minimal impact to productivity. The optimization problem: energy usage over multiple process and equipment.
- **QML Solution**: A QML model that optimizes the energy usage dynamically was implemented using Quantum Variational Algorithms.
- **Implementation Steps**:
  Data Collection: The data includes usage of energy, performance of machines, and environmental conditions using IoT sensors.

Model Training: An energy consumption pattern identification and optimal energy setting recommendation QML model was developed as a hybrid.

Real-Time Adjustments: The system autonomously adjusted equipment settings so that all products could be produced without consuming excessive amounts of energy.

- **Results**:

  Energy Savings: Energy consumption reduced by 20%.

  Emissions Reduction: Lower usage of energy rendered down 15% carbon emission.

  Cost Savings: Overall, $1.2 million in annual energy costs were saved.
- **Key Takeaways**: With regards to energy intensive industries, QML can be very helpful in meeting the energy efficiency and sustainability goals [18].

6.5 Fraud Detection in Financial Transactions

- **Background**: The complexity of fraud patterns was hence making it increasingly difficult for a large financial institution to detect the fraud transactions. However, for reason of the real time business, Classical ML could not handle and process such real time transaction data efficiently.
- **QML Solution**: The Quantum Boltzmann Machine (QBM) was used by the bank for the purpose of fraud detection.
- **Implementation Steps**:

  Data Processing: Quantum states are generated from transaction data describing the customer behaviour patterns and transaction histories.

  Fraud Pattern Detection: Real Time Anomalies and Suspicious Patterns were identified by the QBM.

  Integration: An integrated QML model was applied with existing fraud detection systems with a hybrid approach.
- **Results**:

  Detection Rate: Fraud detection accuracy increased by 25%.

  Response Time: This reduced response times down to 40%.

  Customer Satisfaction: This improved customers' trust on the company in its ability to resolve fraud issues more quickly.
- **Key Takeaways**: QML accelerates the speed and improves accuracy in fraud detection, which is now a crucial necessity in the financial sector's security.

6.6 Autonomous Vehicles in Logistics

- **Background**: An autonomous delivery vehicle logistics company had planned to deploy vehicles but had challenges with navigation and obstacles detection.

- **QML Solution**: An autonomous vehicle was trained using a Quantum Reinforcement Learning (QRL) algorithm that the company integrated into its company.
- **Implementation Steps**:
  Simulated Training: Simulations of autonomous vehicles trained are done on a complex route in a quantum enhanced environment.
  Real-World Testing: In real time the obstacle detection and route planning were optimized by deployment of QRL algorithms.
  Dynamic Learning: The QML vehicles were adapted to new environments and traffic patterns.
- **Results**:
  Efficiency: Reduced delivery times by 18%.
  Safety: Improved obstacle detection accuracy by 22%.
  Scalability: It allowed for deployment of autonomous vehicles in the various environments for deployment.
- **Key Takeaways**: QML is used to develop safe, efficient and innovative autonomous systems, and accelerates the development of automated logistics.

7 Future Outlook of Quantum Machine Learning (QML) in Industrial Automation

Quantum Machine Learning (QML) in industrial automation is a fertile future for quantum computing. At this early stage, QML is likely to bring about a transformation in the way that industries face and overcome the inherent complexity of different markets. It looks into potential advances on the part of QML, how it will affect, applications, and the challenges it will have to face as the coming years progresses.

7.1 Advancements in Quantum Hardware

- **Scalability**: Quantum computers operating now lose qubits and are noisy. Eventually, quantum hardware will become more scalable, more powerful. Noise will be reduced using advanced error correction methods, cutting noise down to the level that QML models became reliable and suitable for industrial use.
- **Quantum Speedup**: Quantum computers will run at speeds that classical systems cannot, and so will cut the time needed to train machine learning models and optimize problems down to size.
- **Hardware Integration**: We will bridge the gap between quantum experimentation and quantum industrial scale by integrating quantum computers with other systems (hybrid systems).

7.2 *Evolution of QML Algorithms*

- **Improved Algorithms**: Future work will be on constructing domain specific quantum algorithms aimed at answering industrial problems in optimization, robotics and predictive analytics.
- **Hybrid Models**: We can expect the use of classical models combined with quantum models (so called hybrid models) to become more common, exploiting the strengths of both technologies to solve problems efficiently.
- **Broader Applications**: More complex, more adaptive automation systems will include quantum versions of deep learning models, reinforcement learning and generative models [19].

7.3 *Impact on Industrial Automation*

- **Smart Factories and Industry 4.0**: Smart factories can be enabled by QML to achieve real time process optimization, minimizing waste and hastening process efficiency. Intelligent decisions will be made by QML powered automation systems according to changing conditions to create fully adaptive production lines.
- **Predictive Maintenance**: Quantum enhanced predictive maintenance systems will analyse massive amounts of sensor data with greater accuracy thereby minimizing unplanned downtime and lowering operational costs.
- **Robotics and AI**: In dynamic environments, faster learning and real time adaptation in the form of QML will revolutionize robotics to make robots more efficient. The increased use of human machine collaboration in industrial environments, will be aided by cobots equipped with QML capabilities.
- **Energy Efficiency**: QML will help make industrial plants, smart grids and factories more energy efficient so industries can meet sustainability goals without the additional cost.
- **Supply Chain Optimization**: By analysing and optimizing complex systems, QML will help with end to end supply chain processes like routing, inventory management and demand forecasting.

7.4 *Applications Beyond Current Horizons*

- **Cybersecurity in Automation**: Quantum-Enhanced Security: With use of QML, cybersecurity threats are detected in industrial automation systems better, with protecting data integrity and shielding from quantum attack.
- **Advanced Human–Machine Interfaces**: Natural Language Processing (NLP): It will improve NLP so that humans and machines can more easily communicate

with each other, and create more intuitive interfaces for industrial automation systems.

- **Autonomous Systems**: Self-Learning Systems: For example, QML will be used by autonomous vehicles, drones and robots to better navigate, make smarter decisions, and adapt to real time changes in industrial operations.

7.5 Industry-Specific Applications

- **Manufacturing**: By using QML, production lines will become on demand and customizable to the minutest detail, playing to changing consumer demands in real time.
- **Healthcare**: QML will be used by pharmaceutical industries for precise control in pharmaceutical production, making quality and regulations' compliance.
- **Energy and Utilities**: In smart grids, QML will equally optimize energy distribution and demand by balancing the two. The benefits will include improved predictive analytics for solar and wind as renewable energy sources, which ultimately means less reliance on intermittent power generation.
- **Transportation and Logistics**: QML models based on real time will optimize transportation routes, thereby decreasing delivery time and fuel consumption. The deployment of QML powered predictive maintenance will help make transportation fleets more efficient and long lasting.

7.6 Key Benefits of QML in the Future

- **Enhanced Productivity**: QML driven automation systems will radically enhance productivity through the optimization of processes, reduced errors, and minimized downtime.
- **Accelerated Innovation**: QML will speed up the pace of innovation of the industry, in that it allows for rapid testing and creation of new automation strategies.
- **Sustainability**: The energy efficient QML systems will aid industries to beat its environmental objectives, reduce waste and decreasing carbon foot prints.
- **Cost Reduction**: QML will save big money by improving efficiency, reducing downtime and optimizing processes [20].

7.7 Challenges to Overcome

- **Quantum Hardware Accessibility**: At first it may be quite the high cost and limited availability of quantum computers which may at least impede its adoption. Nevertheless, cloud-based quantum platforms will resolve this issue [21].

- **Talent Gap**: But because quantum scientists and engineers in QML will be in high demand, supply will not be able to keep up with demand. There are future efforts to be done with education and training.
- **Integration with Legacy Systems**: Integration of QML solutions in existing industrial systems, however, may require substantial upgrades.
- **Standardization**: If there are no standard protocols implemented which allow QML deployment among the industries, there might be a further difficulty in collaboration amongst them.

7.8 Roadmap for the Adoption of QML

7.8.1 Short-Term (1–5 Years)

- We concentrate on hybrid quantum classical systems to solve particular industrial issues.
- Push toward quantum research with all the collaborations between academics, public and private sector.
- Creating industry specific applications and their related QML software tools and frameworks.

7.8.2 Mid-Term (5–10 Years)

- Industrial application such as manufacturing, healthcare, logistics, etc.
- Improvements in quantum hardware scalability and error correction.
- Integration of QML with AI and IoT for comprehensive Industry 4.0 solutions.

7.8.3 Long-Term (10 + Years)

- Full-scale adoption of QML in industrial automation, with quantum systems handling large-scale operations autonomously.
- New quantum-based industries and business models.
- Becoming a standard industrial infrastructure.

8 Conclusion

The introduction of Quantum Machine Learning (QML) into industrial automation represent a major milestone in the ongoing progress of technological evolution. As industrial processes become increasingly complex and there is ever greater need for efficiency, adaptability, and sustainability, QML emerges as a transformative solution. By merging unparalleled computational power of quantum computing with the intuitive power of machine learning, industries are equipped with the combination of computation and decisions and also prediction capabilities to solve problems that were previously intractable. The applications to industrial automation using QML include: Predictive maintenance and supply chain optimization; Quality control; Energy management; Autonomous systems. QML enables processing of large datasets at an unprecedented speed to improve decision making while minimising downtime and reducing operational costs. Because it can model highly complex systems with enhanced accuracy, industries are able to optimise the utilisation of resource, increase productivity and realise substantial reduction in costs.

For addressing sustainability challenges, we believe that QML is one of the most promising aspects. QML assists in reducing carbon emissions, and achieving global environmental goals, by optimizing energy consumption in industrial plants, and by making industrial processes more efficient. Furthermore, its potential applications in renewable energy management and smart grid optimization highlight its importance in the transition toward greener industries. QML is coming because the quantum hardware and algorithms have matured. Broadened QML application will be enabled by the increased scalability and error correction capabilities of the emerging quantum processors. In fact, hybrid quantum–classical models are anticipated to bridge the chasm between the current maturity of technologies and the promise of quantum capabilities, enabling the process of continuous industrial evolution into the quantum era.

QML has huge potential on its side, but it also has several challenges that have to be solved for the adoption to be successful. Barriers that industries have to overcome include limited accessibility to quantum hardware, high costs and a requirement to develop specialized talent. The problem is compounded by the fact that integration with existing legacy systems, and the lack of standardization, make adoption even more difficult. All of this, however, is beginning to create collaborative efforts between governments, academia and private sectors that can help develop these quantum technologies and overcome these barriers.

QML will revolutionize industrial automation, but besides that, QML set to make advances in the healthcare, finance, and logistics areas. In the short term, the deployment of hybrid systems and industry specific applications will drive quantum technology, and in the mid to long term, fully autonomous quantum powered systems will take off. These will be the drivers of innovation cycles, the reshaping of business models and other opportunities to create new economic activity.

Hence, the age of transformation to QML has been characterized as above-faster, smarter and more efficient-than ever. Being able to handle complex systems, optimize operations, and improve sustainability marks it down as one of the critical parts of the industrial spread. However, the continued progress in QML and domain specific quantum optimization will continue to steer a bright future for the field. If industries embrace this technology, they can create unmeasured opportunities, stretching the limits and creating an opportunity for the creation of a smarter and more sustainable world.

References

1. Reddy, G. P., et al.: OPC UA implementation for industrial automation—Part 1: evaluating the performance of PubSub model. In: Proceedings of the 2023 1st International Conference on Circuits, Power and Intelligent Systems (CCPIS), pp. 1–6. IEEE (2023)
2. Szymanski, T.H.: Securing the industrial-tactile internet of things with deterministic silicon photonics switches. IEEE Access **4**, 8236–8249 (2016)
3. Márquez-Vera, M. A., et al. Microcontrollers programming for control and automation in undergraduate biotechnology engineering education. Digit. Chem. Eng. **9**, article no. 100122 (2023)
4. Makmal, A., et al.: Meta-learning within projective simulation. IEEE Access **4**, 2110–2122 (2016)
5. Yang, E.: Retrofitting supervisory control and data acquisition systems with microprocessor-based communications processors. IEEE Trans. Ind. Electron. Control. Instrum. **28**(4), 376–380 (1981)
6. Pakkala, D., et al.: On design of cognitive situation-adaptive autonomous mobile robotic applications. Comput. Ind. **167**, article no. 104263 (2025)
7. Piccialli, F., et al.: Guest editorial: industrial internet of things: where are we and what is next? IEEE Trans. Industr. Inf. **17**(11), 7700–7703 (2021)
8. De Micheli, G., et al.: Advances in quantum computation and quantum technologies: a design automation perspective. IEEE J. Emerg. Select. Top. Circuit. Syst. **12**(3), 584–601 (2022)
9. Umeda, S., et al.: Planned maintenance schedule update method for predictive maintenance of semiconductor plasma etcher. IEEE Trans. Semicond. Manuf. **34**(3), 296–300 (2021)
10. Wang, X., et al.: Managing your supply chain pantry: food waste mitigation through inventory control. IEEE Eng. Manage. Rev. **47**(2), 97–102 (2019)
11. Chencinski, E.W., et al.: Flash storage integration in the IBM system z EC12 I/O drawer. IBM J. Res. Dev. **57**(6), 11:1–11:14 (2013)
12. Elmarady, A.A., Rahouma, K.: Studying cybersecurity in civil aviation, including developing and applying aviation cybersecurity risk assessment. IEEE Access **9**, 143997–144016 (2021)
13. Bakhshi, R., Sandborn, P.A.: A return on investment model for the implementation of new technologies on wind turbines. IEEE Trans. Sustain. Energy **9**(1), 284–292 (2018)
14. Delbaere, M., Ferreira, R.: Addressing the data aspects of compliance with industry models. IBM Syst. J. **46**(2), 319–334 (2007)
15. Devillers, L., Cowie, R.: Ethical considerations on affective computing: an overview. Proc. IEEE **111**(10), 1445–1458 (2023)
16. Rahman, M.S., Reza, H.: A systematic review towards big data analytics in social media. Big Data Min. Anal. **5**(3), 228–244 (2022)
17. Lazzaro, A., D'Addona, D.M., Merenda, M.: A detailed study on algorithms for predictive maintenance in smart manufacturing: chip form classification using edge machine learning. IEEE Open J. Ind. Electron. Soc. **5**, 1190–1205 (2024)
18. Choobineh, M., Mohagheghi, S.: Optimal energy management in an industrial plant using on-site generation and demand scheduling. IEEE Trans. Ind. Appl. **52**(3), 1945–1952 (2016)
19. Munawar, H., Surendro, K.: Utilizing quantum algorithms to achieve carbon neutrality in urban areas: a systematic review. Alex. Eng. J. **108**, 911–936 (2024)
20. Nawaz, S.J., et al.: Quantum machine learning for 6G communication networks: state-of-the-art and vision for the future. IEEE Access **7**, 46317–46350 (2019)
21. Khan, M.W., Martin, J., Lim, M.: Instrumentation of low-cost IoT-agnostic data-capturing platform with enhanced sampling and bandwidth capabilities for subsurface characterization. IEEE Trans. Instrumen. Meas. **73**, 1–10, Art. no. 3002710 (2024)

A Hybrid Quantum–Classical LSTM Approach for Predicting the Stock Market

Sarvesh Arora, Deepika Kumar, and Shilpa Gupta

Abstract Prediction of financial instruments to create enormous wealth has been a source of fascination since the dawn of capitalism in the world. The objective of this study is to improvise over the current research for prediction and forecasting attempts. This paper methodology predicts the close price of National Stock Exchange 50 using quantum machine-learning and hybrid quantum–classical model QLSTM (Quantum Long Short-Term Memory). QLSTM achieves quantum benefits by combining LSTM with Variational Quantum Circuits (VQC). The proposed methodology has been compared with Bi-directional LSTM (BiLSTM), Long Short-Term Memory (LSTM), Gated Recurrent Unit (GRU), and Feedforward Neural Network (FNN). Mean Squared Error (MSE), Root MSE (RMSE), Mean Absolute Percentage Error (MAPE), Weighted MAPE (WMAPE), Symmetric MAPE (SMAPE) and R^2 score has been taken as the evaluation criteria. QLSTM achieved an impressive R^2 score of 0.9804, which has been taken as the primary metric. It also had the best MSE and RMSE errors of 116,461.40 and 341.26 with the lowest MAPE, WMAPE, and SMAPE of 0.0176, 0.0195, and 0.0178, respectively. The proposed approach improved the R^2 Score by 6–7% over the next best model considered in the study. Other metrics as well confirm the advantages of Quantum LSTM.

Keywords Time series forecasting · Price prediction · Stock markets · Quantum machine learning · Quantum LSTM · Variational quantum circuits

---

S. Arora · D. Kumar (✉)
Department of Computer Science and Engineering, Bharati Vidyapeeth's College of Engineering, New Delhi, India
e-mail: deepika.kumar@bharatividyapeeth.edu

S. Gupta
Department of Computer Science and Engineering, JIMS Engineering and Management Technical Campus, Greater Noida, Uttar Pradesh, India

© The Author(s), under exclusive license to Springer Nature Switzerland AG 2025
A. Ghosh et al. (eds.), *Quantum Machine Learning in Industrial Automation*, Information Systems Engineering and Management 65,
https://doi.org/10.1007/978-3-031-99786-0_4

1 Introduction

The stock market is a platform where participants buy and sell equity shares of firms as well as other financial instruments (as buyers and sellers, respectively). Participants might be investors or traders looking for gains in the short-term and long-term. Additionally, the market helps businesses to raise capital, which benefits a nation's economy. As per the efficient market hypothesis, the existing market price signifies all relevant and available information [1, 2]. Any price change is due to new information. Prices are efficient in terms of information, which suggests that stock prices can be predicted using transaction data. This is pretty natural because of the many unpredictable aspects and new information that will begin to affect the prices [3]. It is perplexing to predict how the stock will fare. A variety of factors impact prices, such as international economic conditions, political environments, corporate performance, and financial reports, among others. Other elements influencing predictions include physical and psychological aspects; rational and illogical conduct; etc. All of these elements work together to make stock values dynamic and volatile. This makes forecasting extremely complex and challenging. High accuracy is not necessary to make money in the market; even 60% accuracy can provide significant returns. However, trading or investing money in the market and earning substantial gains is still a challenging endeavour for active fund managers. A statistic from 2008 to 2012 showed that more than 90% of Canadian equity fund managers performed worse than the S&P/TSX Composite Index, and hence, the overall market [4]. Machine Learning (ML) presents a promising avenue for predicting stock prices and other financial assets, potentially outperforming traditional economic methods. Apart from the prospect of wealth creation, it might also help to screen out poor investments and trading decisions that are likely to fail. As per research, ML algorithms have the potential to anticipate the markets more effectively than various economic approaches and methods. It has also been proven that the financial market's effective forecasting is affected by input variables, forecasting periods, forecasting methods, and the maturity of the market [5]. The research also explored the benefits of using ML in the field of quantitative finance [6]. The application of a neural network with economic variables can perform better than a trading strategy and generate better profits [7]. Figure 1 depicts the timeline of various methodologies employed for financial time series:

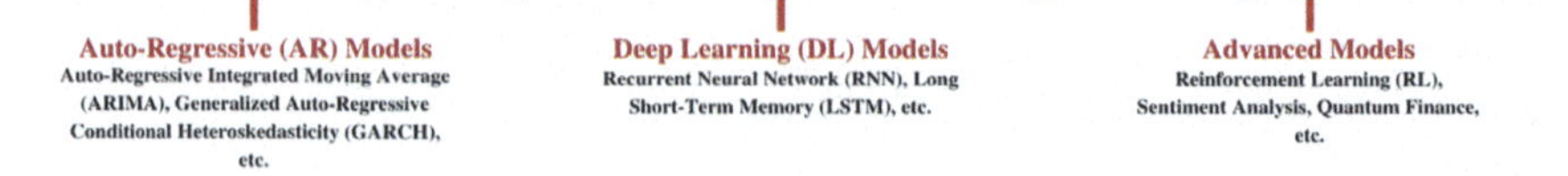

Fig. 1 Timeline of various methodologies employed for financial time series [8]

A study overview traces the evolution of financial time-series modeling approaches, starting with various statistical models like moving averages (MA) and auto-regressive models, based on a linear combination or non-linear combination (like GARCH, etc.) of past values of the variable. The research then evolved to classical ML models (RF, SVR, and so on) and neuron-based deep learning models (examples: RNN, LSTM, etc.). The approach progressed from Deep Learning (DL) to hybrid models, which are a combination of auto-regressive, ML, and DL models such as ARIMA-GARCH, ANN-ARIMA, NN-Fuzzy Systems, etc. The most recent methods involved Reinforcement Learning based on reward functions (such as Self Organizing Maps), Sentiment Analysis (based on news, and financial reports, etc.), and Quantum Finance (Quantum Binomial models, Quantum algorithms, etc.) [8]. Predicting stock prices using classic time series approaches has proven challenging. The dynamic structure makes it difficult to identify consistent patterns. This task is arguably better suited to deep learning as it can deal with complex data, find hidden structures within data and predict their impact on future prices. Furthermore, Neural Networks are favored since no mathematical model assumptions are required before predicting values. A survey of several literature publications demonstrates the popularity of neural networks and DL for stock market prediction, highlighting that neural networks are the most preferred method for this purpose [9]. RNN and LSTM models are much more capable of managing time series data. They are the most preferred deep learning models for all kinds of financial applications, especially in the stock price prediction category [10]. Since the design of the first quantum approach by the father of quantum computing, David Deutsch, the popularity of this field has been ever-growing [11, 12]. There is a major advantage of using Quantum Computing (QC) and Quantum Machine Learning (QML) in machine learning tasks.

The research objectives have been summarized as follows:

Instead of using Classical neural networks in LSTM model to predict the prices, Variational Quantum circuits are used in the quantum–classical hybrid LSTM.

The proposed QLSTM model has been trained individually along with LSTM, BiLSTM, GRU, and FNN for comparison purposes.

A detailed comparison using evaluation metrics such as MSE, RMSE, MAPE, Weighted MAPE, Symmetric MAPE and R-squared have been performed as mentioned in Table 1.

The research paper is structured as: Sect. 2 describes a comprehensive literature review of various techniques or algorithms used for predicting finances. Section 3

Table 1 Comparative analysis of models considered as per various metrics

MODEL	MSE	RMSE	MAPE	WMAPE	SMAPE	R^2 score
LSTM	647,406.03	804.62	0.0446	0.0497	0.0460	0.8910
BiLSTM	479,781.87	692.66	0.0375	0.0420	0.0385	0.9192
GRU	886,424.06	941.50	0.0403	0.0479	0.0420	0.8507
FNN	1,494,794.54	1,222.62	0.0782	0.0843	0.0820	0.7483
QLSTM	116,461.40	341.26	0.0176	0.0195	0.0178	0.9804

elucidates the methods and materials used for conducting this research. Section 4 illustrates the proposed methodology. Section 5 contains the experimental setup which is followed by the experimental results and analysis explained in Sect. 6 which is followed by Conclusion section.

2 Literature Review

Multiple research employing different approaches on various stocks/indices have been studied in the literature. Various research metrics have been studied to predict the prices in the stock market w.r.t various time frames, like intraday, weekly, daily, etc., to make predictions [12, 13]. Also, stock market prediction can be done either by predicting price values/return percentages or by classifying the direction of closing (positive or negative) for the trading days.

Shen et al. [14] utilized Support Vector Machines (SVM) and reinforcement learning to predict NASDAQ, S&P500, and DJIA among others, with a prediction accuracy (consisting of the top 4 features) of 74.4, 76, and 77.6%, respectively. A data set containing daily prices from January 4th, 2000, to October 25th, 2012, has been utilized for experimentation, including stocks, currencies, and commodities. Vijh et al. [15] used Random Forest and ANN models to forecast the next day's close for listed companies: Goldman Sachs, Nike, Johnson & Johnson, Pfizer, and JP Morgan Chase. Open, Close High, Low Adjusted Close, and volume utilizing data from May 4th, 2009 to May 4th, 2019. ANN achieved a value of 0.42 RMSE, 0.77 MAPE, and 0.013 Mean Bias Error (MBE) as its best values among all stocks. Ghani et al. [16] used Linear Regression (LR) along with three Months Moving Averages (3MMA), Exponential Smoothing (ES), and MS Excel for graph and tabular representation for Amazon, Apple, and Google stocks. For a one-month price prediction, the best R^2 score (also called Coefficient of determination or R-Squared) was 71.63% with 4% MAPE. Ismail et al. [17] collected daily closing prices from May 15th, 2000, to March 27th, 2018. Using the Persistent Homology (PH) tool for next-day direction prediction, the SVM performed better as compared to LR, ANN, and RF, with the best average performance of 66.68%. Patel et al. [18] employed ANN, SVM, RF, and Naive-Bayes (Multivariate Bernoulli Process) on ten years of data (2003–12) of CNX Nifty, S&P BSE Sensex, Infosys Ltd., and Reliance Industries. Ten technical indicators/parameters have been utilized for the analysis whereas Naive-Bayes performed best with 90.19% accuracy. Chen et al. [19] utilized high-frequency transaction data in a one-minute timeframe from the Chinese stock market's CSI 300 futures contract ranging from February 20th, 2017–April 20th, 2017 in the train-test ratio of 90:10 to predict the behavior of prices. AutoEncoder (AE) and restricted Boltzmann machine (RBM) outperformed Back Propagation (BP), Radial Basis Function (RBF) and Extreme Learning Machine (ELM). Network in all three ways the data was divided, with the best MAPE value of 0.0001. Pang et al. [20] employed deep LSTM with the embedding layer to predict the stock market by utilizing relevant information from the time series of the stock. A single stock on the Shanghai A-shares composite index ranging from January 1st, 2006, to October 19th, 2016, was considered out of which 10 and 20% were used for validation and test set, respectively. A measurement accuracy of 57.2% was achieved by the best deep LSTM model. Using domestic trends on Google as a barometer of public sentiment and macroeconomic determinants, Xiong et al. [21] used LSTM to model S&P 500 volatility. Training set from October 19th, 2004, to April 9th, 2012, and testing set from April 12th, 2012, to July 24th, 2015, were considered. It resulted in a MAPE value of 24.2%, outperforming other models such as linear Ridge/Lasso and autoregressive Generalized AutoRegressive Conditional Heteroskedasticity (GARCH) by a wide margin. Patel et al. [22] used a two-stage fusion method on CNX Nifty and S&P BSE Sensex data, with Support Vector Regression (SVR) in the first stage and ANN and RF in the second. The SVR-ANN model performed the best at predicting prices for 30 days, with MAPE values of 2.66 and 2.58% for CNX Nifty and BSE Sensex indices, respectively. Using 60 different economic factors and three dimensionality reduction methods, Zhong et al. [23] predicted the S&P Index's direction for the next day. Component Analysis (CA), Kernel Principal CA (KPCA), and Fast Robust PCA (FRPCA), in conjunction with a NN, were utilized on 2518 trading days of data between June 1st, 2003, and May 31st, 2013. The results indicated that PCA was the most accurate, obtaining 57% accuracy. Hao and Gao [24] employed LSTM, SVM, Convolutional Neural Network (CNN), Normalized Fuzzy Neural Network and Multiple Pipeline models on 20 years of S&P 500 stock prices data ranging from January 30th, 1999 to January 30th, 2019. The combined results of multiple LSTMs performed better than other models with over 74.55% accuracy over a month. The authors have adopted a dynamic approach to the LSTM network to forecast the Nifty by utilizing price at the opening, and closing, along with high and low. The algorithm achieved a value of 0.00859 in daily percentage change prediction in terms of RMSE [25]. Gururaj et al. [26] employed LR and SVM individually over The Coca-Cola Company stock data from January 2017 to January 2018. The results were compared on various metrics such as MSE, R^2 score, etc. The SVM outperformed the LR model with an R^2 score of 0.93. Azevedo et al. [27] applied Qubit Neural Networks on three different datasets: DJIA (January 1998 to August 2003), S&P500 (January 1970 to August 2003), and the Brazilian company Petrobras (January 1995 to July 2003). 50% dataset was employed for training, 25% for validating, and the rest 25% for testing dataset on five metrics: MSE, Normalized MSE, Prediction on change in Direction (POCID), MAPE, and Average Relative Variance (ARV). Qubits Neural Network performed better or equivalent to ANN in most cases across the three datasets on various metrics. Gao et al. [28] trained Multilayer Perceptron (MLP), LSTM, and CNN on three different datasets consisting of SP500, CSI300, and Nikkei225 stock indices from July 2008-September 2016. Using 90% dataseta for training, the research resulted in R^2 scores and MAPE values in the range of 0.9759–0.9889 and 0.0067–0.0091 for all three datasets. Budiharto [29] employed LSTM to predict the price of Bank Central Asia Tbk PT with data fetched from Yahoo Finance. With 80% of data used for testing, average accuracy for 1-year and 3-year datasets were in the range of 93.83–94.59% and 78.28–88.47% respectively for various epoch values. Shastri et al. [30] utilized the stock market data available on Twitter using certain keywords for sentiment analysis. This Twitter data, which was stored in Hadoop Distributed File System (HDFS) has been used with data from the Bombay Stock Exchange (BSE). Two different datasets, one consisting of records ranging from 2013 to 2016, and another one of records from the year 2016 are considered with a 75:25 train-test split. With MAPE as the metric, the Multilevel Perceptron algorithm produced an accuracy of 91.8 and 98.42% for the respective datasets implying that the market predictions are more successful for shorter periods. Singh [31] implemented a quantum-based optimization method integrated with a Fuzzy time series (FTS) and performed a comparison with various FTS and non-FTS models on three different datasets (mentioned with dataset record range): daily average temperatures of Taipei (June 1996 to September 1996), TAIFEX index (August 3rd, 1998–September 30th, 1998), TSEC weighted index (1999 to 2004). The approach achieved MAPE (%) of 0.52–1.11 (month-wise), 0.066, and 1.08 respectively for the three datasets. Ghosh et al. [32] suggested an LSTM algorithm to forecast stock values for many Indian banks, including SBI, ICICI and HDFC. The information was obtained from the official BSE website. As training data, two months of data were used. The algorithm forecasted future values for three, six, one, and three years by using their proposed methodology. Over time, the error levels continued to fall dramatically. The average inaccuracy for one month was roughly 232.6, while the average error for three years was 0.89. The study demonstrated that LSTM excelled on forecasting on a considerably broader variety of data. Qiu et al. [33] developed an Attention-based LSTM approach to forecast future-stock values, to compare the findings, three different models were used: LSTM, LSTM with wavelet denoising, and the GRU neural network. The stock data were collected from January 2000 to July 2019. Along with HSI data from January 2002 to July 2019. There were six variables in the data. MSE, RMSE, MAE used to evaluate prediction outcomes. The best R^2 score of 0.947, 0.9621, and 0.8783 for each dataset respectively attained by the Attention-based model. Lu et al. [34] proposed a CNN-LSTM algorithm for forecasting the stock-market, by utilizing stock data from July 1991 to August 2020 of the Shanghai-Composite Index. The experiment revealed that CNN-LSTM model predicted the prices more reliably as compared to conventional models by 2.2, 0.6, 0.5, and 0.2%, respectively. The study by Kurani et al. [35] highlights important aspects of new approaches and technologies whereas ANN was a hybrid model and ANN-MLP, GARCH-MLP, or a combination of Backpropagation and Multilayer Feed-forward networks, delivered improved results. Hafezi et al. [36] presented a new model dubbed the bat-neural network multi-agent system (BNNMAS) designed to forecast Deutscher Aktienindex (DAX) stock indices over the next eight years. MAPE was used as evaluation metric. BNNMAS achieved the best MAPE of 2.84 compared with 3.49 for Genetic Algorithm Neural Network (GANN), 10.75 for General Regression Neural Network (GRNN), and 12.3 for Exact Radial Basis Network (RBE). Selvin et al. [37] compared DL models like RNN, CNN and LSTM compared their effectiveness over ARIMA models using stock data of NSE-Listed firms: Infosys, and Tata Consultancy Service (TCS). Data from July 2014 to October 2014 has been used for training and the prediction been done for October 2014 to November 2014. The research discovered that ARIMA model has a greater error percentage value of 31.91, 21.16, and 36.53% over DL models (error ranges being 2.36–4.18%, 7.65–8.96%, and 3.63–3.83%) for each stock respectively, proving that neural models are more accurate and suited for predicting the effects of excessive volatility on stock markets. Lee [38] presented a Chaotic Type-2 Transient-Fuzzy Deep Neuro-oscillatory Network for global financial data prediction, which includes main cryptocurrencies, major commodities, forex, and various financial indices. The author presented an innovative notion of chaotic neural oscillators, that serve as the DL network's 'transient-fuzzy input neurons.' The results validated the proposed approach's high potential. Zou and Qu [39] proposed daily volumes and prices for top ten S&P 500 companies from 2004 to 2013. MSE was utilized to determine the most effective model among ARIMA, LSTM, Stacked-LSTM, and Attention-LSTM, the later one outperformed in terms of results as compared to others. Due to the time series' long-term dependency model forecasted financial time series substantially better. However, Stacked-LSTM despite having much more complicated structure, could not outperform LSTM. Rundo [40] implemented a supervised deep learning and reinforcement learning method to anticipate the short-term movement in the FOREX market to optimize the return on investment in a high-frequency trend (HFT) algorithm. The author also suggested a grid trading engine for high-frequency operations. The trading system was verified throughout many fiscal years and on the EUR/USD cross to validate the suggested model. The amount of the investment return (98.23%) and the minimized drawdown (15.97%) verified the efficacy of the suggested trading technique.

There is so much research conducted using models to forecast future prices in the stock market. Models based on average prices were found to be the simplest type for prediction but were less accurate. Classical machine learning-based models performed better as compared to others. However, learning complex series and prediction is still a challenge. To learn the complex series of prices and predicting and forecasting future values, neuron-based models were resulted good. LSTM has been discovered as the best and most widely used DL model for this purpose. The hybrid approaches has been utilized for better and more accurate prediction attempts, with the recent advancement in the field of Quantum ML and the introduction of software for the use of quantum hardware, a new horizon opened for advanced research. Utilizing this opportunity, this research aims to implement QLSTM, an approach that combines Variational Quantum Circuits (VQC), and the popular model for prediction, LSTM to successfully predict future stock prices using previous data sequences. This has been implemented with the help of a Quantum simulator by the software framework, PennyLane.

In this study, RNNs like LSTM, BiLSTM, and GRU and simple networks like Feedforward Neural Network (FNN) are employed and compared with a hybrid quantum–classical model, called Quantum LSTM or QLSTM to predict and forecast the stock market prices.

3 Materials and Methods

A significant amount of research has been done using various classical ML and DL models, but only a few have used quantum computing. Here, the authors proposed a hybrid quantum–classical approach, referred to as Quantum LSTM. The details for the methodology has been discussed in the following subsections.

3.1 Dataset Description

For this study the set of data has been taken from Kaggle for training and testing the various models for the financial time-series predictions [41]. It consists of data from 29 indices ranging from November 6th, 1996 (or the date of the inception of that particular index) to January 31st, 2022. It has 8 columns, namely: Index, Date, Volume, Open, High, Low, Close, and Currency. Figure 2 by depicts the available records of their daily closing prices in the dataset.

3.2 Pre-processing

NIFTY 50 index dataset has been utilized for this research. Various algorithms have been applied using its daily open, high, and low values as features to forecast the future close prices of the time series. The volume and currency columns have not been considered. The dataset has been divided into a proportion of 85:15 as training and testing set respectively. The neural networks were trained by using 85 percent of the dataset. This set comprises records of 5550 trading days ranging from November 06, 1995, to February 14, 2018 and 15 percent was used for testing and performance evaluation of the models. This set comprises records of 980 trading days from the period February 15, 2018, to January 31, 2022, a period of roughly 4 years.

3.3 Box-Cox Transformation

The Box-Cox transformation has been utilized on the data values to normalize the data. Box-Cox is a power transformation method specified by a single parameter, lambda (λ). It was selected to address the abnormal distribution of data. It helps best approximate a normal distribution curve by evaluating the optimal value of lambda (λ) for the NIFTY 50 records and utilizing it for the transformation [42]. The following equations represent the Box-Cox transformation, where w represents the transformed variable, t represents the time period, y is the target variable, and λ (lambda) represents the parameter according to which the transformation takes

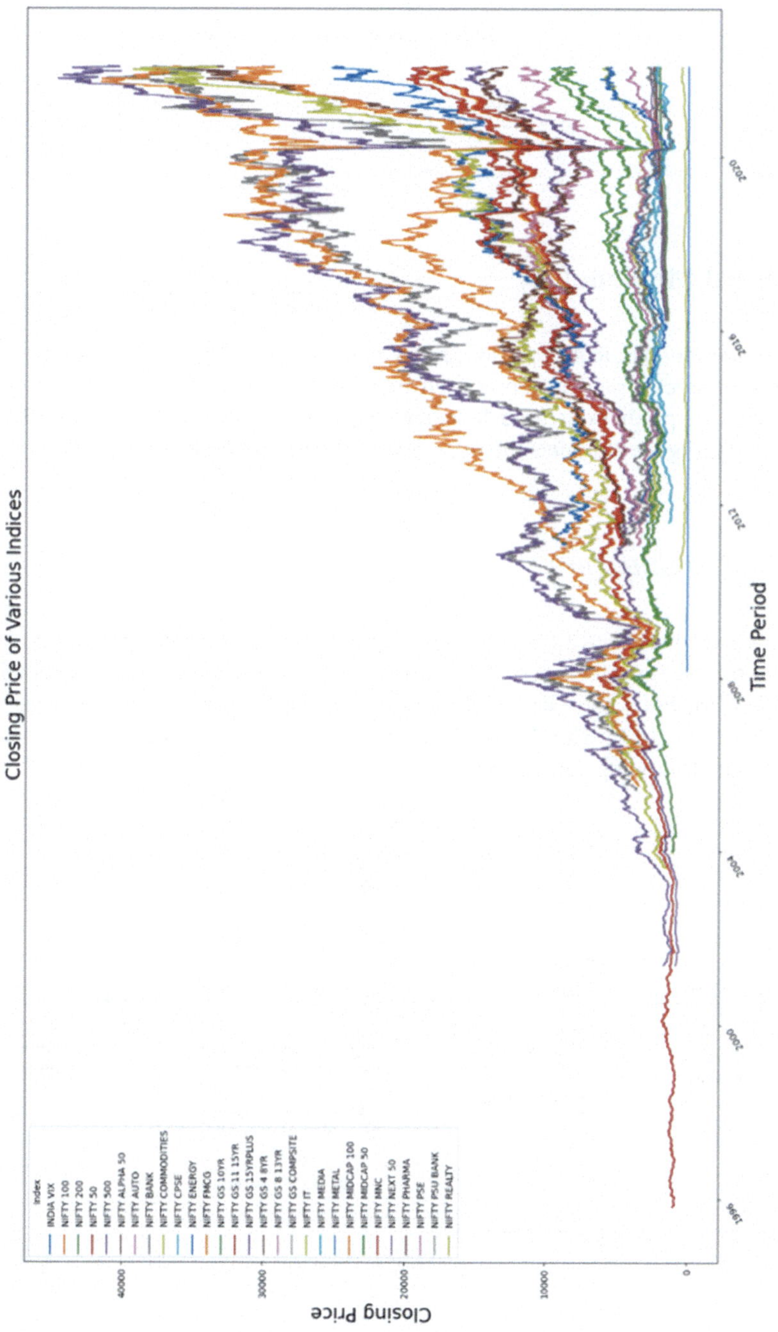

Fig. 2 Close prices of various Indices

place.

$$w_t = \log(y_t) \quad if\, \lambda = 0; \tag{1}$$

$$w_t = \frac{(y_t^{\lambda} - 1)}{\lambda} \quad otherwise; \tag{2}$$

4 Proposed Methodology

A hybrid quantum–classical approach, referred to as QLSTM, has been proposed for forecasting stock prices and proposed methodology have been compared and analyzed with the predictions used for neural networks, such as LSTM, BiLSTM, GRU and FNN. Figure 3 depicts the overview of the research methodology:

4.1 Model Architecture

The authors have proposed LSTM with quantum benefits, referred to as QLSTM. This results for the same has been compared and analyzed with three RNNs namely LSTM, BiLSTM, FNN and GRU. All these models have been implemented using Pytorch framework [43]. The detail of the proposed model architecture has been discussed in the following subsections.

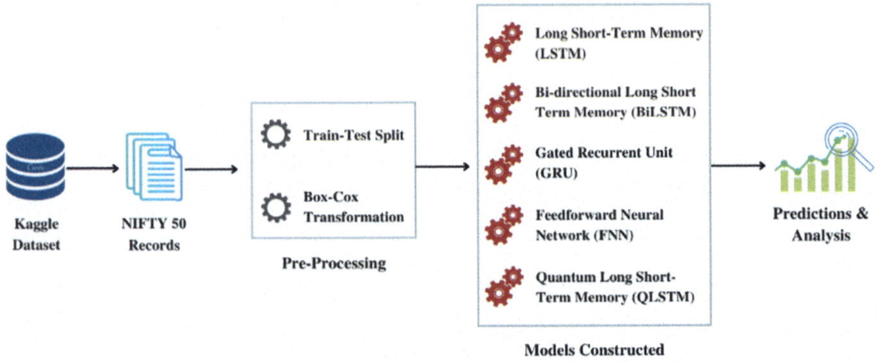

Fig. 3 Overview of the research methodology

4.1.1 Long Short-Term Memory (LSTM)

In sequence prediction tasks, LSTM-RNN effectively capture order dependencies. They feature three gates—input, forget, and output—that enhance their ability to remember information over long sequences. This capability allows LSTM models to excel in predicting data with long-term trends, making them among the most advanced models for forecasting [44–46].

4.1.2 Bi-Directional LSTM (BiLSTM)

BiLSTM processes input from forward (past to future) as well as backward (future to past). Unlike traditional LSTM, which only processes data sequentially in one direction, BiLSTM uses an additional LSTM layer to learn from the reversed input sequence. The outputs from both LSTM layers are combined using methods like mean, addition, multiplication, or concatenation to produce the final results [47].

4.1.3 Gated Recurrent Units (GRU)

GRU can be define as RNN architecture variant that uses gating methods to regulate and govern the flow of information between neural network cells. It consists of two gates: reset (for short-term memory) and update (for long-term memory), which makes it much faster to train. Its structure enables it to capture dependencies from huge data sequences adaptively without rejecting information from previous sections of the sequence [48].

4.1.4 Feedforward Neural Network (FNN)

FNN, or Feedforward Neural Network, processes input data in one direction without cycles among nodes. Unlike recurrent neural networks, which loop back on themselves, FNNs move data from input to output through hidden layers without feedback loops. This classic neural network model is also called a single-layer perceptron in its simplest form [49].

4.1.5 Quantum LSTM (QLSTM)

Quantum LSTM combines quantum computing and machine learning by integrating Variational Quantum Circuits (VQC) into the LSTM structure. VQCs, also known as Parameterized Quantum Circuits (PQCs), replace classical neural networks to enhance data compression and feature extraction capabilities [50]. VQCs are crucial in QLSTM for leveraging quantum advantages, blending quantum computing with

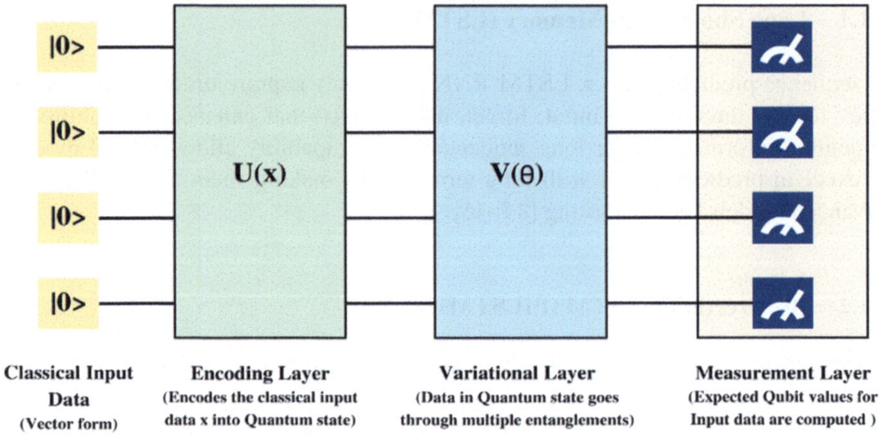

Fig. 4 Architecture of Variational Quantum Circuits (VQC)

machine learning techniques [51]. The architecture typically includes three layers: Encoding Layer, Variational Layer, and Measurement Layer, as illustrated in Fig. 4.

Encoding Layer: This layer encodes the quantum state with the classical data. The classical vector is transformed into quantum amplitudes, which identify the quantum state, using a predetermined process known as the encoding approach. The translation of the classical input vector into rotation angles will direct the single-qubit rotations. The *arctan* function has been used as its domain is R (all real numbers), which matches the range of input values. This layer encodes the classical data into the quantum state of *arcsin* and, *arccos* [52].

Variational Layer: A pipeline of unitary operations utilizing multiple CNOT gates and single-qubit rotation gates will subsequently be applied to the encoded classical data, which is now in a quantum state. CNOT gates are used cyclically to pairs of qubits to create multi-qubit entanglement. The rotation angles for single-qubit gates along the x, y, and z axes are adjusted during iterative optimization using gradient descent. Increasing the levels of entanglements and rotations enhances the layer's depth and the number of variational parameters.

Quantum Measurement Layer: This is the final layer of a VQC block, where the expected values of each qubit are measured in the computational basis. There exist a lot of software frameworks for QC as well as QML such as Google's Tensorflow Quantum (TFQ), Xanadu's Pennylane, etc. The simulation of quantum circuits required for this model utilizes Pennylane software framework [53, 54]. Additionally, Pennylane also provides a user-friendly interface for quantum computers from various firms, including Google, Microsoft, and IBM [55]. Pennylane supports several simulators, including 'default.qubit' (simple state simulator), which was used for the QLSTM model. The output is a fixed-length vector processed on a standard computer. The QLSTM model can be grouped into three layers:

Forget Block: The VQC_1 block uses the sigmoid function to process v_t and produce a vector f_t with values ranging from 0 to 1. This vector decides whether to "forget" or "retain" elements in the previous cell state c_{t-1} by acting element-wise on c_{t-1} (i.e., $f_t{*}c_{t-1}$). A value of 1 means the element is fully retained, while 0 means it is completely forgotten. Typically, the values in f_t are between 0 and 1, indicating partial retention of information. This makes QLSTM effective for learning or modeling temporal dependencies. This can be represented by Eq. (3).

$$f_t = \sigma(VQC_1(v_t)) \tag{3}$$

Here, σ denotes the sigmoid function. VQC represents the VQC circuit block.

Input and Update Block: This section determines what additional information will be incorporated into the cell state. This section has two VQCs.

VQC_2: Processes v_t, and its output is passed through a sigmoid function to decide which values are added to the cell state.

VQC_3: Takes the same concatenated input, and applies *tanh* function to generate a new cell state candidate, C_t.

Finally, the output of VQC_2 is element-wise multiplied by C_t, to update the cell state, c_t. These steps ensure the cell state is updated based on relevant information from v_t and the concatenated input. Mathematically, these can be represented by the following equations.

$$i_t = \sigma(VQC_2(v_t)) \tag{4}$$

$$C_t = \tanh(VQC_3(v_t)) \tag{5}$$

$$c_t = f_t * c_{t-1} + i_t * C_t \tag{6}$$

Here, *tanh* denotes the hyperbolic tangent function. c_t represents the dimensions of the cell state.

Output Block: After updating the QLSTM cell state, the next step involves determining its output. First, VQC_4 examines v_t and applies the sigmoid function to identify which cell state c_t values are significant to the output. This can be mathematically represented as:

$$o_t = \sigma(VQC_4(v_t)) \tag{7}$$

For VQC_1 to VQC_4, the input is the concatenation v_t of the previous time step's hidden state h_{t-1} with the current input vector x_t. Each VQC block produces a vector based on measurements yielding Pauli Z expectation values for each qubit. These values are then processed through nonlinear activation functions (sigmoid and, tanh).

The cell state c_t is passed through the tanh function and then element-wise multiplied by o_t. This value may then be processed further using VQC_5 to obtain the hidden state h_t or VQC_6 to obtain the output y_t. This can be represented by the following equations.

$$h_t = VQC_5(o_t * \tanh(c_t)) \tag{8}$$

$$y_t = VQC_6(o_t * \tanh(c_t)) \tag{9}$$

5 Results and Analysis

The models, namely: QLSTM, LSTM, BiLSTM, GRU, and FNN are trained individually using the training dataset to analyze and predict the price values. These models are then used to forecast NIFTY 50 close price in the test dataset sequentially. This is known as sequential time-series forecasting. It involves predicting future values by analyzing the sequence of historical observations. The trained models are sequentially applied to the test dataset and evaluated as per the metrics. To test the quality and effectiveness of various methodologies, six evaluation metrics are considered. These six metrics MSE, RMSE, MAPE, WMAPE, SMAPE, and Coefficient of Determination (also called R-Squared or R2 score)—are used to evaluate the performance of each approach.

This research has explored the approach of examining the results of employing quantum computing along with machine learning and the improvement it brings over traditional deep learning models for complicated predictions. Each model has anticipated and predicted the price values for the test set, which comprises 980 trading days after being trained on the training set, which itself comprised 5550 trading days of data. The predicted close values for the test set have been contrasted with the original values for the same period, i.e., from February 14th, 2018 to January 31, 2022. Adam Optimizer has been used for tuning the parameters of neural network to enhance the model's performance and reduce the MSE loss, which is used as the criterion. Every model in this research has been evaluated using six different metrics: MSE, RMSE, MAPE, WMAPE, SMAPE, and R2 score (the primary metric for the final comparisons). Also, the predicted values with real values have been visualized in this section. Figures 5a, b represent the performance of various models with MSE and RMSE as metrics.

As illustrated in Fig. 5a, b, the proposed hybrid quantum approach performs the best for the test set, after being trained on the initial 85% of records of the total dataset. The QLSTM had an MSE of 116,461.40. It is a major improvement over the typical LSTM and GRU deep learning models, which obtained an MSE of 647,406.03 and 886,424.06, respectively. When compared to its peer, GRU, LSTM has done substantially better and has claimed the lead by a considerable margin. The performance demonstrates that Bi-directional LSTM or BiLSTM achieved better values than its unidirectional equivalent, LSTM. The MSE value for BiLSTM was

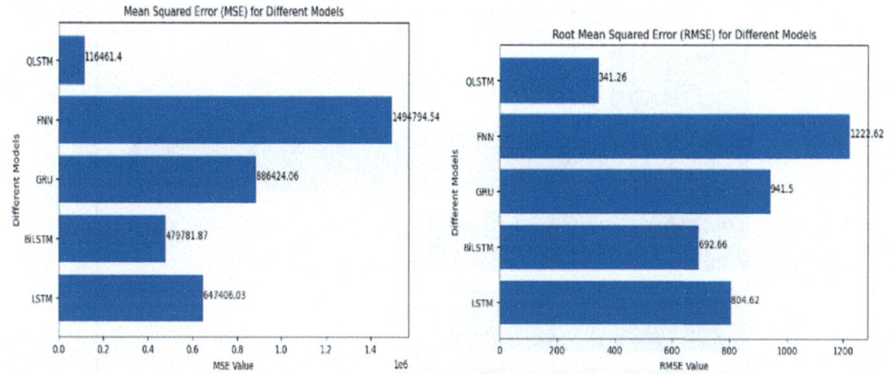

(a) (b)

Fig. 5 **a** MSE, and **b** RMSE, for various models

found to be 479,781.87. It highlights the fact that Bi-directional models are better able to learn complex sequence problems like price predictions and time series due to the advantage of being trained in both forward as well as backward directions, which helps to understand the sequences better. Although none of the models could outperform the proposed model with quantum advantage. GRU's poor performance relative to its other recurrent counterparts can be ascribed to the fact that its structure is substantially simpler than that of different types of LSTM considered in this study. It has two gates, as opposed to three gates in the unidirectional models: QLSTM and LSTM, and the bi-directional variant of LSTM, known as BiLSTM. Not Surprisingly, FNN being a much simpler unidirectional artificial neural network (ANN), did not outperform any other recurrent network. Unlike RNNs, FNN does not have a feedback loop mechanism, which would have been better suited for this type of complex prediction. It was also the only non-recurrent network model in this study. QLSTM outperformed the other recurrent models: LSTM, BiLSTM, and GRU by a difference of 530,944.63, 363,320.47, and 769,962.66 in MSE values, respectively. For the same measure, the difference between conventional unidirectional and bi-directional LSTM models is 167,624.16, with BiLSTM performing better. The MSE values of both classical unidirectional RNNs: LSTM and GRU models differ by 239,018.03. The order of performance is determined to be the same when using the RMSE metric. LSTM and BiLSTM performed similarly, with BiLSTM gaining the lead. They both significantly outperformed GRU and FNN, who themselves had a vast performance gap with GRU dwarfing the FNN model. However, QLSTM excelled over all the other models, with RMSE values being less than half of those of its nearest competitor, BiLSTM. FNN performed inferior to all with 1,494,794.54 and 1,222.62 values for MSE and RMSE metrics, respectively. The hybrid quantum–classical was identified as the only model with MSE and RMSE values less than 250,000.00 and 500.00 limits, respectively, whereas the pair of classical LSTM models, namely LSTM and BiLSTM, were the only classical models with MSE values less than 750,000.00. GRU and FNN were the only models which outdid the 750,000.00 value threshold

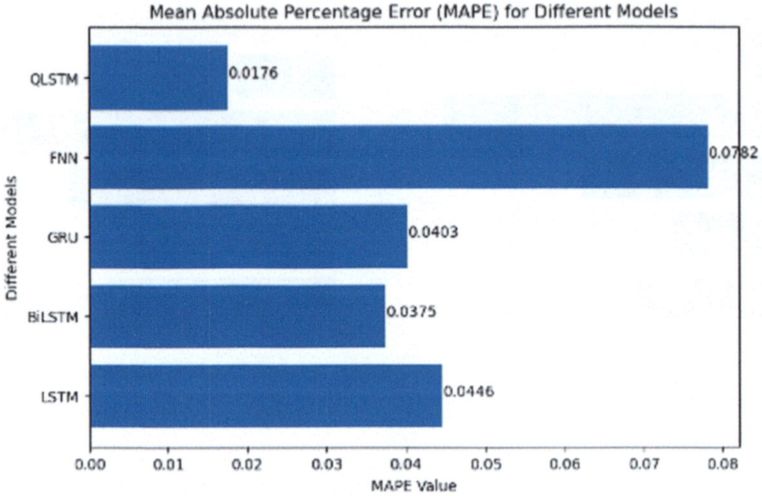

Fig. 6 Mean absolute percentage error (MAPE) for various models

in the MSE metric. All three classical RNNs had RMSE of less than 1,000.00. The most basic model in this research, FNN, had its RMSE value in 4 digits with an MSE value of around 1.5 million. Figure 6 highlights the performance comparison for the MAPE metric.

Furthermore, when MAPE is considered, LSTM and BiLSTM reached values of 0.0446 and 0.0375, respectively. The bi-directional approach performed better than its unidirectional counterpart. Surprisingly, GRU had a MAPE of 0.0403 and displayed better performance for this metric than LSTM. This was not the case when MSE or RMSE were considered. However, BiLSTM still outperformed GRU for the MAPE metric. FNN had a MAPE score of 0.782, which was around double of any other classical model for this criterion. However, QLSTM performed best in this measure as well, with a MAPE value of 0.0176, which is remarkably smaller than any other model for the same metric. The order of performance for additional variants of this measure, namely Weighted MAPE (WMAPE) and Symmetric MAPE (SMAPE), was found to be the same, with QLSTM outperforming every other model. Like in the case of MAPE, GRU outperforms LSTM for WMAPE and SMAPE metrics but it was unable to surpass the performance of the Bi-directional LSTM model. The performance of all the models for WMAPE and SMAPE metrics are illustrated in Fig. 7a, b.

As usual, FNN fared much worse for WMAPE and SMAPE metrics as well, highlighting the difficulty a relatively simple model faces for complex tasks. For FNN, the differences are enormous in terms of performance numbers for all three versions of the MAPE metric, almost double its nearest competitor, LSTM. One thing is common among all the models for the three types of MAPE: they illustrate the trend of MAPE values being the smallest followed by the Symmetric version of

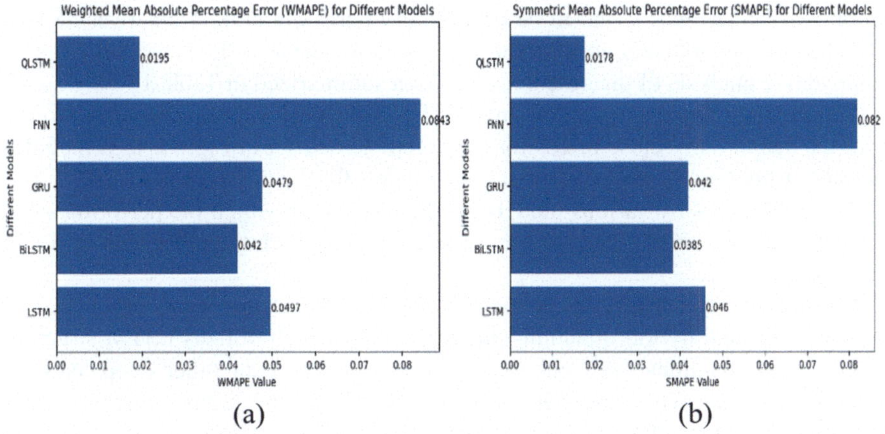

Fig. 7 **a** WMAPE, and **b** SMAPE, for various models

MAPE or SMAPE and Weighted MAPE or WMAPE values. Figure 8 represents the performance of various models with R2 score as the metric.

The QLSTM has achieved an R^2 score of 0.9804, which is the preferred comparison metric among all other metrics. It is a major improvement over the typical LSTM and GRU deep learning models, which obtained 0.8910 and 0.8507, respectively. When compared to its peer, GRU, LSTM has done substantially better and has claimed the lead by a considerable margin. It superseded GRU by around a 4% difference in R^2 score. The performance demonstrates that Bi-directional LSTM or BiLSTM achieved better values than its unidirectional equivalent, LSTM. The R^2

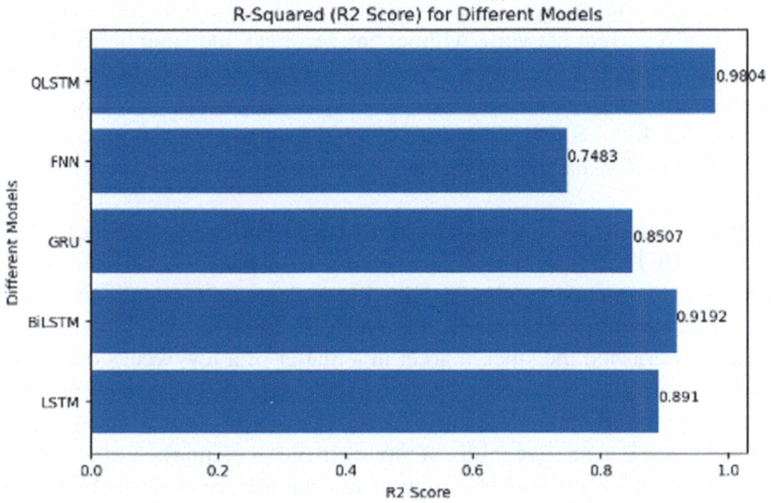

Fig. 8 R^2 score for various models

score of BiLSTM was found to be 0.9192, an improvement of 2–3% over LSTM. FNN had a comparatively poor performance with 0.7483 as its R^2 score. A complete comparative analysis of the outcomes has been summarized in Table 1.

The predicted values by all the models for NIFTY 50 values have been plotted with the actual daily close prices for the complete dataset in Fig. 9a–e. It includes the plot of final predicted values by all the models for the complete dataset.

The predictions by various models for the test set, on which the performance is evaluated as per various metrics, are visualized along with the real close prices in Fig. 10a–e.

Considering the results of all the metrics, the hybrid quantum approach, i.e., the unidirectional hybrid quantum–classical model which utilizes LSTM structure coupled with parameterized circuits, performed better than other deep learning models evaluated. This research shows the benefit of the hybrid quantum–classical model over its classical counterparts. This model performance was followed by BiLSTM, which was the only bi-directional model in this research. BiLSTM was followed by its unidirectional counterpart, LSTM. It emphasizes the fact that bi-directional models are better suited to deal with complex sequence problems such as price forecasting and time series since they are trained in both forward and backward directions, which aids in understanding the sequences. GRU performance was found to be between LSTM and FNN models. Its performance was better than LSTM for some metrics, namely, MAPE, WMAPE, and SMAPE. For other evaluation criteria such as R^2 score, MSE, and RMSE, LSTM achieved better performance than GRU in the order of performance. GRU's relatively poor performance in comparison to its other recurrent counterparts can be attributed to the fact that its structure is significantly simpler with two gates, as opposed to the different types of LSTM considered in this study, which have three gates in each cell. FNN, the unidirectional ANN, however, achieved the last place among all the models considered in this study for every metric by a wide margin. The performance of FNN can be justified by the fact that it is a non-recurrent network, without any feedback loop and hence, faces difficulty in complex predictions. The outcomes of the research highlight the advantage of utilizing Variational Quantum Circuits (VQC) for quantum benefits combined with classical models like LSTM. Considering the main metric of this research, i.e., R^2 score, the final performance rankings are as follows: QLSTM > BiLSTM > LSTM > GRU > FNN.

The market is volatile and complicated as depicted by the exponential change in values of NIFTY 50. Moreover, the period for which the records exist in the test dataset was impacted by the COVID-19 pandemic situation. This devastating once-in-a-blue-moon event disrupted the normal economic cycle due to the global lockdowns and crashed the stock market in a matter of a few weeks due to the fear sentiments in the financial world. Moreover, the actions by various governments and economic institutions for printing large amounts of money and lowering the interest rates caused a massive influx of money in the stock markets around the world including India. This contributed to the swift and massive rise in the financial markets, including the NIFTY 50 index which has been used as the dataset. Every model tried to predict prices in such abnormal and dynamic world conditions. However, a hybrid

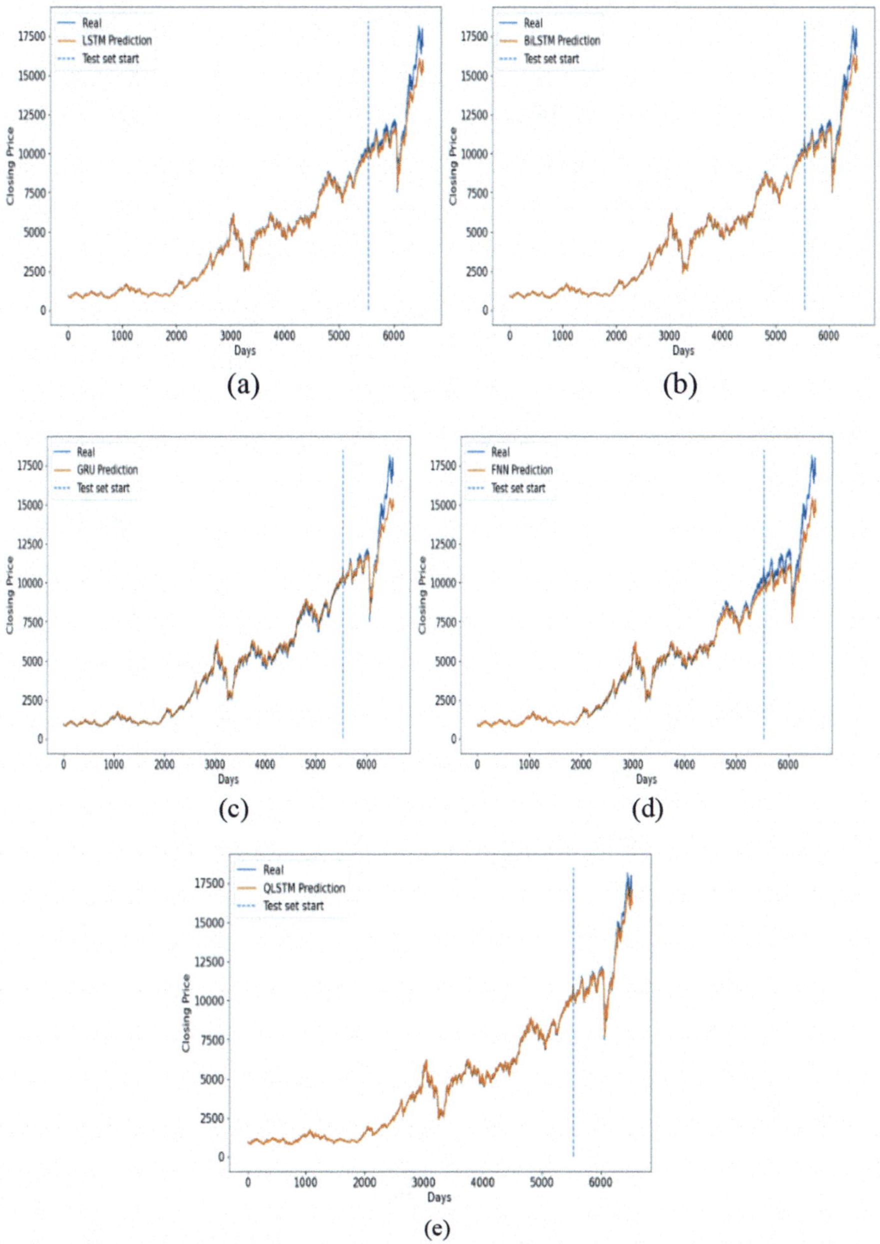

Fig. 9 **a–e** Prediction plots of all models' results for the complete dataset: **a** LSTM, **b** BiLSTM, **c** GRU, **d** FNN, and **e** QLSTM

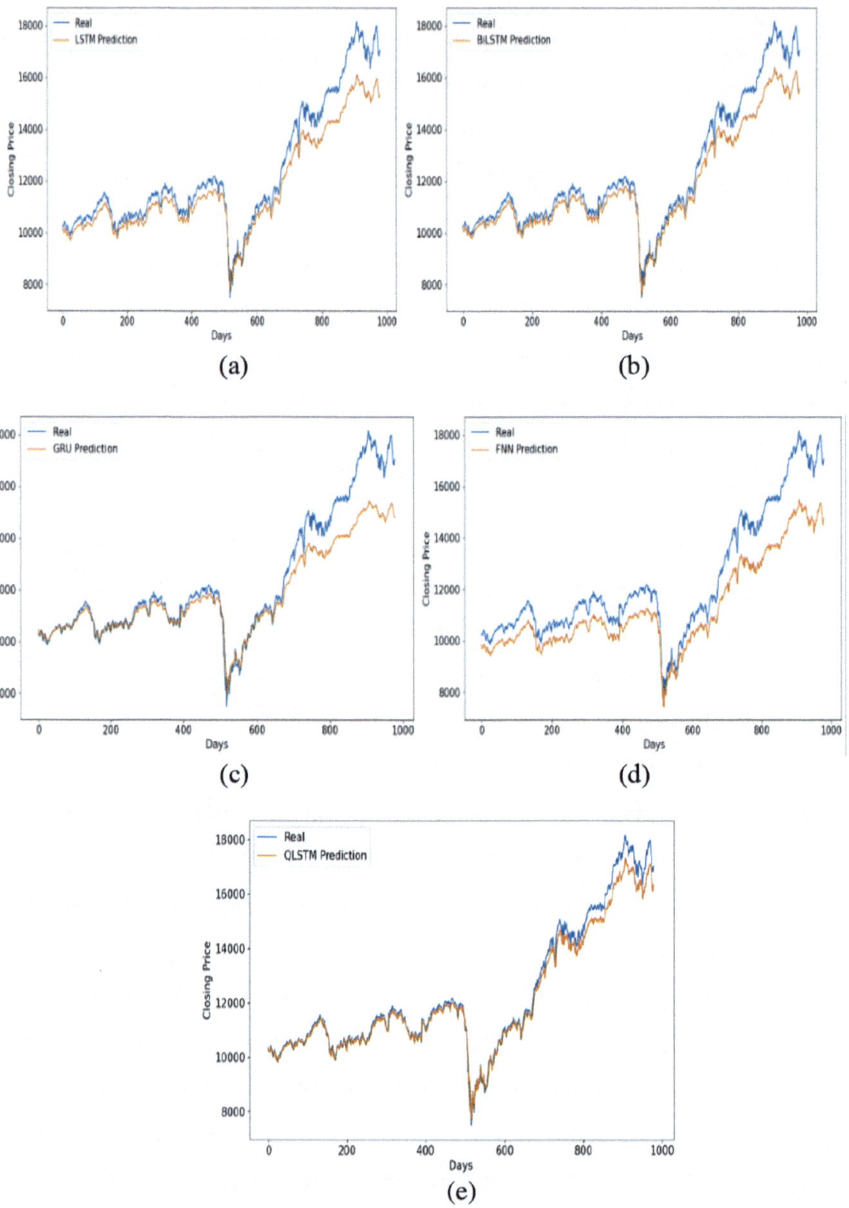

Fig. 10 **a–e** Prediction plots of all models' results for the Test set: **a** LSTM, **b** BiLSTM, **c** GRU, **d** FNN, and **e** QLSTM

quantum–classical approach was much more successful than other models for such a challenging prediction attempt even in an exceptional world scenario.

6 Conclusion and Future Work

Anticipating the stock market is a complex and arduous endeavor. However, the advancements in algorithms and technologies provide the opportunity to attempt this type of research. The paper highlights one such attempt to utilize the power of machine-learning with quantum computing by employing a hybrid quantum–classical model known as QLSTM. The analysis concludes that employing the quantum aspect results in a significant advancement in the prediction over the currently popular classical machine learning and deep learning capabilities. Most importantly, it opens more researchers to the idea of employing quantum machine learning and attempting for better results in various types of research. More complex hybrid approaches can be employed for better predictions. The incorporation of technical indicators like moving averages, RSI (Relative Strength Index), volumes traded, etc. would be very beneficial for accurate predictions. Moreover, fundamental factors such as free cash flow, profits, business growth numbers, etc. should also aid in significant improvement for predicting the prices of stocks, some of which make up the indices such as NIFTY 50. Also, Market sentiments are a very important factor that affects prices. The sentimental aspect can be considered by utilizing the analysis of multiple news, financial reports by various financial advisors, and relevant social media posts. Importantly, in complex and dynamic problems of time-series forecasting, the more training records and the smaller the forecasting period is considered, the better it is likely to perform. As technology advances and quantum devices such as quantum computers become readily available, it would be interesting to see the performance and accuracy advantage they offer over quantum simulators.

References

1. Fama, E.F., Fisher, L., Jensen, M., Roll, R.: The adjustment of stock price to new information. Int. Eco. Rev. **10**(1), 1–21 (1969)
2. Qian, B., Rasheed, K.: Stock market prediction with multiple classifiers. Appl. Intell. **26**(1), 25–33 (2007)
3. Fama, E.F.: Efficient capital markets: a review of theory and empirical work. J. Financ. **25**(2), 383–417 (1970)
4. Hodges, D.: Is Your Fund Manager Beating the Index (2013)
5. Hsu, M.W., Lessmann, S., Sung, M.C., Ma, T., Johnson, J.E.: Bridging the divide in financial market forecasting: machine learners vs. financial economists. Expert Syst. Appl. **61**, 215–234 (2016)
6. Cheng, C., Sa-Ngasoongsong, A., Beyca, O., Le, T., Yang, H., Kong, Z., Bukkapatnam, S.T.: Time series forecasting for nonlinear and non-stationary processes: a review and comparative study. IIE Trans. **47**(10), 1053–1071 (2015)

7. Niaki, S.T.A., Hoseinzade, S.: Forecasting S&P 500 index using artificial neural networks and design of experiments. J. Ind. Eng. Int. **9**(1), 1–9 (2013)
8. Rundo, T., di Stallo, Battiato.: Machine learning for quantitative finance applications: a survey. Appl. Sci. **9**(24), 5574 (2019)
9. Kumar, D., Sarangi, P.K., Verma, R.: A systematic review of stock market prediction using machine learning and statistical techniques. Mater. Today Proc. **49**, 3187–3191 (2022)
10. Huang, J., Chai, J., Cho, S.: Deep learning in finance and banking: a literature review and classification. Front. Bus. Res. China **14**(1), 1–24 (2020)
11. Huang, H.Y., Broughton, M., Mohseni, M., Babbush, R., Boixo, S., Neven, H., McClean, J.R.: Power of data in quantum machine learning. Nat. Commun. **12**(1), 2631 (2021)
12. Deutsch, D.: Quantum theory, the Church–Turing principle and the universal quantum computer. Proc. R. Soc. Lond. A. Math. Phys. Sci. **400**(1818), 97–117 (1985)
13. Jiang, W.: Applications of deep learning in stock market prediction: recent progress. Expert Syst. Appl. **184**, 115537 (2021)
14. Singh, S., Madan, T.K., Kumar, J., Singh, A.K.: Stock market forecasting using machine learning: today and tomorrow. In: 2019 2nd International Conference on Intelligent Computing, Instrumentation and Control Technologies (ICICICT), vol. 1, pp. 738–745. IEEE (2019, July)
15. Vijh, M., et al.: Stock closing price prediction using machine learning techniques. Procedia Comput. Sci **167**(2020), 599 (2020)
16. Umer, M., Awais, M., Muzammul, M.: Stock market prediction using machine learning (ML) algorithms. ADCAIJ: Adv. Distrib. Comput. Artif. Intell. J. **8**(4), 97–116 (2019)
17. Ismail, M.S., Noorani, M.S.M., Ismail, M., Razak, F.A., Alias, M.A.: Predicting next day direction of stock price movement using machine learning methods with persistent homology: evidence from Kuala Lumpur Stock Exchange. Appl. Soft Comput. **93**, 106422 (2020)
18. Patel, J., Shah, S., Thakkar, P., Kotecha, K.: Predicting stock and stock price index movement using trend deterministic data preparation and machine learning techniques. Expert Syst. Appl. **42**(1), 259–268 (2015)
19. Chen, L., Qiao, Z., Wang, M., Wang, C., Du, R., Stanley, H.E.: Which artificial intelligence algorithm better predicts the Chinese stock market? IEEE Access **6**, 48625–48633 (2018)
20. Pang, X., Zhou, Y., Wang, P., Lin, W., Chang, V.: An innovative neural network approach for stock market prediction. J. Supercomput. **76**(3), 2098–2118 (2020)
21. Xiong, R., Nichols, E.P., Shen, Y.: Deep Learning Stock Volatility with Google Domestic Trends (2015). arXiv preprint arXiv:1512.04916
22. Patel, J., Shah, S., Thakkar, P., Kotecha, K.: Predicting stock market index using fusion of machine learning techniques. Expert Syst. Appl. **42**(4), 2162–2172 (2015). https://doi.org/10.1016/j.eswa.2014.10.031
23. Zhong, X., Enke, D.: Forecasting daily stock market return using dimensionality reduction. Expert Syst. Appl. **67**, 126–139 (2017)
24. Hao, Y., Gao, Q.: Predicting the trend of stock market index using the hybrid neural network based on multiple time scale feature learning. Appl. Sci. **10**(11), 3961 (2020)
25. Nguyen, D.H.D., Tran, L.P., Nguyen, V.: Predicting stock prices using dynamic LSTM models. In International Conference on Applied Informatics, pp. 199–212. Springer, Cham (2019, November)
26. Gururaj, V., Shriya, V.R., Ashwini, K.: Stock market prediction using linear regression and support vector machines. Int. J. Appl. Eng. Res. **14**(8), 1931–1934 (2019)
27. Azevedo, C.R., Ferreira, T.A.: Time series forecasting with qubit neural networks. In: Conference: ASC'07 Proceedings of the Eleventh IASTED International (2007, August)
28. Gao, P., Zhang, R., Yang, X.: The application of stock index price prediction with neural network. Math. Comput. Appl. **25**(3), 53 (2020)
29. Budiharto, W.: Data science approach to stock prices forecasting in Indonesia during Covid-19 using Long Short-Term Memory (LSTM). J. Big Data **8**(1), 1–9 (2021)
30. Shastri, M., Roy, S., Mittal, M.: Stock price prediction using artificial neural model: an application of big data. EAI Endors. Trans. Scalable Inform. Syst. **6**(20), e1–e1 (2019)
31. Singh, P.: FQTSFM: a fuzzy-quantum time series forecasting model. Inf. Sci. **566**, 57–79 (2021)

32. Ghosh, A., Bose, S., Maji, G., Debnath, N., & Sen, S.: Stock price prediction using LSTM on Indian Share Market. In: Proceedings of 32nd international conference on computer applications in industry and engineering, vol. 63, pp. 101–110 (2019, September)
33. Qiu, J., Wang, B., Zhou, C.: Forecasting stock prices with long-short term memory neural network based on attention mechanism. PLoS ONE **15**(1), e0227222 (2020)
34. Lu, W., Li, J., Li, Y., Sun, A., Wang, J.: A CNN-LSTM-based model to forecast stock prices. Complexity (2020)
35. Kurani, A., Doshi, P., Vakharia, A., Shah, M.: A comprehensive comparative study of artificial neural network (ANN) and support vector machines (SVM) on stock forecasting. Ann. Data Sci. 1–26 (2021)
36. Hafezi, R., Shahrabi, J., Hadavandi, E.: A bat-neural network multi-agent system (BNNMAS) for stock price prediction: case study of DAX stock price. Appl. Soft Comput. **29**, 196–210 (2015)
37. Selvin, S., Vinayakumar, R., Gopalakrishnan, E.A., Menon, V.K., & Soman, K. P.: Stock price prediction using LSTM, RNN and CNN-sliding window model. In: 2017 international conference on advances in computing, communications and informatics (ICACCI), pp. 1643–1647. IEEE (2017, September)
38. Lee, R.S.: Chaotic type-2 transient-fuzzy deep neuro-oscillatory network (CT2TFDNN) for worldwide financial prediction. IEEE Trans. Fuzzy Syst. **28**(4), 731–745 (2019)
39. Zou, Z., Qu, Z.: Using LSTM in stock prediction and quantitative trading. In: CS230: Deep Learning, Winter, pp. 1–6 (2020)
40. Rundo, F.: Deep LSTM with reinforcement learning layer for financial trend prediction in FX high frequency trading systems. Appl. Sci. **9**(20), 4460 (2019). https://doi.org/10.3390/app9204460
41. Kaggle: Available Online: https://www.kaggle.com/datasets/andrewmvd/india-stock-market?select=nse_indexes.csv. Accessed on 10th Aug 2022
42. Box, G.E., Cox, D.R.: An analysis of transformations. J. Roy. Stat. Soc.: Ser. B (Methodol.) **26**(2), 211–243 (1964)
43. Paszke, A., Gross, S., Massa, F., Lerer, A., Bradbury, J., Chanan, G., Chintala, S., et al. (2019). Pytorch: An imperative style, high-performance deep learning library. In: Advances in Neural Information Processing Systems, p. 32.
44. Graves, A.: Long short-term memory. In: Supervised Sequence Labelling with Recurrent Neural Networks, pp. 37–45 (2012)
45. Sutskever, I., Vinyals, O., Le, Q.V.: Sequence to sequence learning with neural networks. In: Advances in Neural Information Processing Systems, p. 27 (2014)
46. Hochreiter, S., Schmidhuber, J.: Long short-term memory. Neural Comput. **9**(8), 1735–1780 (1997)
47. Yang, M., Wang, J.: Adaptability of financial time series prediction based on BiLSTM. Procedia Comput. Sci. **199**, 18–25 (2022)
48. Chung, J., Gulcehre, C., Cho, K., Bengio, Y.: Empirical Evaluation of Gated Recurrent Neural Networks on Sequence Modeling (2014). arXiv preprint arXiv:1412.3555
49. Sazli, M.H.: A brief review of feed-forward neural networks. Phys Sci Eng. (2006)
50. Chen, S.Y.C., Yoo, S., Fang, Y.L.L.: Quantum long short-term memory. In: ICASSP 2022–2022 IEEE International Conference on Acoustics, Speech and Signal Processing (ICASSP), pp. 8622–8626. IEEE (2022, May)
51. Benedetti, M., Lloyd, E., Sack, S., Fiorentini, M.: Parameterized quantum circuits as machine learning models. Quantum Sci. Technol. **4**(4), 043001 (2019)
52. Mitarai, K., Negoro, M., Kitagawa, M., Fujii, K.: Quantum circuit learning. Phys. Rev. A **98**(3), 032309 (2018)
53. Bergholm, V., Izaac, J., Schuld, M., Gogolin, C., Alam, M.S., Ahmed, S., Killoran, N., et al.: Pennylane: Automatic Differentiation of Hybrid Quantum-Classical Computations (2018). arXiv preprint arXiv:1811.04968

54. Pennylane-Qulacs Plugin. https://github.com/PennyLaneAI/pennylane-qulacs. Originally written by Steven Oud (@soudy)
55. Geoffrey, A.S.B.: Ab Initio Calculations of Dissociation Energy of Water Molecule on Near-Term Quantum Devices Using the Pennylane Interface to Quantum Hardware (2021)

Quantum Machine Learning Applications

Priyanka Jammwal

Abstract One of the most advanced fields to date, Quantum Machine Learning (QML), balancing the power of quantum computing and artificial intelligence, has an application base across the industries and is definitely at the top of the line. This chapter of the book presents the real-world effect of QML, by the use of which it has changed the healthcare, finance, and the supply chain management fields and thus transformed them. QML achieves this nadir of computing power by employing quantum principles like superposition, entanglement, and quantum gates that drive the generation of algorithms outpacing classical methods in computational science and math. By means of QML, pharmaceuticals have been cut length of time needed for molecular simulations, which has led to an improvement in financial risk analysis through portfolio optimization. Elaborating on this, QML is already a big player in the whole industry. In supply chain management, it automates logistics and optimizes resource distribution, which enhances the efficiency of operations. A comprehensive study of quantum algorithms embracing Quantum Support Vector Machines (QSVM) and Vibrational Quantum Algorithms (VQA) further illustrates their successful application. QML is a technology with a big future but still facing obstacles, for example, hardware constraints, data preparation intricacies, and scalability hurdles. For that reason, hybrid quantum–classical systems are emerging as a means of addressing the currently existing technological divide. QML promises to be a game-changer in Industry 4.0, with its novel advancements likely to affect many different areas. This chapter is a detailed study of QML applications, illustrating the connection of quantum computing theory with real-world implementation. It closes with case studies and future research directions, delivering invaluable information on the ever-changing state of quantum machine learning.

Keywords Quantum machine learning · Quantum computing applications · Quantum algorithms · Industry 4.0 · Healthcare · Finance · Supply chain optimization · Variation quantum algorithms

P. Jammwal (✉)
Department of CSE, Chandigarh University, Mohali, Punjab, India
e-mail: priyankaarajput210@gmail.com

© The Author(s), under exclusive license to Springer Nature Switzerland AG 2025

A. Ghosh et al. (eds.), *Quantum Machine Learning in Industrial Automation*,
Information Systems Engineering and Management 65,
https://doi.org/10.1007/978-3-031-99786-0_5

1 Introduction to Quantum Machine Learning Applications

1.1 Overview of Quantum Machine Learning (QML)

Quantum Machine Learning or QML is a cross-disciplinary area of research. It is a diversified field that combines the ideas of quantum computing with machine learning [1]. By merging the different properties of quantum mechanics such as superposition, entanglement as well as the quantum parallelism, QML overcomes the limitations of the classical machine learning algorithms. QML reduces the complexity of the classical model and is much more scalable. It has the potential to address the problems more efficiently than the ongoing conventional techniques. Quantum computers compute the information based on the quantum bits or the quits. These are capable of representing and computing a set of states in parallel which allows for the complex computational tasks at rates impossible for the classical systems [2]. QML algorithms which uses these quantum features have the potential to transform industries by offering quicker insights, low computational expenses, and allowing for the analysis of enormous datasets.

1.2 Relevance and Scope in Modern Industries

Quantum Machine Learning has many applications in the industrial world. High computational intensity as well as the high data complexity industries are likely to gain the most from the QML [3]. It can also help in the pharmaceutical and the healthcare sector by speeding up the drug delivery and facilitating the personalized medicine through faster analysis of the genomic data. In case of finance QML improves the fraud detection system and optimize the trading strategies on the basis of quantum enhanced models. In artificial intelligence, QML notably improves deep learning algorithms by use of quantum-improved optimization, and it further speeds up natural language processing and recommendation systems. Supply chain as well as logistics activities may be optimized via detailed routing along with inventory control solutions, while the energy domain may be assisted through optimized renewable energy modelling in addition to grid optimization. QML's applicability can be greatly extended into each field with a need for wide-ranging computation. As a result, it is an indispensable technology throughout the future.

1.3 Differences Between Classical and Quantum Machine Learning Applications

Traditional computational models set the pillars of classical machine learning, where algorithms are limited by the machine and classical hardware, hindering their potential however; quantum machine learning takes it up a level bringing all the paradigms that reform computation [4]. Traditional systems represent data in binary state (0 and 1 s) while quantum systems use quits with many states at once thanks to superposition at the level of data representation. For instance, traditional systems do all the work with huge volumes of data, which can also take a considerable amount of time and resources, while quantum algorithms such as Grover's search and the Harrow–Hassidim–Lloyd (HHL) algorithm can complete certain tasks exponentially faster than the best classical algorithms. Traditional models struggle with problems like optimization on a big scale and high dimensional analytics to solve complex problems. But quantum models do well because they exploit entanglement and quantum parallelism. In addition, classical machine learning employs traditional algorithms such as gradient descent, support vector machines, and neural networks, whereas QML methods use quantum-based algorithms like quantum neural networks, vibrational quantum circuits, and quantum kernel methods. Despite being in its nascent state, QML is yet to gain full traction, despite the immense benefits that it has to offer [5]. Such challenges of hardware limitations, error rates in quantum systems, and hybrid designs in quantum classically are being fixed through rigorous R&D efforts.

Literature Review

Maheshwari discussed how Quantum technologies have grown to be useful resources across different fields, including chemistry, agriculture, natural language processing, and medicine [22]. Progresses in computing capacity and machine learning algorithms facilitated the advent of Quantum Machine Learning (QML)—an area of increasing size that integrates traditional data handling with quantum computation approaches. QML has generated great interest in medicine, particularly for data processing and classification tasks. Due to its disruptive power in biomedical applications, researchers have extensively studied numerous QML algorithms and quantum circuits applied to health data.

Gowda provides ways in which quantum algorithms could make imaging technologies much less risky and therefore more accurate in diagnostic tests. QML is a revolution that sweeps away all the work of drug discovery. The tremendous potential of computation that quantum systems offer might significantly make the synthesizing and screening of drugs to take much lesser time. It will have the possibility to walk through huge databases of chemicals much more quickly than could be done through normal systems. Scanning detailed molecular shape and quantum-level molecular interactions, a new therapeutic practice may be found for diseases where no treatments presently exist. In this chapter, the promising future of QML applications in medicine is discussed, and the current boundaries and barriers facing integration are investigated [28]. Of primary focus are the susceptibility of quantum computing hardware to external interference and the early stages of quantum algorithms specifically designed for healthcare data. The convergence of quantum computing and healthcare will have fundamental impacts on patient treatment, medical research, and the pharmaceutical development business. This information here provides a general overview of the existing situation, possible advantages, and limitations of using QML in drug development and medical diagnostics. Additionally, it represents a bright future for medical research.

2 Key Principles of Quantum Computing in Machine Learning

2.1 Quantum Superposition and Entanglement

Several principles inherent to quantum computing violate the classical paradigms associated with them. Superposition is a quantum system's ability to occupy multiple states simultaneously—unlike classical binary systems that can only be in one state at any single moment. This factor gives machine learning, with its datasets, never before seen parallel processing powers. A quantum bit, or quit, encodes 0 and 1 simultaneously, offering exponential processing power well beyond the borders of classical computing [6].

Entanglement is an even more profound quantum phenomenon, in which several quits become inextricably linked, so that the state of one quit instantly affects another quit far away, regardless of physical distance. This phenomenon allows quantum systems to do complex computational tasks with levels of efficiency never seen before.

2.2 Quantum Gates and Circuits for Machine Learning

Quantum gates serve as the fundamental computational units in quantum circuits, analogous to logic gates in classical computing. These gates manipulate quit states through precise quantum transformations, enabling complex computational operations. Unlike classical gates that perform binary operations, quantum gates can rotate, flip, and entangle quit states with remarkable precision [7]. This allows for the creation of quantum circuits that can perform sophisticated machine learning transformations that would require exponentially more computational resources in classical systems.

The design of quantum machine learning circuits involves carefully constructed sequences of quantum gates that can perform specific computational tasks. These circuits can implement quantum versions of classical machine learning algorithms, such as support vector machines, clustering algorithms, and neural network architectures. The key innovation lies in the ability to perform high-dimensional linear algebra operations with unprecedented efficiency. Quantum circuits can perform matrix transformations and eigenvalue decompositions that would be computationally prohibitive for classical systems.

Quantum circuit design for machine learning requires deep understanding of both quantum mechanics and computational theory. Researchers develop specialized quantum algorithms that can efficiently prepare, manipulate, and measure quantum states to solve specific machine learning problems [8]. These algorithms often leverage quantum phenomena like interference and entanglement to create computational shortcuts that dramatically reduce computational complexity. The potential applications range from accelerated training of neural networks to more efficient optimization of complex multidimensional problems.

The practical implementation of quantum machine learning circuits faces significant challenges. Quantum systems are inherently fragile, requiring extremely low temperatures and sophisticated error correction mechanisms. Current quantum computers have limited quit counts and high error rates, which restricts their practical applicability. However, ongoing research continues to push the boundaries of quantum computing technology, with major technological companies and research institutions investing heavily in developing more stable and scalable quantum computational platforms.

Quantum machine learning represents a convergence of quantum physics, computer science, and artificial intelligence. The potential applications are profound, ranging from drug discovery and material science to cryptography and complex system modelling. By exploiting quantum superposition and entanglement, these approaches offer the potential to solve computational problems that are fundamentally impossible or prohibitively expensive using classical computing methods.

The future of quantum machine learning depends on continued technological advancement in quantum hardware, error correction techniques, and algorithmic innovation. As quantum systems become more sophisticated, they promise to revolutionize computational approaches across multiple scientific and technological domains. Researchers continue to explore the theoretical foundations and practical implementations of quantum computational techniques, pushing the boundaries of what is computationally possible.

3 Applications of Quantum Machine Learning in Key Domains

3.1 Healthcare: Drug Discovery and Precision Medicine

With unparalleled power to compute molecular models and create custom treatment vaccines, quantum machine learning is transforming medicine. For instance, quantum algorithms can simulate complicated interactions between molecules in drug discovery with unparalleled accuracy, exponentially shortening the timelines from years to months [9]. So instead of modelling molecular interactions using huge amounts of computational power, quantum systems can evaluate multiple molecular conformations at the same time.

Quantum machine learning has yet another important application in precision medicine, which is where quantum machine learning has strong revolutionary potential. Quantum algorithms used for large genomic datasets can identify subtle genetic patterns that accurately predict the risk of future disease.

Quantum-computing-based machine learning models can additionally revolutionize medical imaging analysis. They have the ability to analyse complex medical imaging data, identifying features and potential anomalies that are sometimes impossible even for trained eyes to see. These advanced methodologies have a profound impact on early disease diagnosis, particularly in complex scenarios such as cancer, in which identifying minor structural changes can meaningfully improve patient prognosis. Quantum-enhanced image identification could have the ability to detect precise changes invisible, or unrecognizable by a human audience, or standard computational techniques.

3.2 Finance: Risk Analysis and Portfolio Optimization

The area of finance has increasingly leveraged quantum machine learning to solve complicated optimization problems that are integral to risk management and investment strategy. Quantum algorithms excel at addressing high-dimensional optimization problems that are intractable through classical computational approaches. Although, classically, portfolio optimization is a computationally intensive problem, quantum computational methods can radically redefine our approach, examining a huge number of potential investment scenarios at once.

Quantum machine learning also offers outstanding potential in risk analysis. Through analyses involving wide-ranging financial data sets, accompanied by simulations around detailed market interactions, quantum algorithms can create improved predictive models regarding exactly how markets act [10]. These specific, strong calculation methods are helpful in observing many financial elements within current market conditions. These methods also let finance experts get more thorough risk analyses about possible results, too. Quantum systems' skill within analyses of substantially large datasets as well as searches in the middle of many possible outcomes simultaneously gives a definitive edge above usual computational methods.

Quantum machine learning may be used inside of fraud detection systems for betterment of them through examination of multiple advanced transactional patterns as well as detection of several subtle anomalies. Since quantum solutions can automatically change and spot multiple patterns of advanced actions connected to fraud, fraud detection systems often depend on many pre-set rules [11]. These systems, when handling many data streams simultaneously and spotting detailed links, can yield more effective and responsive financial security systems.

3.3 Supply Chain: Optimization and Resource Allocation

Supply chain management is a complex optimization problem that has a natural fit with quantum algorithms based on machine learning [12]. Quantum algorithms can run many different logistical scenarios in parallel—discovering optimal routing, inventory control and resource allocation strategies.

These computational approaches perform well on large sets of available information while considering multiple variables simultaneously, and provide more comprehensive and effective solutions than traditional computational approaches.

Quantum-driven optimization has the potential to greatly improve inventory control with more accurate models of demand forecasting. Quantum machine learning can also process multiple history datasets, accounting for complex interdependencies to form more accurate predictive models. These algorithms can better pick up subtle shifts in markets, seasonal variations and emerging trends that traditional methods might miss, providing supply chain managers with more accurate decision-making tools by processing vast troves of data and testing multiple potential scenarios at once. The logistics route optimization is another integral field of the quantum machine learning purpose.

Quantum algorithms can consider an array of potential transportation routes in parallel, weighing distance, fuel consumption, traffic patterns, and delivery restrictions. Quantum systems can potentially solve such complex optimization problems exponentially faster than conventional digital computing approaches and create optimal and cost-effective logistical pathways.

Quantum machine learning has broad potential across many use cases; it has the potential to disrupt computational approaches to everything from healthcare to finance to supply chain [13]. As quantum computing technologies mature over time, these advanced computational approaches will increasingly provide unique insights and optimization capabilities, carrying transformative implications for how multidimensional complex challenges are tackled and resolved.

4 Quantum Algorithms in Machine Learning Applications

4.1 *Quantum Support Vector Machines (QSVM)*

Quantum Support Vector Machines can provide an entirely new perspective for classification problems and utilize the rules of quantum computation to far surpass traditional machine learning approaches [14]. Although classical SVMs do not manage well with high-dimensional data, QSVM can easily work with complex, high dimensional feature space due to the quantum superposition and entanglement. At the heart of the breakthrough are quantum kernel methods with the remarkable speed to embed input data in higher-dimensional spaces.

The benefit of the approach in terms of computation is particularly evident when it comes to complex classification problems with considerable datasets. Support vector machined become exponentially computationally costly when applied to high-dimensional spaces, whereas their quantum reciprocals can span multiple feature dimensions simultaneously [15].

Such algorithms leverage quantum circuits to compute kernel functions, making possible the execution of implicit non-linear transformations that would be prohibitively expensive for classical systems. This enables a far more powerful pattern recognition capability, especially in domains like biological data analysis, picture recognition, and the ability to recognize complicated patterns.

Quantum kernel estimation allows for a more complex feature mapping, discovering subtleties in patterns that traditional algorithms may miss. Yet, "quantum" information could also be used in the definition of a kernel: the quantum computational framework is capable of delivering kernel matrices directly as a result of measuring quantum circuits, providing a new kind of insight into the relationships among the data. This performs more complex similarity calculations that can identify complex, nonlinear relationships of classes in the input data.

4.2 *Vibrational Quantum Algorithms (VQA)*

Vibrational Quantum Algorithms are a mixed approach where quantum and classical calculation techniques are joined, creating a useful structure for optimizing hard optimization problems. The algorithms use parameterized quantum circuits, along with the ability to be readily optimized to address multiple computational issues. These circuits can also be dynamically optimized [16]. The innovation is based purely on employing classical optimization methods specifically for quantum circuit parameter optimization in order to maximize synergies deeply within the computational process.

The flexibility with VQAs is applicable among a variety of machine learning tasks, for example, optimization, generative modelling, as well as function approximation for complex functions. Through mindfully designing quantum circuits with many learnable parameters, scientists can develop, for themselves, multiple flexible computational models that can be trained for fitting multiple problem spaces.

The hybridity within these algorithms enables them to take advantage of the strengths within quantum as well as classical computing, solving computational problems that neither model could solve by itself.

VQAs' practical uses tend to include iterative optimization procedures in which quantum circuits produce candidate solutions while classical optimization methods improve upon these solutions. This allows for computational approaches that are more resilient and flexible. This is especially true in areas involving complex, high-dimension optimization problems. The dynamic adjustment within quantum circuit parameters offers absolutely unmatched flexibility throughout solving notably detailed computational problems.

4.3 Quantum Reinforcement Learning

Quantum reinforcement learning is a new method for building flexible learning systems able to explore complicated decision-making areas. By fully harnessing quantum computational concepts, these algorithms are able to explore an entire collection of possible action routes in parallel, greatly speeding up the learning process.

Quantum systems are able to represent as well as handle policy and value functions with impressively greater computational efficiency, which makes it possible to explore especially hard decision spaces in further detail [17].

Quantum reinforcement learning's key advantage exists in that it may explore throughout potential strategies by means of quantum superposition concurrently. Usually, reinforcement learning algorithms explore action spaces in an iterative way, but quantum methods are able to look at many potential trajectories at the same time. This can allow for greater detailed exploration within difficult decision environments, especially in scenarios with many interactive variables and indeterminate outcomes.

Quantum reinforcement learning has many practical applications within a variety of fields. Furthermore, it has applications within robotics, calculated decision making, and optimization of complex systems. These methods may result in the development of more clever and adaptive decision-making systems by enabling a more powerful investigation into decision spaces. The use of quantum computing techniques for processing state spaces that are complex and high-dimensional offers a novel potential regarding the construction of more powerful systems of artificial intelligence.

Active progress within quantum algorithms for machine learning shows a main change inside computation. Quantum approaches, by radically changing how we picture computing tools, could help resolve undergoing computing issues typically considered too difficult [18]. As quantum computing technologies advance on a constant basis, these algorithmic methods in an advanced state will gradually offer computational capabilities that are unmatched for the tackling of detailed computational challenges across a wide variety of areas.

5 Challenges in Realizing Quantum Machine Learning Applications

5.1 *Quantum Hardware Limitations*

Quantum computing hardware poses the biggest challenge to realizing advanced machine learning applications. Current quantum systems are troubled with physical constraints, greatly limiting their use in practice. Qubits represent greatly delicate computational elements. Such elements call for close to absolute zero temperatures and also extraordinary shielding facing ecological noise [19]. Many current quantum processors are able to maintain coherent quantum states for microseconds, posing large obstacles to advanced computational procedures.

The current generation of quantum hardware is characterized through its limited qubit counts and non-negligible error rates. However, common computers can do many operations with almost full consistency, but quantum computers battle to be computationally reliable. Each additional quit markedly increases the system's quantum coherence complexity. The susceptibility of the system's quantum coherence is also greatly increased by each added quit. This basic limit implies that even quantum algorithms, advanced as they are, are limited by the physical constraints from current quantum hardware.

Varied quantum computing architectures pose special challenges for people. Superconducting qubits, along with trapped ion systems as well as topological qubits, each have their own respective strengths in addition to weaknesses. Superconducting qubits compute with speed, yet they exhibit a propensity toward errors. Even though systems using trapped ions have quantum states with more stability, they function with decreased speeds. Presently, finding a quantum computing platform that works for everything is a remaining problem.

5.2 *Data Pre-processing for Quantum Systems*

Preparing classical data for processing by quantum computers presents a challenge that is difficult as well as subtle. Quantum algorithms demand certain data encoding methods, and classical information can be effectively cast onto quantum states through those methods. Classical data representations may be unable for a direct translation into quantum computational models, thereby prompting the need for certain advanced data transformation methods.

Quantum state preparation, in principle, is a process involving preparing quantum circuits so they encode classical information in a form for compatibility with quantum computational principles [20]. Data dimensionality, feature scaling, and computational efficiency must be into taken consideration of. For the sake of preserving necessary information, researchers must create new encoding methods while conforming to the odd computational nature of quantum systems.

The entire process of encoding such data should also consider something. That something is the naturally probabilistic nature of quantum computation. Quantum representations of data, as opposed to standard binary representations, use complex probability amplitudes, which can considerably change the computational method. It involves creating new machine learning approaches that can adequately manage quantum probability distributions and deal with the natural uncertainties of quantum computing systems.

5.3 Scalability and Error Correction

Scalability is certainly among prominent challenges in quantum machine learning. As quantum hardware grows by adding qubits, keeping its computational accuracy is a challenge. Error correction methods are quite necessary to build trustworthy quantum computational structures; however, they induce large computational overhead within them. Quantum error correction necessitates the use of extra qubits for detecting and correcting computation errors. This can potentially lower the effective computational power of the system by orders of magnitude.

Quantum error correction methods involve creating extra quantum data that can find, as well as fix, computer mistakes without destroying the original quantum state. The methods do generally require multiple physical qubits to provide one logical qubit for improved computation stability. The cost of correcting mistakes can be meaningful, and creating a single reliable logical qubit might require many physical qubits.

The making of truly fault-tolerant quantum computer systems remains as an active research issue [21]. Approaches focus on creating better quantum error-correcting codes and seeking new quantum computer designs that are naturally more stable. A way in which to address scalability issues involves topological quantum computing, relying on unusual quantum states more resistant by nature to outside disturbance.

The route toward the practical applications absolutely depends upon parallel advances. These advances occur across several different fields. Progress within quantum hardware, in data encoding, as well as through error correction techniques must, in conjunction, develop to achieve the ultimate potential within quantum computational methods. Even though there are many challenges on the horizon, active study regarding this subject improves quantum computational capabilities, offering promising understandings into what future advanced machine learning technology may resemble.

6 Future Trends in Quantum Machine Learning Applications

6.1 Hybrid Quantum–Classical Systems

The most promising short-term strategy for implementing quantum machine learning is to use hybrid quantum–classical computational systems. These systems use classical computing infrastructure for general system management and coordination while leveraging quantum processors for certain computational subtasks [23]. The core innovation is the development of computational frameworks that optimize performance across complicated problem domains by dynamically allocating computational resources between quantum and conventional systems. The computational approach entails pinpointing particular computational problems where quantum methods offer observable benefits. Complex linear algebra operations, high-dimensional data transformations, and some optimization tasks that are computationally expensive for classical systems are all solved well by quantum systems. Through meticulous design, Scientists can create computational models that take advantage of the unique strengths of both quantum and traditional computational models through the use of hybrid architectures.

The parameterized quantum circuits employed in today's hybrid systems can be dynamically adjusted using the application of standard optimization methods. This provides a means by which more flexible and dynamic computational methods can evolve according to specific issue requirements. The approach provides an ability to phase in quantum computational capability gradually, providing a useful path to evolving more sophisticated machine learning systems.

6.2 Quantum Advancements in Artificial Intelligence

By making more sophisticated computational methods possible for troublesome learning tasks, quantum computing can potentially revolutionize artificial intelligence. Of particular promise are quantum neural networks, capable of processing high-dimensional data with levels of computational efficiency previously unknown [24]. Quantum neural structures could simultaneously follow multiple paths of computation through quantum superposition compared to linearly constrain classical neural networks.

More sophisticated pattern recognition abilities can be created through machine learning models fuelled by quantum computing methods. Quantum computers, by simultaneously studying a number of potential paths to compute, can explore complex, high-dimensional data. This enables more comprehensive data analysis methods that can identify subtleties in patterns and relationships that conventional computational processes might overlook.

The aim of creating quantum-improved machine learning methods is to construct more intelligent and adaptive computational systems. For instance, quantum reinforcement learning methods have the potential to explore multiple probable decision pathways simultaneously, making it possible to apply more intricate adaptive learning strategies. Such innovative computing techniques have the promise to create artificial intelligence systems with greater reasoning capabilities.

6.3 Integration into Industry 4.0

Industry 4.0 is a technology revolution with regard to intelligent, networked industrial systems. Quantum machine learning technologies are poised to play an instrumental role in enabling more sophisticated industrial optimization methods. Manufacturing, logistics, and complex system design represent some of the fields in which quantum computational approaches can potentially work on multiple variables within a system in parallel and yield optimal configuration methods.

Predictive maintenance is among the most critical use cases in which quantum machine learning can have immense benefits. Processing vast quantities of sensor data while simultaneously probing into different potential modes of failure, quantum algorithms will be able to construct more accurate predictive models. These advanced computer techniques will recognize subtle patterns indicative of imminent piece of equipment breakdowns in advance, considerably reducing maintenance costs as well as boosting operational effectiveness.

Supply chain optimization is another life-changing domain where quantum machine learning can be provided [25]. Quantum computational techniques are able to handle multiple logistical factors simultaneously, generating more sophisticated routing and resource planning strategies. Through the processing of complex, multi-dimensional data with unprecedented computational velocities, these technologies can allow companies to construct stronger and more adaptive industrial systems.

The integration of quantum machine learning within Industry 4.0 is not merely an improvement in technology—it is a reimagining of computational responses to industrial issues at its heart. As further developments in quantum technologies take hold, they will be capable of enabling more responsive, intelligent, and efficient industrial systems that dynamically adapt to high-level operational problems.

7 Case Studies of Quantum Machine Learning Applications

7.1 Pharmaceutical Drug Discovery

Quantum machine learning has demonstrated revolutionary potential in drug research, particularly molecular modeling and drug discovery procedures [26]. Researchers at IBM and Google have succeeded in building quantum algorithms that can simulate complex molecular interactions with unprecedented precision. These computational techniques can analyze the molecular geometries of possible drug molecules simultaneously, considerably reducing the time and computational resources required for developing drugs.

A notable case study is the application of quantum computational techniques in COVID-19 treatment research. Quantum algorithms were successful in simulating protein interactions and potential drug binding profiles, which led to the identification of potential molecular compounds faster than traditional computational methods. By leveraging quantum superposition, researchers were able to examine numerous molecular configurations simultaneously, accelerating the discovery of potential therapeutic agents during the global pandemic.

Quantum machine learning's computational ability enables researchers to examine complex molecular data sets that are computationally intractable by classical systems [27]. These techniques can even evaluate thousands of potential molecular structures simultaneously, picking up on subtle patterns of interaction that might be missed by classical computational techniques. Such a capability represents a revolutionary advance in computational drug discovery.

7.2 Financial Risk Assessment

Banks have also begun exploring quantum machine learning for more advanced risk analysis and portfolio optimization methods. Goldman Sachs and JPMorgan Chase have made significant investments in quantum computational research, developing extremely sophisticated algorithms that can compute intricate financial data sets and analyze multiple possible market scenarios at once.

Quantum algorithms demonstrate exceptional performance in portfolio optimization by taking into account several investment alternatives simultaneously in high-dimensional, intricate financial landscapes. These computational techniques are able to process large amounts of historical market data, identifying subtle patterns and risk factors with high accuracy. The ability to consider multiple potential scenarios simultaneously provides financial analysts with more sophisticated risk assessment tools.

One of the strongest case studies is constructing quantum-strengthened fraud detection platforms. Since quantum machine learning algorithms can process several transactional data sets in parallel, they can identify advanced patterns of fraudulent activity that traditional systems cannot. Such high-level computation provides financial institutions with more potent security tools that can adapt to shifting fraudulent techniques.

7.3 Climate Modelling and Environmental Research

Quantum machine learning is increasingly contributing to climate science by making more sophisticated environmental modelling approaches possible. Researchers in national laboratories and universities are developing quantum computational Methods that can manipulate complex climate information with unparalleled computation speed.

Climate modeling entails the calculation of vast, multi-dimensional data sets of atmospheric conditions, oceanic currents, and geologic interactions [29]. Quantum algorithms can search many possible climate scenarios simultaneously, generating more accurate models of long-term environmental change. These computational techniques can identify the subtle interactions between different environmental systems that classical modeling techniques might miss.

A good example of significant implementation is quantum-enhanced climate prediction models that are capable of processing satellite data, geological data, and atmospheric measurements simultaneously. Such advanced computational techniques allow scientists to better understand complex environmental systems, making climate change mitigation more effective.

7.4 Ongoing Research and Future Perspectives

There is also active research on creating more robust quantum computational structures and further expanding practical application domains. Major technology companies and research institutions continue to fund the quantum computing infrastructure, realizing the revolutionary potential these new computational techniques hold.

Cooperative research initiatives between technology companies and universities are accelerating innovations in quantum machine learning capacity at an exponential rate. The initiatives aim to break the current limitations of quantum hardware, develop more sophisticated quantum algorithms, and expand practical application strategies in many scientific and industrial applications.

Future quantum machine learning promises the ability to increasingly sophisticated computational tools capable of handling complex problems across disciplines. As quantum hardware technologies become more sophisticated, the advanced computational tools will provide capabilities to solve intractable complex computational challenges beyond the scope of current classical systems.

8 Conclusion and Future Outlook

Quantum machine learning is a ground-breaking computational paradigm that has the potential to transform computational approaches in a wide range of scientific and industrial applications. The integration of quantum computation concepts with advanced machine learning methodology has unparalleled potential for surmounting challenging computational issues that traditional systems deem insoluble.

The strength of quantum machine learning in itself is its ability to examine all possible paths of computation at the same time through quantum superposition and entanglement. Unlike classical processing techniques limited by linear processing, quantum systems can potentially process high-dimensional data with unimaginable efficiency. The capacity to do so enables more sophisticated pattern recognition, optimization, and predictive modeling across many areas of application.

A few of the notable research landmarks have demonstrated the potential of quantum machine learning in important areas of drug research, risk management for finance, climatology, and optimization problems. Quantum algorithms successfully resolved computational challenges in drug development, portfolio management, and ecological prediction, revealing glimpses into the revolutionary power of these pioneering computational techniques.

Substantial challenges remain for the realization of full-scale quantum machine learning applications. Current quantum hardware limitations, including qubit numbers and high error rates, are significant barriers to universal adoption. Researchers continue to work on more fault-tolerant quantum computational architectures, focusing on hybrid quantum–classical systems that are capable of leveraging the strengths of both computational frameworks.

Guidelines for future research will understandably move toward a few prominent areas. Better quantum hardware with improved qubit coherence is the first priority of research. Better error correction techniques and more fault-resistant quantum computational structures will be needed to construct viable quantum machine learning structures. Researchers will continue to explore new quantum algorithms that can effectively encode classical information into quantum computational frameworks.

Quantum machine learning integrated into Industry 4.0 and next-generation technology systems has the capability to create more intelligent and responsive computing systems. With the evolution of quantum technologies, these are likely to enable more sophisticated methods of solving complex optimization problems, predictive modeling, and creating artificial intelligence.

Collaborative work between research institutions, technology companies, and universities will be central to the future of quantum machine learning. Interdisciplinary approaches that combine quantum physics, computer science, and domain expertise will be crucial to realizing the potential of these ground-breaking computational techniques.

Long-term quantum machine learning vision is far beyond incremental advances in technology. These new approaches to computation promise to radically shift the way we process complex information, optimize, and design better computational systems. As quantum technology continues to improve, it will likely create entirely new computational models that lie outside our present technology imagination.

References

1. Biamonte, J., Wittek, P., Pancotti, N., Rebentrost, P., Wiebe, N., Lloyd, S.: Quantum machine learning (2016). https://doi.org/10.1038/nature23474
2. Huggins, W., Patil, P., Mitchell, B., Whaley, K., Stoudenmire, E.: Towards quantum machine learning with tensor networks (2018). https://doi.org/10.1088/2058-9565/aaea94
3. Havlíček, V., Córcoles, A., Temme, K., Harrow, A., Kandala, A., Chow, J., Gambetta, J.: Supervised learning with quantum-enhanced feature spaces (2018). https://doi.org/10.1038/s41586-019-0980-2
4. Allcock, J., Zhang, S.: Quantum machine learning (2018). https://doi.org/10.1093/nsr/nwy149
5. Benedetti, M., Lloyd, E., Sack, S., Fiorentini, M.: Parameterized quantum circuits as machine learning models (2019). https://doi.org/10.1088/2058-9565/ab4eb5
6. Carrasquilla, J.: Machine learning for quantum matter (2020). https://doi.org/10.1080/23746149.2020.1797528
7. Khan, T., Robles-Kelly, A.: Machine learning: quantum vs classical (2020). https://doi.org/10.1109/ACCESS.2020.3041719
8. O'Quinn, W., Mao, S.: Quantum machine learning: recent advances and Outlook (2020). https://doi.org/10.1109/MWC.001.1900341
9. Buffoni, L., Caruso, F.: New trends in quantum machine learning (2020). https://doi.org/10.1209/0295-5075/132/60004
10. Hassanzadeh, P.: Towards the quantum-enabled technologies for development of drugs or delivery systems (2020). https://doi.org/10.1016/j.jconrel.2020.04.050
11. Li, W., Deng, D.: Recent advances for quantum classifiers (2021). https://doi.org/10.1007/s11433-021-1793-6
12. Batra, K., Zorn, K., Foil, D., Minerali, E., Gawriljuk, V., Lane, T., Ekins, S.: Quantum machine learning algorithms for drug discovery applications (2021). https://doi.org/10.1021/acs.jcim.1c00166
13. García, D., Cruz-Benito, J., García-Penalvo, F.: Systematic literature review: quantum machine learning and its applications (2022). https://doi.org/10.1016/j.cosrev.2024.100619
14. Wang, X., Lin, Z., Che, L., Chen, H., Lu, D.: Experimental quantum-enhanced machine learning in spin-based systems (2022). https://doi.org/10.1002/qute.202200005
15. Marshall, S., Gyurik, C., Dunjko, V.: High dimensional quantum machine learning with small quantum computers (2022). https://doi.org/10.22331/q-2023-08-09-1078
16. Simoes, R., Huber, P., Meier, N., Smailov, N., Füchslin, R., Stockinger, K.: Experimental evaluation of quantum machine learning algorithms (2023). https://doi.org/10.1109/ACCESS.2023.3236409
17. Manjunath, T.D., Bhowmik, B.: Quantum machine learning and recent advancements (2023). https://doi.org/10.1109/AISC56616.2023.10085586
18. Zaman, K., Marchisio, A., Hanif, M. A., Shafique, M.: A survey on quantum machine learning: Current trends, challenges, opportunities, and the road ahead (2023). https://doi.org/10.48550/arXiv.2310.10315
19. Sarkar, S.: Quantum machine learning (2023). https://doi.org/10.22214/ijraset.2023.49421
20. Liu, J., Liu, M., Liu, J., Ye, Z., Alexeev, Y., Eisert, J., Jiang, L.: Towards provably efficient quantum algorithms for large-scale machine-learning models (2023). https://doi.org/10.1038/s41467-023-43957-x
21. Olivera-Atencio, M., Lamata, L., Casado-Pascual, J.: Benefits of open quantum systems for quantum machine learning (2023). https://doi.org/10.1002/qute.202300247

22. Bhatia, A., Saggi, M., Kais, S.: Quantum machine learning predicting ADME-Tox properties in drug discovery (2023). https://doi.org/10.1021/acs.jcim.3c01079
23. Melnikov, A., Kordzanganeh, M., Alodjants, A., Lee, R.K.: Quantum machine learning: from physics to software engineering (2023). https://doi.org/10.48550/arXiv.2301.01851
24. Lamata, L.: Quantum machine learning implementations: proposals and experiments (2023). https://doi.org/10.1002/qute.202300059
25. Avramouli, M., Savvas, I., Vasilaki, A., Garani, G.: Unlocking the potential of quantum machine learning to advance drug discovery (2023). https://doi.org/10.3390/electronics12112402
26. Wong, Y., Zhou, Y., Liang, Y., Qiu, H., Wu, Y., He, B.: The new answer to drug discovery: quantum machine learning in preclinical drug development (2023). https://doi.org/10.1109/PRML59573.2023.10348356
27. Ullah, U., Garcia-Zapirain, B.: Quantum machine learning revolution in healthcare (2024). https://doi.org/10.1109/ACCESS.2024.33534617
28. Gowda V.D., Kumar, A., Ananth, B., Vasanthakumar, G., Singh, M.: 3 Quantum machine learning in healthcare: diagnostics and drug discovery. In: Raj, P., Song, H., Le, D., Vyas, N. (eds.) Quantum Machine Learning: Quantum Algorithms and Neural Networks, pp. 39–64. De Gruyter, Berlin, Boston (2024). https://doi.org/10.1515/9783111342276-003
29. Tomar, S., Tripathi, R., Kumar, S.: Comprehensive Survey of QML: From Data Analysis to Algorithmic Advancements (2025). https://doi.org/10.48550/arXiv.2501.09528

Understanding Quantum Computing: From Qubits to Quantum Algorithms

Vikash Kumar, Ankush Joshi, Gesu Thakur, Naman Chauhan, and Ritik Kumar Singh

Abstract Quantum computing is a revolution in computing. This chapter outlines some basic concepts governing the principles of quantum computers and introduces quantum systems. The chapter explains qubits, Quantum computing utilizes superposition and entanglement as data structures, making the differences between quantum data handling and traditional data handling. Fundamental mathematical and physical principles of the quantum state are about to be addressed, explaining how quantum algorithms can be constructed. The chapter also introduces some major Quantum algorithms, such as Grover's search algorithm and Shor's factoring algorithm, play a significant role in the field of quantum computing to show how quantum computing can completely change the fields of data science, cryptography, and optimization. Quantum circuits, error correction, and quantum gates—essential tools for constructing practical quantum computers—are also discussed in this chapter. Moreover, the chapter discusses the status of quantum computing technology today, with emphasis on progress in theory and practice. It also discusses the challenges to be developing scalable quantum systems, such as hardware constraints, noise, and quantum decoherence. Finally, this chapter establishes a firm ground for anyone looking to grasp the principles, possibilities, and problems of quantum computing, and prepares the ground for a more thorough investigation of this revolutionary technology. As this area continues to grow, quantum computing knowledge will become an integral component in defining the future of computational science and technology.

V. Kumar (✉) · A. Joshi · G. Thakur · N. Chauhan · R. K. Singh
COER University, Roorkee, India
e-mail: vikashpaper1991@gmail.com

A. Joshi
e-mail: ankushjoshi1987@gmail.com

G. Thakur
e-mail: drgesuthakur@gmail.com

N. Chauhan
e-mail: namanc206@gmail.com

R. K. Singh
e-mail: ritiksingh0101@gmail.com

© The Author(s), under exclusive license to Springer Nature Switzerland AG 2025

A. Ghosh et al. (eds.), *Quantum Machine Learning in Industrial Automation*, Information Systems Engineering and Management 65,
https://doi.org/10.1007/978-3-031-99786-0_6

Keywords Quantum algorithms qubits and superposition · Quantum entanglement · Grover's and Shor's algorithm · Quantum decoherence · Error correction

1 Introduction

Nobel Laureate Richard Feynman first put forth the concept of a quantum Calculator in 1981. Helium renowned that amp standard calculator would not work fit to accurately and expeditiously Representation measures automatic systems just that amp green case of twist amp calculator itself built of measure automatic elements which obey measure automatic torah would get yet work fit to Representation measure systems in effect. Because the quantity of Information needed to fully represent a quantum system grows exponentially classical Calculators are unable to mimic such a system utilizing sub-exponential time and space Complicated. Not with standing away attractive employ of the non-classical characteristics of the measure systems they are founded along measure Calculator get work exponentially big amounts of information inch good polynomial sentence. So, this Method Ing skill could very use for best solve other problems outside the domain of quantum physics. Mechanization and suitable arsenic level computation swelled much faster than its initial establishment of optional Newtonian mechanization imitations. The curriculum of measuring Procedures has now greatly extended into modelling measure natural systems for determining and comprehensive application set, including mathematical Information theory, cryptography, and speech theorization. Quantum computing thereby enters significant grounds within computational paradigms by leveraging measurement information in ways posited as unachievable through standard computers. Since quantum bits (qubits) can be in a superposition of states, sometimes also entangled with each other, they can perform complicated calculations very efficiently, as opposed to the classical bits where phenomena take place only in one of the two states 0 and 1. Herein, this provides a grounded overview of some principles of quantum computing and also describes what one does with them when modelling a large scale quantum computer [1]. Table 2 shows different Applications Area of Quantum Computing.

2 Fundamentals of Quantum Computing

2.1 The Basic Unit of Quantum Information

Quantum bits or qubits are like the bits found in classical computers.

$$0 \rightarrow |0\rangle$$

$$1 \rightarrow |1\rangle$$

Despite its limited number of possible units, Information of the fundamental unit of quantum Computing.

$$|Y\rangle = x_1|0\rangle + x_2|1\rangle$$

Specifies the quantum phase of a particular qubits. The term qubit is used to describe a quantum information bit. When objects need exist in either state at the same time they are called qubits. Bits can be either 0 or 1, unlike traditional computers where each bit has no capacity. The variables x_1 and x_2 fictitiously (complex probability amplitudes) assume values. Blurred (complex probability amplitudes) are signified by the complex variables x_1 and x_2 (beta) associated with a square modulus (in the entire range) of them. Successive performances of the operator can eventually come to the steady states represented by the individual states $|0\rangle$ and $|1\rangle$.

The probabilities always add up to one:

$$|x_1|^2 + |x_2|^2 = 1$$

In this equation x_1 and x_2 stand for complicated coefficients that meet the normalization condition (2.2), which makes physical sense for a quantum state. In difference to bits or qubits are 2D vectors i.e. reside in the Hilbert space. Their physical realizations are one atom, one electron, one photon, or one superconducting circuit in the cooled state with moving electrons [2].

A qubit measurement probability distribution graph visually depicts how likely a qubit is to collapse to a certain state when it is measured. A qubit collapses when it is measured, which means it can shift to one of two classical states, $|0\rangle$ or $|1\rangle$, with a certain chance based on its quantum state. This occurs because of the superposition of qubit states. Qubit Measurement Probability Distribution is shown in Fig. 1.

The two possible states are represented in the graph description.

$|0\rangle$ (blue bar): Likelihood that measuring the qubit causes it to collapse to state $|0\rangle$.

$|1\rangle$ (red bar): The chance that the qubit will end up in the $|1\rangle$ state when we measure it. The measurement probability, or the Y-axis, shows the odds of the qubit landing in either state, with values ranging from zero to one [3].

Superposition and Measurement

Superposition enables a qubit to run in more than one phase at the same time. A characteristic that classical bits lack. Upon measurement, the qubit settles on $|0\rangle$ or $|1\rangle$ based on probabilities $|\alpha|^2$ and $|\beta|^2$. The Bloch sphere presents the approach of visualizing the quantum country of a unmarried qubit in three dimensions. It affords an intuitive manner to recognize the lifestyles of superposition and the motion of

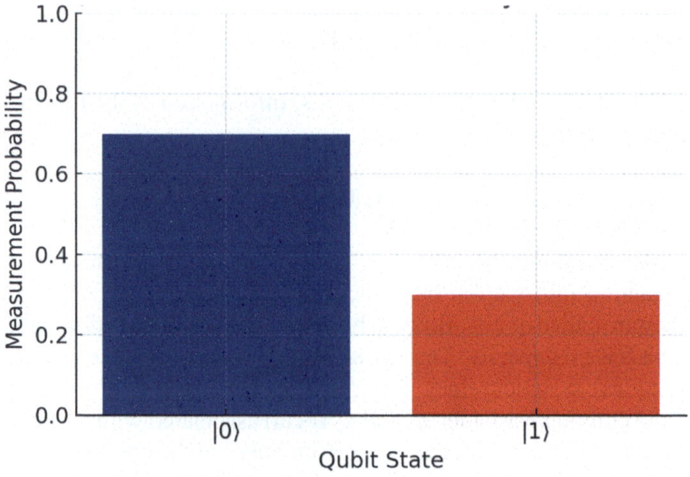

Fig. 1 Qubit measurement probability distribution

quantum gates. Gives one a clearer view of how qubits are manipulated. Demonstrates the rotational use of quantum gates on qubits. Helps to illustrate the difference between a quantum computer and a classical one which operates with bits in the state $|0\rangle$ or $|1\rangle$ [4].

Superposition: An Essential Quantum Idea

A bit in classical computing may be Binary value. In contrast, a qubit in quantum computing exist in a superposition of both phase at the identical time. This manner that a qubit represents a mixture of 0 and 1, each with certain probabilities, as opposed to being confined to simply one of these values. When measured, these probabilities dictate whether the qubit will crumble to 0 or 1. Quantum computers are a good deal more effective than conventional computers for precise obligations because of their potential to exist in more than one states right now, letting them system good sized quantities of facts in parallel. Thanks to superposition, quantum computer systems can carry out many operations simultaneously, and the Bloch sphere affords a useful visualization of the way qubits function. Understanding these concepts is essential to grasp quantum computing and its potential to real world problem-solving in areas like artificial intelligence, shorting and cryptography.

Bloch Sphere Qubit Visualization

The phase of a qubit is show in three dimensions by the Bloch sphere. It offers a natural approach to comprehend quantum operations and superposition. Qubits can reside anywhere on the sphere, in contrast to classical bits, which have defined locations (either 0 or 1).

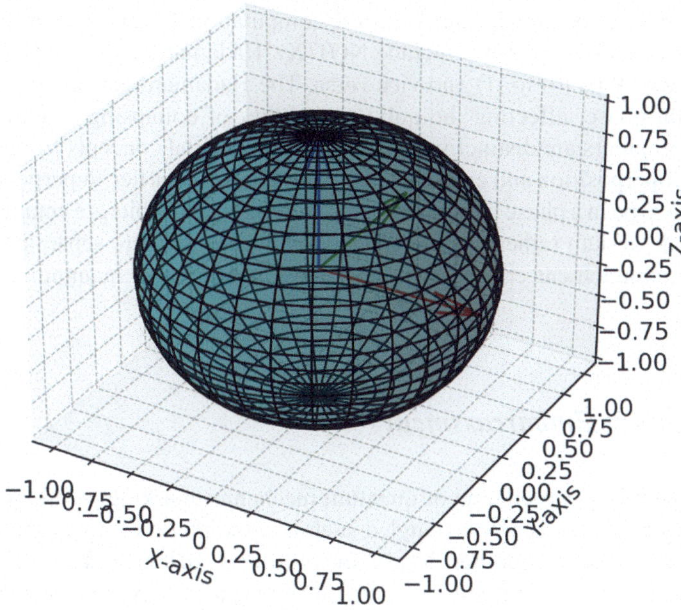

Fig. 2 Bloch sphere: this shows a qubit in visual

The Bloch sphere facilitates the visualization

Superposition: Qubits are found across the sphere, not only at the poles that stand for 0 and 1.

Quantum Gates: Quantum gates change the state of qubits without measuring them by rotating them around various axes of the sphere (Fig. 2).

Phase Difference: By displaying qubits' phase, a special quantum characteristic that affects processing capacity, the sphere helps demonstrate how qubits vary from classical bits [4].

2.2 *Quantification*

A qubit may be existed in a superposition value of $|0\rangle$ and $|1\rangle$. Measurement causes the qubit to collapse to either state $|0\rangle$ or $|1\rangle$, but it does not predict which it will be. It does, however, allow calculation of the probability i.e. A qubit is collapse to a particular phase. Among other models, such as the quantum Turing machine and the adiabatic quantum computing and quantum circuit model is by far the most popular one used for quantum computing. This model utilizes quantum gates to manipulate the input qubit, like how classical computers operate. By using a series of gates, complex calculations can be performed. To fully characterize a quantum gate, one can describe how it transforms the computational basis, such as $|0\rangle$ or $|1\rangle$. The most frequently used gates include NOT(X), Hadamard, and CNOT(CX). The basic NOT gate flips $|0\rangle$ to $|1\rangle$ and vice versa. The Hadamard gate creates an equal superposition phase, affecting the qubit probability of collapsing to either phase. Quantum computers are distinctive due to two fundamental concepts from quantum mechanics superposition and entanglement. As mentioned earlier, superposition will allow a qubit to be in a phase between $|0\rangle$ and $|1\rangle$. When two qubits are entangled, they act as a single system rather than independently even if they are separated by large distances. Entanglement enhances the information density of quantum computers [5].

2.3 *Quantum Embarrassment*

Embarrassment is a cornerstone of quantum mechanics marking an anomaly when more than two qubits become entangled such that the phase of one qubit instantly references the phase of another despite everything of distance. This implies that knowing the phase of one qubit measures of the right of way. The Mechanism of Entanglement Think about two qubits in the Bell state or another entangled state:

$$|\Phi^+\rangle = \frac{1}{\sqrt{2}}(|00\rangle + |11\rangle)$$

- Until they are measured, the two qubits in this instance lack distinct individual states.
- The second qubit must likewise be $|0\rangle$ if the first qubit is measure and determined $|0\rangle$.
- Other qubit will be $|1\rangle$ if the first qubit is measured as $|1\rangle$. Even if there are significant distances between the qubits, this correlation still holds true [6].

2.3.1 Important Entanglement Properties

- **Non-locality**: Even over great distances, the state of the entangled partner can be instantly ascertained by measuring one qubit.
- **Quantum Correlation**: Entanglement correlations, in contrast to classical correlations, cannot be accounted for by prior knowledge.
- **Utilized in Quantum Computing**: Entanglement plays a decisive role in quantum cryptography Rules like Quantum important Distribution (QKD) and enables techniques such as superdense coding and quantum teleportation [7].

3 Quantum Algorithm with Multiple Qubits

Overview

A quantum computer is based on the principles of superposition and Embarrassment where qubits are used to perform computations rather than the traditional bits. Complex computations and processes can be performed using multi-qubit systems (quantum gates, entanglement, quantum parallelism). Multi-qubit algorithms enable powerful applications like Grover's explore and Shor's Factoring and reside at the heart of quantum computing. Before you can leverage the capabilities of quantum computing, you need a clear understanding of the fundamentals of entanglement, gate operations, and measurement [8].

3.1 *Multiple-Qubit Display*

The representation of a system with n qubits is as follows:

$$|Y\rangle = \sum (x_i|i\rangle) \text{ for } i = 0 \text{ to } 2^{n-1}$$

Multi-qubit Algorithm

Step 1: Initialize the Qubits

All qubits are initialized in the |0 state:

$$|Y\rangle = |00\ldots0\rangle$$

Put the Hadamard gates to each qubit to generate a superposition phase.

$$h^{(\otimes N)}|00\ldots0\rangle = \left(1/\sqrt{(2^n)}\right)\sum |i\rangle \text{for } i = 0 \text{ to } 2^{n-1}$$

Step 2: Utilize Quantum Gates in Step Two

- Entanglement is accomplished by the CNOT Gate.
- **Pauli Gates (x, y, z)**: Execute phase shifts and bit flips.
- **Phase Gates (S, T)**: Incorporate regulated phase changes.

Step 3: Quantum Fourier Transform (QFT)

QFT is essential quantum algorithms and is applied as:

$$\text{QFT}|x\rangle = \left(1/\sqrt{N}\right)\sum e^{(2\pi_i xk/N)}|k\rangle \text{for } k = 0 \text{ to } N-1$$

Step 4: Measurement

Collapse the quantum state into classical bits by measuring the qubits.

4 Shor's Factorization Algorithm

Shor's algorithm is a quantum computing that presents an important challenge to traditional encryption systems, such as RSA, by allowing the efficient factoring of large numbers. Utilizing the principles of quantum computing, it solves the integer factorization problem significantly quicker than any other classically best-known algorithms. This problem is almost insurmountable for classical computers; thus, Shor's algorithm showcases what quantum computing can do. Of course, its potential implications in the field of cryptography make it is most important algorithms of Quantum Computing [9].

4.1 *Shor's Algorithm Steps*

Estimating Quantum Phases to Assess Periodicity

The objective is to identify the non-trivial elements of a big composite number N.

A random integer and is selected so that

$$\gcd(a, N) = 1 \text{ and } 1 < a < N.$$

The function has periodic behaviour with interval.

Note that $f(x) = a^x \bmod N$ means there is an integer period r, s.t.

$$f(x + r) = f(x)$$

With the help of quantum Fourier transform (QFT) and quantum phase estimation, we can efficiently compute period r on a quantum computer; our measurement collapses the state into an approximation of after modular exponentiation and QFT on a superposition over x [10].

Determining Factors by Using the Greatest Common Divisor (gcd)

After determining the time r, the following calculation must be done:

$$\gcd\left(a^{(r/2)} - 1, N\right)\text{not equal to 0 mod N } \left(a^{(r/2)} - 1\right)\left(a^{(r/2)} + 1\right) = 0$$

If the r value is even, then $a^{(r/2)}$ will be congruent to 1 mod N. The gcd operation will yield non-trivial factors of N.3. Deriving Prime Factors using Classical Post-processing.

- To determine the prime factors, the gcd is calculated using the Euclidean technique.
- The process is repeated if the obtained factor is trivial, such as 1 or N.

- The factorization process is finished once non-trivial factors have been identified [11].

4.2 *Importance of the Shor's Algorithm*

RSA encryption is considered secure primarily because all the traditional factorization algorithms, especially the GNFS, can only factor large numbers in subexponential time. In stark contrast, Shor's algorithm runs exponentially faster than any classical algorithm and operates in polynomial time $O((\log N)^3)$. If scalable quantum computing is ever developed, it would suffice to say that RSA encryption would be dramatically weakened, thereby emphasizing the importance of finding new cryptographic algorithms resistant to quantum computers [12].

5 VQAs

A subclass of divisional quantum–classical algorithms known as variational quantum algorithms (VQAs) takes advantage of the strengths of both quantum computing and traditional computers. These algorithms are primarily designed for tasks in machine learning and optimization, utilizing parameterized quantum circuits that are fine-tuned with classical methods to achieve desired outcomes. They are particularly useful in noisy intermediate-scale quantum (NISQ) systems, where practical quantum error correction is not yet feasible [13].

5.1 *Important Algorithms for Variational Quantum*

Variational Quantum Eigen solver (VQE): VQE is a hybrid quantum classic approach that is very useful for simulation of quantum chemistry, as it is mostly used to determine the energy of the ground state of molecular systems.

VQE operates: Initialization: To represent the quantum phase Measurement of quantum circuit to build.

Quantum Execution: Using the selected ansatz, the quantum computer creates and measures quantum states.

Classical Optimization: To reduce the energy expectation value, the parameters are updated by a classical optimizer using the data from the quantum observations.

Iteration: This procedure is repeated until the system Hamiltonian's lowest eigenvalue is reached by the ideal parameters.

Applications of VQE: VQE include the identification of molecules' ground and excited states in quantum chemistry. Electronic structure simulation is a material science application. Resolving intricate many-body problems in nuclear physics [14].

5.2 *Algorithm for Quantum Approximate Optimization (QAOA)*

QAOA is targeted at solving combinatorial optimization problems (e. g., in machine learning, finance, and logistics). QAOA works as follows, and is inspired by adiabatic quantum computing:

Problem Mapping: The optimal solution is encoded by the ground state of a cost encods the solution to reduces problem that is mapped onto the blinding problem.

Set of parameters Quantum Circuit: A circuit with adjustable parameters is built utilizing alternating layers of cost and mixing Hamiltonians.

Quantum Measurement: The cost function's expected value is quantified.

Classical Optimization: To minimize the cost function, a classical optimizer modifies the parameters.

Iteration: This procedure keeps going until the most accurate approximation is discovered.

Applications of QAOA: Applications include graph partitioning issues. Optimization of routing and scheduling. Financial portfolio optimization [15].

5.3 *Benefits of VQA's*

Hybrid Approach: It refines quantum computations using classical optimizers.

NISQ-Compatible: Compatible with the quantum hardware of today.

Scalability: As quantum technology develops, it can be used in larger systems [16].

VQAs' Barren

Plateaus Challenges: There may be areas in optimization landscapes where gradients disappear, which makes training challenging (Table 1).

Hardware Noise: Decoherence and noise are problems with current quantum devices.

Ansatz Selection: The key to success is selecting an efficient parameterized circuit [17].

Table 1 Challenges in quantum computing

Challenge	Description	Possible solutions
Decoherence and noise	Quantum states are fragile and affected by the environment	Develop better error correction codes like Surface Code
Scalability	Increasing the number of stable qubits is difficult	Research in topological qubits and superconducting qubits
Hardware limitations	Qubit coherence times and gate fidelity need improvement	Improve materials and cooling techniques
Algorithm efficiency	Not all problems benefit from quantum speed-up	Develop hybrid quantum–classical algorithms
Quantum network security	Current cryptographic systems are in danger of being harmed by quantum computers	Implement quantum key distribution (QKD) for secure communication

Table 2 Different applications area of quantum computing

Application area	Description	Impact
Cryptography	Quantum computing can break RSA encryption using Shor's algorithm	Calls for quantum-safe encryption methods
Drug discovery and material science	Creating quantum simulations of molecular interactions	Faster drug development and new material discovery
Financial modelling	Optimization of portfolios, risk assessment, and fraud detection	More accurate predictions and better investment strategies
Logistics and supply chain	Solving optimization problems like the traveling salesman problem	Cost reduction and improved efficiency
Artificial intelligence and machine learning	Quantum algorithms for faster training of AI models	Improves pattern recognition and decision-making

6 Quantum Error Correction (QEC) Algorithms

Algorithms for correction of quantum errors (QEC) and other sources of quantum noise cause quantum computers to be highly susceptible to errors. Correction codes of quantum error (QEC) are required to ensure computing correctness. QEC detect and correct errors, without disturbing quantum information. Fault-tolerant quantum computation needs to be supported by Quantum Error Correction (QEC). Surface Code, Steane Code and Shor Code all do the jobs of reducing quantum errors, but with different trade-offs. We will soon achieve fault-tolerant quantum computing through QEC advancements which would enable more complex and useful quantum applications [18].

6.1 Shor Code

The first quantum error correcting algorithm ever developed, known as the Shor algorithm, protects qubits against phase-flip as well as against bit-flip errors. To get nine physical qubits, one logical qubit is produced from one. To protect against bit-flip errors, each of these three-qubit codes further encodes the eight states into another three-qubit code to guard against phase-flip errors. Total number of nine physical qubits are used to show one logical qubit. A logical phase may be probed and corrected for errors without disturbing a quantum state thanks to syndrome measurements. Measuring a qubit destroys its quantum state. This determines that built in quantum memories need to refrain from directly measuring the stored information on which operations such as syndrome measurement rely [19, 20].

Errors in Binary Digit-Flip and Phase-Flip

Binary Digit-Flip Error: A Bit-Flip Error occurs when a computer's quantum transforms from |0 to |1 and vice versal.

Phase-Flip Error: This error affects the phase but not the bit value when a qubit in the superposition $(|0\rangle + |1\rangle)/\sqrt{2}$ changes to $(|0\rangle - |1\rangle)/\sqrt{2}$.

Step 1: Guarding Against Bit-Flip Mistakes

For every logical qubit $|\psi\rangle = \alpha|0\rangle + \beta|1\rangle$, three qubits are encoded:

$$|1\rangle \rightarrow |111\rangle, |0\rangle \rightarrow |000\rangle$$

This encoding aids in bit-flip error detection. For instance, majority voting can be used to fix an error that causes $|000\rangle$ to change to $|010\rangle$ (if two bits are 0, we believe the fault occurred in the third bit and repair it) [21].

Step 2: Guarding Against Phase-Flip Mistakes

Each of the three qubits from Step 1 is then encoded using an additional layer of encoding to fix phase-flip errors:

$$1\rangle \rightarrow (|000\rangle - |111\rangle)/\sqrt{2}|0\rangle \rightarrow (|000\rangle + |111\rangle)/\sqrt{2}$$

This guards against phase faults since a comparable majority voting technique can be used to identify and fix an issue that flips the phase in one qubit. The need to identify and fix errors with syndrome measurements is fundamental in quantum mechanics error correction, as it would require the detection and resolution of any deviations caused by direct measurement of the quantum state [22].

6.2 *Surface Code*

The most popular candidate for quantum error correction codes that allows for scalable quantum computing is the Surface Code. It's got several good properties because it's a 2D lattice of physical qubits. It's got ancilla qubits and data qubits in a planar configuration, errors are detected and corrected without disturbing the quantum state with stabilizer measurements and has a very high threshold for fault tolerance, high enough that errors can be controlled even at high rates. It allows you to build large arrays of physical qubits to create a logical qubit and practical error solving techniques. The Surface program is a candidate for large scale quantum computing because it allows bit flip and phase flip errors. Because of decoherence and other external perturbations, the quantum system is highly subject to errors. Classical computers can use simple redundancy to correct errors because quantum error correction needs to preserve quantum coherence. This is made possible by the Surface Code, which stores a logical qubit over many physical qubits and uses stabilizer measurements to detect and correct errors [23, 24].

Surface Code's Organization

Two different kinds of physical qubits make up the 2D grid on which the Surface Code is implemented:

Qubits of data: The quantum information is stored in these checkerboard-patterned qubits.

Qubits for Syndrome (Ancilla): These qubits are positioned in between data qubits and are employed for error detection [25].

Operation of the Surface Code

By using stabilizer measurements to identify system changes without measuring the quantum state, the Surface Code fixes mistakes. The procedure entails:

Step 1: Error Detection Using Measurements of Stabilizers
Errors can be found using syndrome qubits without interfering with the logical qubit that is encoded. To quantify parity variations brought on by bit-flip or phase-flip errors, each stabilizer qubit communicates with its nearby data qubits.

Step 2: Correcting Errors Applying an Algorithm for Decoding
A decoding algorithm (like a minimum-weight perfect matching method) identifies the most likely error sites after an error is identified and conducts correction operations appropriately.

Step 3: **Protection of Logical Qubits**
Using different number of the physical qubits to encode a logical qubit and ongoing syndrome measurements in Surface Code guarantees.

Error Correction Without Measuring Directly

The Surface Code employs indirect measurements to identify mistakes because measuring a qubit directly destroys its quantum state. Parity checks are used to retrieve the error syndrome information, guaranteeing error detection without interfering with the coherence of the logical qubit [26].

Benefits of the Surface Code

Scalability: Surface Code is appropriate for large-scale quantum computers due to its tremendous scalability.

Significant Fault Tolerance: Even in situations where physical qubits have significant mistake rates, it can effectively rectify defects.

Nearest-Neighbour Interactions: This simplifies hardware implementation by requiring only local qubit interactions.

Error Correction Threshold: The Surface Code can withstand more mistakes than many other quantum error correction codes because of its comparatively high threshold (~ 1%) for quantum error correction.

6.3 Steane Code

The Steane program is a Seven-qubit quantum error solving code from the Calderbank_Shor-Steane family. Classical coding theory is used to correct single qubit errors. It combines the dual of the Hamming (7,4) code for phase flip errors with the code for bit flip errors. 7 physical qubits map to 1 logical qubit. The stabilizer measurements allow to identify and correct the errors. It's a widely used QEC technique in theory and practice of quantum computing as it provides a good trade-off between fault tolerance and efficiency. The Steane Code is a big step forward in quantum error correction as it makes quantum computation fault tolerant. Because it's based on classical Hamming codes it inherently corrects against binary digit flip and phase flip errors [27].

Steane Code's Structure

A quantum information equivalent to the classical [7,4,3] Hamming code is known as the Steane code. The technology enables error detection and correction by splitting one logical qubit into seven physical qubits.

Encoding Procedure

$$\text{In a } 7-\text{qubit state, the logical qubit}|\mathrm{Y}\rangle = \mathrm{x}_1|0\rangle + \mathrm{x}_2|1\rangle \text{ is encoded.}$$

The encoding guarantees that parity checks can be used to identify problems. Syndrome measurements are mainly used to repair binary digiti -flip and phase-flip mistakes.

Error Identification and Fixing

By applying parity tests to ascertain whether an error has occurred, the Steane Code enables error detection. Measurements of syndrome aid in determining the precise kind of problem and implementing the required fix [28].

6.4 Multi-qubit System Difficulties

In quantum computing, decoherence, error correction, hardware constraints, and scaling problems pose serious problems for multi-qubit systems. When qubits interact with their surroundings, decoherence occurs, which results in the loss of quantum information and interferes with calculations. Long-term maintenance of stable quantum states is therefore challenging. Errors can be detected and corrected without collapsing the quantum state thanks to quantum error correction techniques like surface codes and the Shor code. These methods, however, increase computing complexity by requiring more qubits. Scalability is further complicated by hardware constraints, since dependable operations need optimization of qubit coherence time, gate fidelity, and communication between qubits. Topological qubits promise inherent error resistance but are still in the early experimental stages; photonic qubits excel in quantum communication but are challenging to implement for large-scale computation; superconducting qubits enable fast gate operations but require ultra-low temperatures; and trapped ions provide high coherence times but operate slowly. Quantum computing progress must overcome the challenges posed by qubit stability, error reduction, and scalable quantum architectures. This is especially true for these technologies that are already in development [29, 30].

7 Future Research Directions

Creating Quantum Computers That Can Handle Failures: Potential Research Directions The future of multi-qubit quantum computing depends on developing fault-tolerant quantum computers and implementing scalable error correction methods [31].

Developing Quantum Algorithms: Improving algorithms for optimization, AI, and cryptography.

Improving Quantum Hardware: Improving gate fidelities and coherence times are two ways to improve quantum hardware.

Investigating Quantum Machine Learning: Blending AI methods with quantum computers [32].

8 Conclusion

As to the summary, multi-qubit quantum algorithms use quantum physics to address a range of issues, including cryptography and search optimization. However, resolving issues with scalability and error correction is necessary to achieve workable quantum computing solutions. The report emphasizes how areas like artificial intelligence, materials research, and secure communication would be greatly impacted by continued developments in quantum hardware and algorithms. Continuous advancements in these fields will influence the development of quantum computing in the future by propelling advancements in artificial intelligence, materials science, and secure communication.

References

1. Strubell, E.: An introduction to quantum algorithms. COS498 Chawathe Spring **13**, 19 (2011)
2. Hayashi, M., et al.: Introduction to quantum information science. Springer (2014)
3. Korbicz, J.K., et al.: Generalized spin-squeezing inequalities in N-qubit systems: theory and experiment. Phys. Rev. A Atom. Mol. Opt. Phys. **74**(5), 052319 (2006)
4. Aberg, J.: Quantifying superposition. arXiv preprint quant-ph/0612146 (2006)
5. Hassija, V., et al.: Present landscape of quantum computing. IET Quantum Commun. **1**(2), 42–48 (2020)
6. Wootters, W.K.: Quantum entanglement as a quantifiable resource. Phil. Trans. Royal Soc. London. Ser. A Math. Phys. Eng. Sci. **356**(1743), 1717–1731 (1998)
7. Życzkowski, K., et al.: Dynamics of quantum entanglement. Phys. Rev. A **65**(1), 012101 (2001)
8. Farhi, E., et al.: Quantum algorithms for fixed qubit architectures. arXiv preprint arXiv:1703.06199 (2017)
9. Kumar, M., Mondal, B.: Study on implementation of shor's factorization algorithm on quantum computer. SN Comput. Sci. **5**(4), 413 (2024)
10. Phoenix, S.J.D., Knight, P.L.: Periodicity, phase, and entropy in models of two-photonresonance. J. Opt. Soc. Am. B **7**(1), 116–124 (1990)
11. Bhat, B.V.R.: On greatest common divisor matrices and their applications. Linear Algebra Appl. **158**, 77–97 (1991)
12. Hayward, M.: Quantum computing and shor's algorithm, 1. Macquarie University Mathematics Department, Sydney (2008)
13. Cerezo, M., et al.: Variational quantum algorithms. Nat. Rev. Phys. **3**(9), 625–644 (2021)
14. Stęchły, M.: Introduction to variational quantum algorithms. arXiv preprint arXiv:2402.15879 (2024)
15. Zhou, L., et al.: Quantum approximate optimization algorithm: performance, mechanism, and implementation on near-term devices. Phys. Rev. X **10**(2), 021067 (2020)
16. Zakari, R.Y., et al.: VQA and visual reasoning: an overview of recent datasets, methods and challenges. arXiv preprint arXiv:2212.13296 (2022)
17. Haug, T., Kim, M.S.: Optimal training of variational quantum algorithms without barren plateaus. arXiv preprint arXiv:2104.14543 (2021)

18. Devitt, S.J., Munro, W.J., Nemoto, K.: Quantum error correction for beginners. Rep. Prog. Phys. **76**(7), 076001 (2013)
19. Napp, J., Preskill, J.: Optimal bacon-shor codes. arXiv preprint arXiv:1209.0794 (2012)
20. Huang, S., Brown, K.R., Cetina, M.: Comparing Shor and Steane error correction using the Bacon-Shor code. Sci. Adv. **10**(45), eadp2008 (2024)
21. Chen, Z., et al.: Exponential suppression of bit or phase flip errors with repetitive error correction. arXiv preprint arXiv:2102.06132 (2021)
22. Li, K., Chaudhary, V., Chowdhury, K.R.: BiP: bit-phase-flip error mitigation in quantum communications. In: ICC 2023-IEEE International Conference on Communications. IEEE, USA (2023)
23. Fowler, A.G., et al.: Surface codes: towards practical large-scale
24. Fowler, A.G., et al.: Surface code quantum communication. Phys. Rev. Lett. **104**(18), 180503 (2010)
25. Taylor, J.R., Van Every, E.J.: The emergent organization: communication as its site and surface. Routledge (1999)
26. Horsman, D., et al.: Surface code quantum computing by lattice surgery. New J. Phys. **14**(12), 123011 (2012)
27. Shaw, B.A.: Quantum steganography and quantum error-correction. University of Southern California (2010)
28. Steane, A.M.: Simple quantum error-correcting codes. Phys. Rev. A **54**(6), 4741 (1996)
29. Ayanzadeh, R., et al.: Multi-qubit correction for quantum annealers. Sci. Rep. **11**(1), 16119 (2021)
30. Chauhan, N., et al.: Ethical decision-making in artificial intelligence: frameworks, challenges, and future directions. In: Responsible AI for Digital Health and Medical Analytics, pp. 267–292. IGI Global Scientific Publishing (2025)
31. Preskill, J.: Reliable quantum computers. In: Proceedings of the Royal Society of London. Series A: Mathematical, Physical and Engineering Sciences, vol. 454, no. 1969, pp. 385–410 (1998)
32. Daniel, M.: Quantum computing AI frontiers-people, technology and economics. In: Proceedings of the International Annual Conference of the American Society for Engineering Management. American Society for Engineering Management (ASEM) (2024)

Quantum Computing Overview

Osama Babakhel, Mujahid Tabassum, and Tripti Sharma

Abstract The following chapter serves as a thorough introduction into the quantum computing world. It starts with the genesis of quantum theory, detailing seminal experiments and breakthroughs such as Planck's quantization of energy and the double-slit experiment, and traces the historical evolution of computation from classical binary systems to quantum computation, including pivotal ideas like Feynman's proposal of quantum simulation and Deutsch's universal quantum computer. It emphasizes the revolutionary nature of quantum mechanics—not just in physics but also in its synthesis into the concept of computation. This chapter sheds light upon the core principles that introduce a beginner to quantum computing—superposition, entanglement, and interference—elucidating these concepts using simple analogies, thought experiments, and mathematical visualizations such as the Bloch sphere and tensor product spaces. Moreover, apart from defining necessary terminology and parameter settings, this chapter delves deeper into the transformational and transcendental potential of quantum technologies, with specific examples including Shor's algorithm for factoring, Grover's search for unstructured databases, and hybrid algorithms like the Variational Quantum Eigensolver (VQE) and Quantum Approximate Optimization Algorithm (QAOA) for complex system simulations and optimization. Classical computing limitations are compared in detail with the emerging capabilities of quantum systems, discussing topics like exponential state spaces and quantum parallelism, thus allowing readers a clear understanding of exactly not just why, but how quantum computing is poised to reimagine computational ecosystems. Furthermore, this chapter provides an overview of current quantum hardware platforms—including superconducting circuits, trapped ions, and the emerging topological qubits—and addresses in detail the technological, economical, and importantly,

O. Babakhel
Savonia University of Applied Sciences, Kuopio, Finland

M. Tabassum (✉)
South East Technological University, Waterford, Ireland
e-mail: Mujahid.tabassum@SETU.IE

T. Sharma
University of Technology and Applied Sciences, Muscat, Oman

© The Author(s), under exclusive license to Springer Nature Switzerland AG 2025
A. Ghosh et al. (eds.), *Quantum Machine Learning in Industrial Automation*, Information Systems Engineering and Management 65,
https://doi.org/10.1007/978-3-031-99786-0_7

the ethical challenges that must be discussed with all stakeholders to fully, equitably realize the ‘quantum advantages.’ Lastly, it lays a comprehensive foundation for the in-depth, technical exploration of subsequent rigorous chapters, while inspiring an appreciation for the profound implications of quantum technology on both science and society.

Keywords Quantum computing and algorithm Shor’s algorithm · Hybrid quantum–classical algorithms · Quantum error correction · Combinatorial optimization · Machine learning

1 Why Quantum Computing? a Historical Perspective

Imagine you lost a key in a huge, messy warehouse and have to find it. The normal (classical) computer would start to check every spot one-by-one, which obviously could take forever. But imagine a computer that could check many places at once, allowing you to find your key faster. A lot faster! This indeed is what quantum computers can do: It isn’t just faster; it completely changes how we solve problems.

We got here, but how? And what discoveries happened that led to quantum computing?

1.1 Historical Background: From Quantum Theory to Quantum Computing

Quantum computing didn’t just magically appear in isolation. It took many crucial discoveries over the last century to come this far. Scientists tried first to deeply understand how tiny particles behaved, and later, they recognized that these discoveries could change computing much later [1] (Table 1).

Table 1 Historical background

Year	Discovery	Why it matters
1900	Planck’s quantum hypothesis	Energy exists in very small packets called quanta. This prompted the beginning of the field of quantum physics
1905	Einstein’s photoelectric effect	It showed that light behaves like both particles and waves. Einstein was awarded the Nobel Prize for this discovery
1927	Heisenberg’s uncertainty principle	Explained categorically that we just cannot know exactly where particles are and how fast they are moving simultaneously, introducing probability into physics

Initially, these discoveries were just interesting ideas about innocuous, very unrelated things, like atoms. Later, scientists began asking whether quantum physics could help us build better computers. And with this chapter, we show you how it does!

1.2 The Birth of Quantum Computing

In the late 1900s, scientists started thinking about applying quantum physics ideas to computing [1]:

- In **1981**, physicist Richard Feynman stated that ordinary (classical) computers can't efficiently simulate quantum systems. He suggested that quantum computers could be perfect for this job, sparking the start of lots of new research.
- In **1985**, David Deutsch introduced the idea of a Quantum Turing Machine. It ended up being the first theoretical model proving quantum computers could be able to solve certain problems significantly faster than classical computers.
- In **1994**, Peter Shor developed the seminal algorithm that showed quantum computers could factor large numbers much quickly. This discovery was enormous as it demonstrated how quantum computing could break modern encryption. This resulted in making cybersecurity a big topic in research circles.

1.3 How is Quantum Computing Different from Classical Computing?

Before we proceed, let's compare classical and quantum computing to better understand their differences to keep up [2] (Table 2).

We can imagine the difference for classical computing as reading one book, page by page, while quantum computing is more like reading multiple books, multiple pages all at once. Not only is this approach a difference of kind, and not degree,

Table 2 Difference between quantum and classical computing

Type	Basic unit	How it works	Limitations
Classical	Bit (0 or 1)	Checks at the rate of one item at a time, methodically (step by step)	Can gets extremely slow with complex problems as factoring large numbers or the simulation of molecules
Quantum	Qubit (can be 0, 1, or both simultaneously)	Is able to process multiple possibilities simultaneously because of quantum properties (superposition and entanglement)	Extremely sensitive to errors, needs specialized conditions

but explains why quantum computing can be orders of magnitude (OOM) faster and more powerful for certain specialized tasks.

1.4 Why Quantum Computing Matters

It should be understood that quantum computing is simply not 'just cool technology'—it has potential to change our lives significantly [3]:

- **Cryptography**: Quantum computers could crack current encryption methods with incredible ease, so we need to reimagine ways to protect data, whether it is sensitive or not.
- **Medicine**: Quantum computers could help create new medicine OOM's faster by accurately simulating molecules.
- **Optimization and AI**: Quantum computers could quickly solve complex, dynamic problems in real-time like planning efficient delivery routes or improving AI systems.
- **Material Science and Environment**: Quantum simulations could help discover or synthesize new materials currently a bottleneck in energy storage, helping fight climate change.

In short, quantum computing is about more than speed; it is a technology to fundamentally transform science, industry, and society.

1.5 Chapter Roadmap

Here's what you get to learn throughout the rest of this chapter [3]:

- **Quantum Bits (Qubits)**: What exactly are qubits, and what makes them special?
- **Quantum Algorithms**: How quantum algorithms can solve problems radically faster than classical methods.
- **Quantum Hardware**: What challenges and bottlenecks currently exist in building real quantum computers?
- **Ethical, Economic, and Technological Issues**: What potential real-world, societal and commercial problems or challenges does quantum computing bring?

1.6 Summary and What's Next?

Quantum computing is fast becoming a reality. It's set to redefine how we solve complex problems, protect data and information, and innovate across various industrial sectors. In the sections ahead, we'll explore quantum computing in detail. Ready to learn more about this exciting technology? Let's dive in!

2 Quantum Bits (Qubits): the Building Blocks

2.1 *What Exactly is a Qubit?*

In classical computing, information is stored using a unit called bits (binary digit). It can either be 0 or 1—like an on and off switch. This seemingly simple, binary approach has allowed us to build all the incredible technology around us: from smartphones to powerful supercomputers and nuclear reactors. However, this same classical computing faces limitations of capacity: when problems become extremely complex, involving many possibilities.

Quantum computing introduces a radical new idea: the quantum bit or "qubit." Unlike classical bits, qubits can be 0 or 1, or even both at the same time! This strange and powerful feature is called superposition, which is a complete game changer [4].

2.2 *Superposition: More Than Just 0 and 1*

Flip a coin and it goes up! While it's spinning in the air, you know it's neither heads nor tails—it's sort of both, until it lands. Only once it does, then do you know for sure. Similarly, a qubit remains in multiple states simultaneously until what we call 'an act of measurement', and then we know!

You're right to wonder: "How can the value of something be two different things at once?" Doesn't this idea go against our normal, everyday experiences? It sure does, but in the quantum world, that's the way it is. This one ability allows quantum computers to explore multiple possibilities simultaneously, or rather allows for the difference of kind from other classical computers that depend on computational ability to be increased proportional to the problem at hand. *Example:* One can express a qubit mathematically as [4]:

$$|\varphi = \alpha|0\rangle + \beta|1\rangle$$

2.3 *How Powerful is Superposition? an Example*

Let's get a real-world problem scenario: you have to solve a maze. A classical computer would test one path at a time, from start to finish. Guess what? That could take a very long time based on the computational ability. A quantum computer on the other hand, could simultaneously explore multiple paths due to superposition. This significantly reduces the time and effort it takes to find the correct path.

The number of possibilities a quantum computer can handle simultaneously grows exponentially with each added qubit. How much, let's see [1, 2] (Table 3).

Table 3 No. of qubits and possible states

Number of qubits	Possible states (simultaneously)
1	2
2	4
3	8
10	1024
50	Over 1 quadrillion
300	More states than atoms in the observable universe!

This exponential increase is exactly what allows quantum computers to tackle certain complex problems OOM's faster than classical computers [4].

2.4 Entanglement: Qubits Connected in a Mysterious Way

Another fundamental idea in quantum computing is entanglement. Imagine two magical dice, connected by something you cannot understand, that always land on the same number, even if they were rolled far apart. This strange connection is like quantum entanglement.

When two qubits are entangled, the state of one instantly influences the state of the other, no matter how far apart they are. This connection is not just an inexplicable phenomenon, it has practical uses. For example, entangled qubits can help quantum computers perform multiple linked operations simultaneously, which greatly enhances their computational power [3].

2.5 Interference: Enhancing Correct Answers

Quantum computing also uses the concept called interference, pretty similar to how noise-canceling headphones work. The headphones cancel out unwanted noise by generating their own sound waves that interfere destructively with outside noise. Quantum computers similarly use interference to amplify correct answers while reducing the chances of error with incorrect answers [5].

2.6 Challenges with Qubits: Fragile and Sensitive

Despite the aforementioned impressive capabilities, qubits are extremely fragile. Their quantum properties (like superposition and entanglement) can easily break down when exposed to disturbances from the environment—like heat, electromagnetic radiation, or vibrations. This problem is known as decoherence, and it's one of the biggest challenges and bottlenecks in quantum computing [2].

The best and brightest among researchers are actively trying different methods to minimize decoherence. One of the ways is to keep qubits extremely cold (close to absolute zero) and isolate them from external disturbances as much as possible. Yet still, managing decoherence remains difficult and is a key area of ongoing R&D.

2.7 *The Bloch Sphere: Visualizing Qubits*

To help understand qubits better, scientists use a tool called the Bloch sphere; a 3D representation of a qubit's state. If the North Pole represents a qubit in state $|0\rangle$ |0\rangle $|0\rangle$, and the South Pole represents state $|1\rangle$|1\rangle $|1\rangle$. Anywhere else on the sphere is a superposition a combination of both states. This visual tool helps scientists understand and manipulate qubits more effectively [4]. *Equation Example:* A general qubit state on the Bloch sphere is written as:

$$|\varphi\rangle = \cos\frac{\theta}{2}|0\rangle + e^{i\Phi}\sin\frac{\theta}{2}|1\rangle$$

2.8 *Quantum Measurement: From Superposition to Reality*

A qubit exists in superposition only until we measure it. After the act of measurement, its superposition "collapses," becoming either 0 or 1. This measurement is, although random, but based on probabilities set by the superposition state.

For example, a qubit may have a 70% chance of being 0 and a 30% chance of being 1. As we measure it, it will "collapse" into either state, with the probabilities corresponding to the percentages [1]. *Equation Example*:

$$|\varphi\rangle = \sqrt{0.7}|0\rangle + \sqrt{0.3}|1\rangle$$

Then the probability of measuring |0 is 0.7 and 1 is 0.3.

2.9 *Why is Measuring Qubits Tricky?*

Unlike classical bits that are static, measuring qubits changes their states. Once the act of measurement is done, qubits lose their quantum properties. This makes designing algorithms tricky because quantum information needs to be kept preserved until the precise moment it is useful to collapse it.

2.10 *Why Are Qubits Difficult to Work With?*

Because qubits are extremely fragile and can lose their quantum properties quickly, practical quantum computing remains challenging at scale. Scientists need cutting edge techniques and R&D to stabilize qubits, reduce errors, and maintain quantum coherence for longer durations. Current quantum computers can only reliably use a finite number of qubits, but research is rapidly advancing to overcome these obstacles, hopefully sooner than later!

2.11 *Summary and What's Next?*

You now know why qubits are essential and how their unique quantum properties like superposition, entanglement, and interference make quantum computing incredibly powerful and very exciting for you students. But how do researchers and scientists put these qubits to work to solve practical, complex problems? That's what quantum algorithms are for, which we will explore further in the next section. Ready to observe these quantum bits in action? Let's move forward!

3 Quantum Algorithms: How Quantum Computers Solve Problems

Quantum computers are incredibly powerful because they can run special algorithms designed to use superposition and entanglement. Let's look closely at some seminal quantum algorithms and why they matter.

3.1 *Shor's Algorithm: Cracking Codes with Quantum Computers*

If we were to unlock a safe by guessing the combination, there are 2 ways in the light of our comparisons. A classical computer would have to test every possible combination one-by-one, as previously established, making it incredibly slow obviously. Peter Shor's algorithm however, developed in 1994, changed everything. It showed how a quantum computer could quickly find the prime factors of very large numbers—something classical computers struggle with.

Here's what it looks like:

1. **Random Selection and GCD Check**
   Choose a random integer *a* (with 1 < a < N) and compute gcd(a, N) [7].

- If gcd(a, N) $\neq$ 1, then you've found a nontrivial factor.
- If gcd(a, N) = 1, proceed.

2. **Period Finding via Quantum Fourier Transform**
   Define the function

$$f(\mathrm{x}) = \alpha^{\mathrm{x}} \text{mode } N$$

This function is periodic with period *r* (the smallest positive integer such that [refer to 4]).

$$\alpha^{\mathrm{x}} = 1(\text{mode } N)$$

Use a quantum circuit to prepare a superposition of states.

- Apply modular exponentiation to compute f(x) for all x simultaneously [this step is implemented efficiently on a quantum computer].
- Perform the Quantum Fourier Transform (QFT) on the superposition. The QFT transforms the state such that measurements will likely yield values related to multiples of L/r (with L being the size of the register), from which r can be determined [4, 6].

$$|x\rangle \rightarrow \frac{1}{\sqrt{L}} \sum_{y=0}^{L-1} \mathrm{e}^{2\pi ixy/L} |y$$

Which concentrates the probability distribution near multiples of L/r.

3. **Extracting Factors**
   Once you have r:

- Ensure r is even.
- Compute

$$\mathrm{y} = \alpha^{\mathrm{r}/2} \text{mode N}$$

If y $\equiv -1$ (mod N), the attempt fails, and you need to choose another a.

- Otherwise, compute gcd (y $-$ 1, N) and gcd (y $+$ 1, N). One of these will be a nontrivial factor of N [6].

Why does this matter? Most modern encryption (like RSA encryption used by banks and secure websites) depends on the difficulty of factoring large numbers. Not that it is impossible, just takes an ungodly amount of time. Shor's algorithm can break this encryption rather quickly. That means quantum computers could disrupt our current systems of security and privacy with incredible ease, forcing us to create new, quantum-safe encryption methods.

3.2 *Grover's Algorithm: Searching Faster*

Imagine a needle in a haystack situation: searching for your name in a huge, unordered list of a million names. Standard algorithms would have a classical computer check about half a million names on average before finding yours. Grover's algorithm, developed by Lov Grover in 1996, speeds this up with unreasonable efficiency using quantum computing. Instead of half a million steps, Grover's algorithm would only take about a thousand steps.

Here's how the Grover's Algorithm does it:

- **Problem Setup**

  Suppose you have an unsorted database of N items and you want to find the unique item that satisfies a certain condition (the "marked" item). Classically, you'd need O(N) queries; Grover's algorithm reduces this to approximately $O(\sqrt{N})$ (see [5]).
- **Initialization**

  Start with n qubits (where $N = 2^n$) all initialized to the state $|0\rangle$. Apply the Hadamard gate H to each qubit to create an equal superposition of all possible N states:

$$|S\rangle = \frac{1}{\sqrt{N}} \sum_{x=0}^{N-1} |x$$

- **Oracle Operation**

  Define an oracle O that flips the phase of the state corresponding to the marked item. For any basis state $|x\rangle$:

$$O|x\rangle = \{-|x\rangle\, , \textit{if x is the target};\ |x\, , \textit{otherwise}$$

  if x is the target, otherwise. This step marks the solution by inverting its amplitude.
- **Diffusion (Amplitude Amplification)**

  After the oracle, the algorithm applies the diffusion operator D, also known as "inversion about the mean." This operator amplifies the amplitude of the marked state while reducing those of all others. Mathematically, it is given by:

$$D = 2|s\rangle\langle s| - I$$

where I is the identity matrix. Essentially, it reflects the state vector through the average amplitude [4].

- **Iteration**
  The combination of the oracle and the diffusion operator constitutes one Grover iteration. Repeating this iteration approximately $\frac{\pi}{4}\sqrt{N}$ times increases the amplitude of the marked state so that a measurement will yield the target with high probability.
- **Measurement**
  Finally, measuring the qubits collapses the state. With high probability, the measurement returns the marked item.

How does it do this? By using quantum superposition and interference, Grover's algorithm quickly narrows down possibilities, increasing the chance of finding the right answer in fewer tries. Although Grover's speedup isn't as dramatic as Shor's algorithm, it still significantly reduces the search time for very large databases.

3.3 Real-Life Applications of Quantum Algorithms

Quantum algorithms aren't merely theoretical—they could and are revolutionizing real-world problems (Table 4).

These examples show just a few ways quantum algorithms could transform industries.

Table 4 How quantum computing related with industry applications

Industry	Quantum algorithm	Application
Cybersecurity	Shor's algorithm	Breaking encryption, forcing us to develop new secure communication standards [6]
Logistics and transportation	Grover's algorithm	Quickly finding optimal routes and solutions, improving efficiency radically [5]
Pharmaceuticals and chemistry	Variational quantum eigensolver (VQE)	Simulating molecules to create new medicines and materials [7]
Finance	Quantum approximate optimization algorithm (QAOA)	Optimizing investment portfolios and financial modeling [8]

3.4 *Variational Quantum Eigensolver (VQE): a Hybrid Approach*

Since current quantum computers are physically limited in qubits and prone to errors, hybrid quantum–classical algorithms like VQE are also being developed. These algorithms deploy quantum computers for parts of the calculation they handle best and classical computers for the rest.

For example, pharmaceutical companies could use quantum simulations through VQE to accurately model complex molecular structures. Quantum computers could efficiently explore potential molecular configurations, while classical computers would handle calculations and optimizations. This combination could significantly speed up drug discovery processes. Here's an example in action: (Add VQE) [7].

3.5 *Quantum Approximate Optimization Algorithm (QAOA): Solving Optimization Problems*

Many practical, real-world issues, such as finding the best routes for delivery vehicles or optimizing investment portfolios, involve numerous possibilities and combinations. Classical computers struggle with these problems as they scale up because they must test each option sequentially. Computation easily becomes the bottleneck. Quantum computers, using QAOA, leverage superposition and entanglement to evaluate many possibilities simultaneously, radically speeding up these optimization tasks. Financial institutions and logistics companies are particularly excited about this technology due to its potential to save both time and money and computational hardware problems. Here's an example in action: (Show QAOA example) [8].

3.6 *Challenges in Quantum Algorithms*

- Error Rates: Quantum computers make mistakes due to the fragile nature of qubits.
- Hardware Limitations: Current quantum computers have relatively few qubits, limiting their practical use.
- Algorithm Complexity: Designing quantum algorithms is difficult and requires deep knowledge of both quantum mechanics, computer science and mathematics [2, 4].

3.7 *Summary and What's Next?*

Quantum algorithms have incredible potential, but we still need better or rather, different hardware to fully benefit from them. Next, we'll explore the hardware that makes quantum computing possible, looking at current technologies and the challenges the best and brightest among scientists are working hard to overcome. Let's dive into the exciting world of quantum hardware!

4 Quantum Hardware: Building the Quantum Machines of Tomorrow

Quantum computers hold amazing potential because of the power of qubits. But making quantum computers practical and widely available is easier said than done. This section explores the hardware—the physical machines—that quantum computing relies on. We'll be learning about the major technologies, understanding their challenges, and see how scientists are working hard to overcome them [4].

4.1 *Why is Quantum Hardware Hard to Build?*

Building quantum computers is challenging because qubits are extremely delicate. They easily lose their quantum properties when disturbed by their environment (which can be in any way), causing errors in calculations. Scientists call this issue decoherence [2].

Imagine meticulously building a house of cards— but one tiny vibration, breeze, or bump, and it all falls apart. Qubits behave similarly, needing extremely stable and controlled environments to work properly. Researchers must keep qubits very cold, nearly at absolute zero (− 273 °C) and isolate them from external disturbances. This makes building practical quantum computers extremely challenging for the market and scaling challenges.

4.2 *Leading Quantum Technologies Today*

Scientists and companies worldwide are developing different types of quantum hardware. Let's explore some of the leading approaches:

Superconducting Qubits

Used by companies like Google and IBM, superconducting qubits involve tiny circuits made of superconducting materials. These qubits must operate at ultra-low temperatures, near absolute zero, to maintain their quantum states [9, 10].

- Pros: Fast operations and relatively well-developed technology.
- Challenges: Requires very expensive cooling systems, vulnerable to noise causing errors.

Trapped Ion Qubits

Companies like IonQ and Honeywell use this technology. Here, ions (charged atoms) are trapped using electromagnetic fields, and lasers are used to control their quantum states [11].

- Pros: Longer coherence times (they stay quantum longer), better and precise operations.
- Challenges: More difficult to scale (adding more qubits makes control complex), operations are slower than superconducting qubits.

Photonic Qubits

Photonic quantum computing, developed by Xanadu, uses particles of light (photons) as qubits. Quantum operations are done through devices called beam splitters and phase shifters [13, 14].

- Pros: Can be operated at room temperature, easier to integrate into existing optical communication systems.
- Challenges: Requires extremely precise photon generation and detection, making scaling to many qubits complex.

Topological Qubits

Microsoft is currently pursuing a completely different approach—topological qubits. These use special particles known as 'anyons', existing in unique quantum states that make them resistant to decoherence [12].

- Pros: Potentially fewer errors and naturally resistant to environmental noise.
- Challenges: Currently still mainly theoretical and more difficult to build reliably; no practical topological quantum computers exist yet (Table 5).

4.3 Why is Quantum Hardware so Difficult to Scale?

Today's quantum computers typically have fewer than 100 qubits. However, to solve real-world complex problems, we will likely need thousands, even millions, of qubits. However, the more qubits added, the harder it becomes to control and maintain them.

Table 5 Comparison of leading quantum technologies

Technology	Pros	Challenges	References
Superconducting qubits	Fast operations; well-developed technology	Requires cryogenic cooling; noise sensitivity	[9, 10]
Trapped ion qubits	Long coherence times; high gate fidelity	Scalability; slower operation speeds	[11]
Photonic qubits	Room temperature operation; integration with optics	Precise photon generation/detection; scaling issues	[13, 14]
Topological qubits	Inherent error resistance	Mainly theoretical; experimental realization pending	[12]

Imagine trying to coordinate a small group of people versus trying to do the same with a huge stadium crowd—it's much more complex and chaotic to a point of impossibility. Similarly, scaling quantum hardware means managing exponentially more qubits without losing their quantum properties, which is extremely difficult, if not yet impossible.

4.4 Quantum Error Correction: Protecting Quantum Information

Since qubits are susceptible to making mistakes due to their fragile quantum states, researchers use something called quantum error correction (QEC). Instead of relying on one qubit, they can use multiple qubits to represent a single piece of information. This redundancy helps detect and fix errors without losing quantum information [15].

Popular quantum error-correcting methods include:

- Shor's Code: Uses 9 qubits to protect 1 qubit.
- Surface Code: Considered the most promising and scalable method for practical quantum computers, but it requires many qubits.

4.5 Engineering Challenges: Cooling and Energy Demands

Many quantum systems, especially superconducting qubits, need extremely cold temperatures to function—mere fractions above absolute zero. Keeping large numbers of qubits cold requires massive and energy-intensive refrigeration systems, adding significant complexity and cost with no reliable scaling systems yet.

Researchers are exploring alternative solutions:

- Developing room-temperature quantum technologies (like photonic qubits).

Table 6 Significant milestones of quantum computing

Year	Breakthrough	Why it matters
2019	Google's quantum supremacy	Demonstrated quantum computers could perform certain tasks radically faster than classical computers [16]
2022	IBM's 127-qubit processor (eagle)	Largest superconducting quantum processor available, showcasing quantum technology scalability potential [14]
2023	IonQ's 29-qubit system	Achieved high fidelity (accuracy), demonstrating trapped-ion advances in quantum computing [17]

- Improving cooling technologies to become more efficient and cost-effective [see research on advanced cryogenic techniques and materials science innovations].

4.6 Recent Quantum Hardware Breakthroughs

Despite these challenges, quantum hardware technology is advancing rapidly (Table 6).

4.7 What's Next in Quantum Hardware?

Quantum hardware is still in the early stages, but progress is not slow. Over the next few years, we can expect to see:

- Increasing Qubit Counts: Quantum processors achieving 1000+ qubits.
- Improved Error Correction: Better methods and processes to protect quantum information.
- Hybrid Quantum–Classical Systems: Combining quantum and classical computing for practical, immediate uses [2].

4.8 Summary and Moving Forward

Quantum hardware faces critical challenges and bottlenecks: decoherence, error correction, scalability, and cooling. But researchers worldwide are working tirelessly to overcome these hurdles. The incredible potential of quantum computing, luckily, continues to drive innovation and investment.

5 Ethical, Economic, and Societal Implications of Quantum Computing

Quantum has profound impacts on society, and is no longer just theory. With new technologies come new responsibilities and challenges. In this section, let's explore the ethical, economic, and societal implications of quantum computing and understand why these issues matter [2].

5.1 Quantum Computing and Cybersecurity

One of the biggest consequences quantum computing could have is in cybersecurity. Most of our current encryption methods (such as RSA) rely on mathematical problems that classical computers find extremely difficult to solve, but not yet impossible. Quantum computers, by virtue of qubits, could solve these problems quickly using algorithms like Shor's algorithm [6]. This poses a serious risk:

- **Risk to Data Privacy**: Personal data, government secrets, and business information could become vulnerable with quantum computers in the wrong hands [18].
- **Urgency for Quantum-Safe Encryption**: There is a critical need to develop and implement encryption methods secure against quantum attacks [18].

5.2 Quantum-Safe Cryptography: Preparing for a Quantum Future

To prevent security risks, researchers are working to develop quantum-safe cryptographic methods. These are encryption systems that even quantum computers would find difficult to break. Promising solutions include:

- **Lattice-Based Cryptography**: Uses very complex mathematical structures that quantum computers struggle to solve efficiently [19].
- **Quantum Key Distribution (QKD)**: Uses quantum mechanics itself to create secure communication channels, making interception nearly impossible [20].

These technologies are imperative for protecting sensitive data in the future quantum era.

Table 7 Economic impact comparison

Sector	Potential benefit	Challenges
Pharmaceuticals	Faster drug discovery via quantum simulations	High initial investment, specialized expertise
Finance	Optimized portfolios and risk management	Need for new quantum-safe security protocols
Logistics and transportation	Efficient routing and supply chain optimization	Integration with existing classical systems
General IT And cybersecurity	Threat to current encryption, need for re-skilling	Job displacement and need for retraining in quantum skills

5.3 Economic Impacts: Winners and Losers

Quantum computing is likely to significantly affect global economics, creating opportunities and challenges:

Industries Set to Benefit

- **Pharmaceuticals and Healthcare**: Incredibly faster drug discovery through quantum simulations [7].
- **Finance and Investment**: Enhanced portfolio optimization and improved risk management for firms using quantum computers [8].
- **Logistics and Transportation**: Radically optimized delivery routes and supply chains using quantum algorithms [5].

Potential Economic Challenges

- **Quantum Divide**: Quantum computing could widen the gap between rich and poor countries because it is expensive and technologically complex. The early adopters will benefit much more than the ones who may not be able to [see recent economic analyses on quantum investments.
- **Job Market Changes**: Many traditional computing and security jobs could become obsolete if not retrained in quantum computing technologies [refer to workforce studies in quantum technology education (Table 7).

5.4 Ethical Concerns and Quantum Computing

New technology almost always brings ethical concerns, and quantum computing will be no exception. Some critical ethical issues include:

- **Privacy**: Quantum computers would easily break current encryption, posing risks to personal and private information [21].
- **Quantum AI (Artificial Intelligence)**: Quantum computing could seriously enhance AI capabilities, leading to more powerful, autonomous decision-making

systems, unlike the current advances in AI. This raises critical issues about accountability, transparency, and potential misuse of technology.
- **Fairness and Accessibility**: Ensuring somehow that quantum technology benefits everyone, not just wealthy nations or corporations.

5.5 *Global Security and Geopolitical Challenges*

Quantum computing could reshape global power dynamics:

- **Quantum Arms Race**: Countries investing heavily in quantum technology might achieve significant advantages in intelligence, cybersecurity, and economic strength but race to develop it in not friendly ways against adversaries [22, 23].
- **International Cooperation**: There's a strong need for global agreements, safeguards, guardrails, and watchdogs on quantum cybersecurity and ethical use to prevent misuse or conflicts.

5.6 *Preparing the Workforce: Education and Training*

As quantum computing develops more and more into the mainstream, there will be a growing need for people trained in quantum technologies:

- **Quantum Engineers and Developers**: Specialists who understand both quantum mechanics and quantum software development.
- **Quantum Cybersecurity Experts**: Professionals skilled in developing and maintaining quantum-safe encryptions.
- **Educational Initiatives**: Universities, governments, and companies must invest and collaborate in quantum computing education programs to prepare the workforce adequately [13, 14].

5.7 *How Should Governments and Businesses Respond?*

To navigate quantum computing's implications, benefits and challenges effectively, governments and businesses should:

- **Invest in Quantum Research and Infrastructure**: Fund R&D to keep pace with rapid advancements.
- **Develop Quantum-Safe Policies**: Prepare and collaborate on cybersecurity strategies that protect critical infrastructure and information.
- **Promote Public–Private Partnerships**: Encourage collaboration between industries, academia, and governments to maximize benefits and manage risks but also share with the third world [18].

5.8 Individual Actions: What Can You Do?

As quantum computing approaches mainstream adoption, individuals can:

- **Stay Informed**: Easiest is to understand quantum technology basics and implications [recommend following resources from IBM, Google, and leading universities].
- **Engage in Learning Opportunities**: Utilize online courses, papers, webinars, and resources offered by organizations like IBM, Google, and universities.
- **Advocate for Responsible Use**: Encourage discussions around ethical standards and responsible technology use within your townhalls, communities, and civil society.

5.9 Summary and What's Ahead

Quantum computing's ethical, economic, and societal impacts are vast and complex yet promising. Addressing these challenges proactively ensures that quantum technology benefits society ethically, equitably, and responsibly. By understanding these implications, we can better navigate the quantum future and harness its incredible potential and hopefully not hurt the world more.

In the next section, we'll outline a roadmap for quantum computing's future: highlighting key milestones, breakthroughs, and the exciting possibilities ahead. Let's explore where quantum computing might take us!

6 Roadmap to the Quantum Future: What Comes Next

Quantum computing technology is evolving rapidly and we are accelerating the advance. Understanding deeply what lies ahead helps us prepare for the incredible changes quantum technology will bring. This section provides a clear, actionable roadmap, highlighting expected milestones and breakthroughs in quantum computing over the next several decades [2].

6.1 The Quantum Computing Landscape Today (2025)

Right now, quantum computers operate in what's called the "Noisy Intermediate-Scale Quantum" (NISQ) era. They can perform certain tasks better than classical computers, but they're still limited in the following ways (Table 8).

Table 8 Current NISQ landscape

Metric	Typical value	Reference
Qubit count	< 500	[9, 10]
Error rates	~ 1–5% per gate operation	[2]
Commercial availability	Experimental/Prototype	[13, 16]

- **Limited Qubits**: Most quantum computers today have fewer than 500 qubits. [For example, current IBM and Google processors typically have 50–100 qubits [9, 10].
- **High Error Rates**: They often and frequently make errors due to qubit fragility [2].
- **Specialized Applications**: Quantum computers today are mostly experimental or used in highly specialized tasks and not yet commercial.

Recent notable achievements include:

- Google's demonstration of quantum supremacy [16].
- IBM's development of quantum processors with over 100 qubits [13, 14].
- IonQ's high-fidelity quantum systems with trapped ions [17].

6.2 Short-Term Milestones (2025–2030)

Over the next five years, we expect several important advancements (Table 9).

- **Scaling Up**: Quantum computers will grow to 1000 or more qubits, allowing for more complex calculations.
- **Improved Error Correction**: Significant progress will be made in quantum error correction, reducing error rates and improving reliability in the fragile space [15].
- **Practical Applications**: Early commercial use of quantum computing in areas like specialized drug discovery, financial modeling, and logistics optimization [15].

Table 9 Mid-term quantum milestones

Milestone	Description	Expected timeframe
Fault-tolerant systems	Error-corrected quantum computers operating reliably	2030–2040
Commercial integration	Widespread use in industries such as finance and healthcare	2030–2040
Hybrid quantum–classical systems	Seamless integration for enhanced computational workflows	2030–2040

6.3 Mid-term Milestones (2030–2040): Achieving Quantum Advantage

Between 2030 and 2040, quantum computing will surely become significantly more practical and widespread:

- **Fault-Tolerant Quantum Computing**: Quantum computers will achieve stability, dramatically reducing errors and making complex calculations not just possible but also feasible.
- **Expansion of Commercial Uses**: Quantum computing will become commonplace in key industries like finance, pharmaceuticals, and transportation.
- **Quantum Integration**: Quantum computing will begin integrating with classical computing infrastructure, creating hybrid systems that leverage strengths from both computing types, much like distributed systems today. This integration is essential for solving large-scale problems in a practical timeframe.

6.4 Long-Term Vision (2040 and Beyond): Quantum Ubiquity

Looking further ahead, quantum computing has the potential to become as transformative as the internet revolution:

- **Universal Quantum Computers**: Quantum computers will become powerful enough to solve currently impossible problems across all scientific and applied disciplines.
- **Quantum Internet**: Quantum communication networks will emerge, providing ultra-secure and fast data transfer globally.
- **Quantum AI**: Quantum-enhanced artificial intelligence will revolutionize fields such as healthcare, climate modeling, material science and technology.

6.5 Challenges that Could Delay Progress

Several hurdles could slow down quantum computing development potentially (Table 10).

Table 10 Key challenges to quantum advancement

Challenge	Description
Engineering/Technical	Scaling qubits, maintaining coherence, reducing error rates
Economic/Funding	High cost of research, development, and infrastructure
Ethical/Societal	Privacy risks, equitable access, and responsible use of quantum technology

- **Engineering and Technical Challenges**: Developing large-scale, error-free qubits is an immense challenge in technical, talent and operational levels.
- **Economic and Funding Limitations**: Quantum computing development requires astronomical investment, potentially limiting rapid global adoption currently.
- **Ethical and Societal Concerns**: Privacy, security, and ethical use of quantum technologies must be addressed ahead of time to gain widespread acceptance and trust [22, 23].

6.6 How to Prepare for the Quantum Future

Preparation is key to successfully navigating quantum computing advancements:

For Students and Professionals

- **Learn Quantum Fundamentals**: Take full advantage of free courses provided by platforms like IBM Quantum Experience, Google AI Quantum, and online university programs. [Refer to IBM's and Google's educational resources.]
- **Engage with Quantum Programming**: Practice writing quantum programs using simulators or available quantum computers online and leveraging research papers and open source platforms.

For Businesses

- **Stay Informed and Agile**: Keep up-to-date with quantum computing advancements and start experimenting with quantum solutions in beta programs.
- **Develop Quantum Strategies**: Plan how to integrate quantum computing into business operations, especially in industries like finance, logistics, and pharmaceuticals.

For Governments

- **Support Quantum Research and Education**: Provide funding and policy support to quantum science and education.
- **Establish Quantum Ethical Guidelines**: Work towards international standards for ethical and secure quantum technology use.

6.7 Summary and Final Thoughts

The journey to practical, widespread quantum computing is filled with exciting milestones, transformative potential, and challenging obstacles. By staying informed, investing in quantum literacy, and preparing strategically, we can harness quantum computing's power responsibly and effectively [2].

References

1. Von Neumann, J.: Mathematical Foundations of Quantum Mechanics. Princeton University Press (1955)
2. Preskill, J.: Quantum computing in the NISQ era and beyond. Quantum **2**, 79 (2018)
3. Aspect, A., Dalibard, J., Roger, G.: Experimental test of Bell's inequalities using time-varying analyzers. Phys. Rev. Lett. **49**, 1804–1807 (1982)
4. Nielsen, M.A., Chuang, I.L.: Quantum Computation and Quantum Information. Cambridge University Press, Cambridge (2010)
5. Grover, L.K.: A fast quantum mechanical algorithm for database search. In: Proceedings of the 28th Annual ACM Symposium on Theory of Computing (1996)
6. Shor, P.W.: Algorithms for quantum computation: discrete logarithms and factoring. In: Proceedings of the 35th Annual Symposium on Foundations of Computer Science (1994)
7. Peruzzo, A., et al.: A variational eigenvalue solver on a photonic quantum processor. Nat. Commun. **5**, 4213 (2014)
8. Farhi, E., Goldstone, J., Gutmann, S.: A Quantum Approximate Optimization Algorithm. arXiv preprint arXiv:1411.4028 (2014)
9. Barends, R., et al.: Superconducting quantum circuits at the surface code threshold for fault tolerance. Nature **508**, 500–503 (2014)
10. Kelly, J., et al.: State preservation by repetitive error detection in a superconducting quantum circuit. Nature **519**, 66–69 (2015)
11. Monroe, C., et al.: Scaling the ion trap quantum processor. Science **339**, 1164–1169 (2013)
12. Nayak, C., et al.: Non-Abelian anyons and topological quantum computation. Rev. Mod. Phys. **80**, 1083–1159 (2008)
13. IBM Quantum Experience and educational initiatives (IBM, Google, etc.)
14. IBM Research (2022). Eagle: IBM's 127-Qubit Processor
15. Fowler, A.G., et al.: Surface codes: towards practical large-scale quantum computation. Phys. Rev. A **86**, 032324 (2012)
16. Arute, F., et al.: Quantum supremacy using a programmable superconducting processor. Nature **574**, 505–510 (2019)
17. IonQ (2023) (Refer to IonQ press releases or research papers on their 29-qubit system)
18. Chen, L.K., et al.: Report on Post-Quantum Cryptography. NISTIR 8105 (2016)
19. Micciancio, D., Regev, O.: Lattice-based Cryptography. In: Post-Quantum Cryptography (2009)
20. Bennett, C.H., Brassard, G.: Quantum cryptography: public key distribution and coin tossing. In: Proceedings of IEEE International Conference on Computers, Systems and Signal Processing (1984)
21. Harari, Y.N.: 21 Lessons for the 21st Century (2018)
22. Brookings Institution reports on quantum technology and geopolitics
23. Mathew, K., Tabassum, M., Ramakrishnan, M.: Experimental Comparison of Uninformed and Heuristic AI Algorithms for N Puzzle Solution

Quantum Computing Algorithm

Waqar Hassan, Mujahid Tabassum, and Mohammad Asif Nawaz

Abstract Quantum computing, a paradigm-shifting technology, leverages the principles of quantum mechanics to perform computations far beyond the capabilities of classical systems. This chapter provides a comprehensive overview of quantum computing algorithms, emphasizing their theoretical foundations, design, and practical applications. It begins by exploring fundamental quantum principles, such as superposition, entanglement, and quantum interference, which serve as the basis for quantum computation. Key algorithms are analysed, including Shor's algorithm for integer factorization, Grover's algorithm for database search, and the Quantum Approximate Optimization Algorithm (QAOA) for solving combinatorial problems. The chapter also introduces the growing field of hybrid quantum–classical algorithms, highlighting their potential for addressing real-world optimization, machine learning, and cryptography challenges. Implementation challenges, such as error correction and the scalability of quantum hardware, are discussed to provide a balanced perspective on the current state of the field. Finally, the chapter will discuss future trends in quantum algorithm research, including advancements in quantum error mitigation and the potential integration of quantum algorithms into existing computational frameworks. This chapter will serve as a resource for researchers, practitioners, and students aiming to understand and contribute to the evolving landscape of quantum computing algorithms.

Keywords Quantum computing and algorithm Shor's algorithm · Grover's algorithm · Quantum approximate optimization algorithm (QAOA) · Hybrid quantum–classical algorithms · Quantum error correction · Combinatorial optimization · Machine learning · Cryptography

---

W. Hassan
Arid Agriculture University, Burewala Campus, Punjab, Pakistan

M. Tabassum (✉)
South East Technological University, Waterford, Ireland
e-mail: Mujahid.tabassum@SETU.IE

M. A. Nawaz
Vibrant Hospitality, Charleston, SC, USA

© The Author(s), under exclusive license to Springer Nature Switzerland AG 2025
A. Ghosh et al. (eds.), *Quantum Machine Learning in Industrial Automation*,
Information Systems Engineering and Management 65,
https://doi.org/10.1007/978-3-031-99786-0_8

1 Introduction

Quantum computing makes a revolutionary change in computational systems through quantum mechanical principles that execute tasks which classical machines cannot achieve. Quantum computers use phenomena including superposition and entanglement with quantum interference to achieve better performance on complicated problems in comparison to conventional processors. The chapter incorporates complete information about quantum computing algorithms which integrates theoretical base elements alongside architectural systems and recent application development.

1.1 Fundamental Principles of Quantum Computing

Quantum computing systems function through qubits which serve as the quantum digital alternatives to classical binary units. Qubits differ from classical bits because they maintain their position between both values while representing states 0 and 1 simultaneously [6]. Quantum computers can simultaneously process an extensive number of solutions owing to their special ability. Through entanglement, qubits create links that make the state of one qubit affect another regardless of geographical distance. Quantum interference technology performs two functions by boosting specific computational paths through amplification while bringing unwanted paths to mathematical nullity for maximizing correct answers [29].

1.2 Pioneering Quantum Algorithms

Developers created multiple algorithms that exploit quantum properties to solve problems which Classical computers found unbreakable.

- **Shor's Algorithm**: Shor developed a quantum algorithm named after him in 1994 which factors large integers much more speedily than existing classical algorithms can. The cryptographic scheme-breaking capability of this algorithm creates significant impacts on the security of sensitive data.
- **Grover's Algorithm**: The quantum algorithm from Lov Grover in 1996 delivers unstructured search problem speedup at a quadratic rate thus enabling faster database searches and optimization tasks.
- **Quantum Approximate Optimization Algorithm (QAOA)**: a quantum approximate optimization algorithm aims to solve combinatorial optimization problems through approximate solutions because obtaining exact answers would be computationally unfeasible. The algorithm finds multiple uses in resolving logistics challenges as well as machine learning needs.

1.3 Hybrid Quantum–Classical Algorithms

Research groups created hybrid algorithms which unite quantum processors with classical processing components because quantum hardware faces present performance limitations. These computational methods distribute specific subtasks between quantum processors alongside classical system operations for other processes. Quantum algorithms prove advantageous for subproblems in fields such as optimization, machine learning and cryptography since they solve these portions better than classical processing methods on their own [25] (Fig. 1).

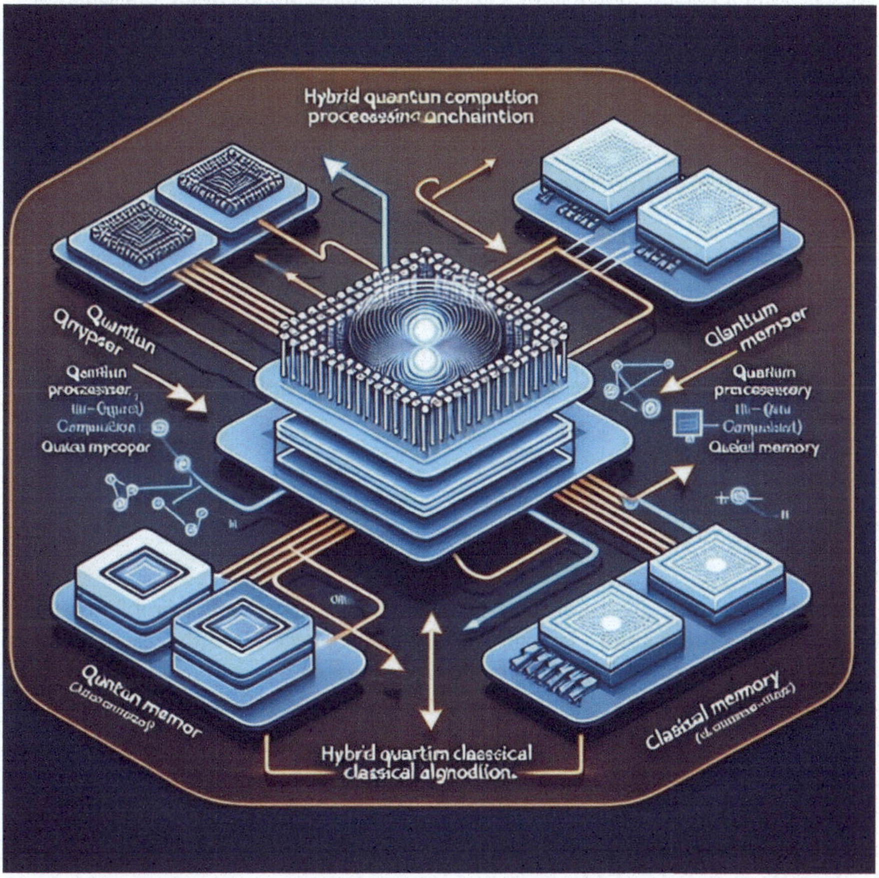

Fig. 1 Hybrid quantum–classical algorithms

1.4 Implementation Challenges

Error correction: The current deployment of quantum computing faces multiple obstacles which block its general use.

Hardware scalability: The sensitivity of quantum systems arises from their inclination toward errors which stem from decoherence together with environmental disturbances. Successful error correction techniques for quantum computation must be developed because they ensure computational precision [25].

Resource Constraints: Developing quantum hardware requires overcoming the major technical barrier of expanding systems through reliable qubit construction.

The number of qubits as well as the coherence times in existing quantum systems severely restricts the complexity range of problems they can solve.

1.5 Current Applications and Future Trends

New early applications start to appear despite the ongoing development of fully fault-tolerant quantum computers. Both IBM and IonQ unite with industries to create quantum algorithms aimed at enhancing artificial intelligence technology as well as managing weather-related risks [22]. The joint efforts of these organizations work to exploit quantum computing capabilities that solve complex practical issues better than standard approaches.

1.6 Conclusion

Quantum computing algorithms use revolutionary solutions for problem-solving which stem from quantum mechanical principles. These algorithms will begin to reveal fresh computational abilities by expanding research and breaking technological obstacles which will result in changes to scientific research and industries [31]. This section provides fundamental knowledge about the existing quantum computing algorithm state along with their projected developments for the future.

2 Fundamentals of Quantum Computing

Quantum computing implements quantum bits (qubits) while classical systems use bits consisting of either 0 s or 1 s to achieve its powerful new algorithm capabilities. The fundamental principles of quantum computing are presented through discussion of qubits and fundamental quantum mechanics principles as well as the procedural aspects of quantum circuits.

2.1 Quantum Bits (Qubits): Concept and Differences from Classical Bits

Every quantum computing system contains qubits as its base element. Qubits surpass ordinary bits because they exist in multiple states at once as opposed to binary 0 or 1 states. A key quantum computer capability stems from having multiple calculations running simultaneously because this leads to exponential power improvement for particular classes of problems [7] (Fig. 2).

- **Superconducting circuits**: Science can accomplish the creation of qubits through multiple physical substance categories.
- **Trapped ions**: The technology known as superconducting circuits forms the basis of quantum processors operated by both IBM and Google.
- **Photonic qubits**: The technology of trapped ions serves as the basis for quantum systems operated by IonQ and Honeywell.
- **Topological qubits**: The utilization of photonic qubits serves optical quantum computing systems. Microsoft conducts research on topological qubits due to their resistance to faults.

Multiple-state existence of qubits serves as the source of quantum computing parallel processing while maintaining distinction from traditional computing methods.

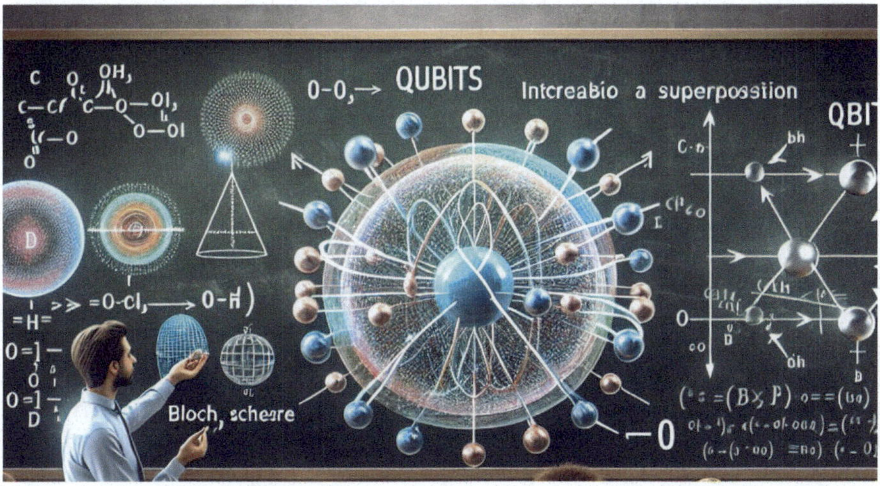

Fig. 2 Quantum bits processing

2.2 *Key Quantum Principles*

2.2.1 Superposition

Superposition represents the fundamental aspect of qubits because it enables them to occupy several states at once. A qubit state can be mathematically described through a linear combination of traditional classical states:

$$|\varphi\rangle = \alpha|0\rangle + \beta|1\rangle$$

where α and β are probability amplitudes such that:

$$|\alpha|^2 + |\beta|^2$$

Quantum algorithms can process multiple problem solutions concurrently due to the superposition feature. Shor's algorithm uses superposition to divide large numbers at an exponential speed which exceeds traditional classical algorithms [12].

2.2.2 Entanglement

Quantum systems exhibit entanglement as a phenomenon that links multiple qubits in a way which makes their states dependent on each other without considering their spatial relationship [9]. The instantaneous information exchange capability of quantum communication and computing depends on such features because it violates classical understanding.

The mathematical description of two-qubit entanglement exists as:

$$|\varphi\rangle = \frac{1}{\sqrt{2}}(|00\rangle + |11\rangle)$$

Measuring one qubit will reveal the state of the other instantaneously through a process that enables high efficiency for quantum computing and protocols such as quantum teleportation and quantum key distribution (QKD).

2.2.3 Quantum Interference

Qubits function through quantum interference because they exhibit wave properties. Quantum computers use probability amplitude manipulation to build better desirable outcomes while minimizing access to undesired possibilities. Quantum interference plays a key role in Grover's search algorithm because it enhances the correct search result probability while decreasing the required number of queries in comparison to traditional methods [32] (Fig. 3).

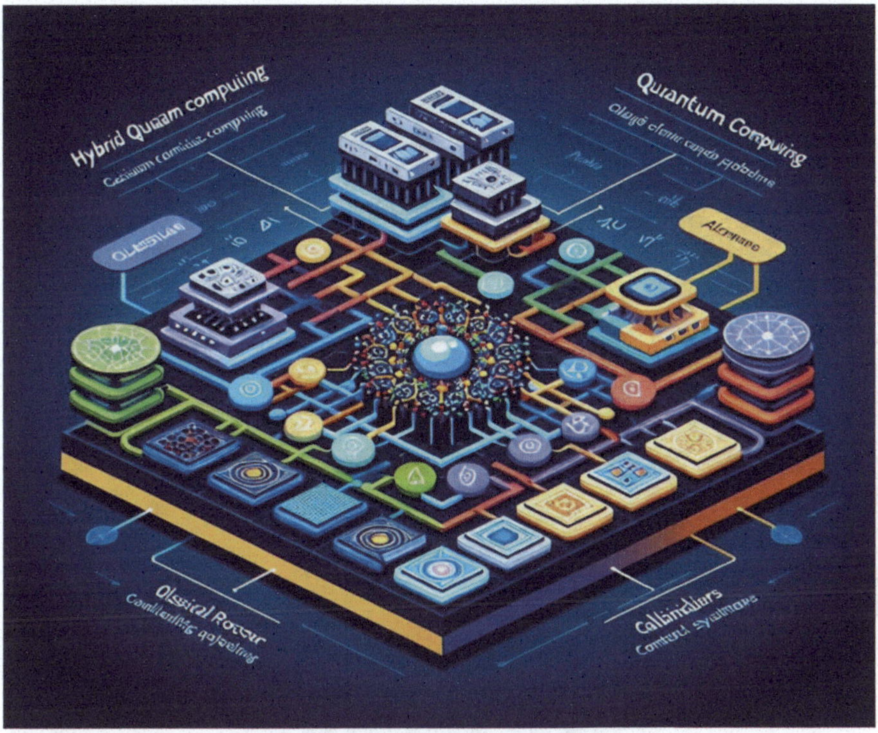

Fig. 3 Quantum processing

Quantum circuits use unitary operations to control interference patterns for creating advantageous computational results by managing constructive and destructive interference effects.

2.3 Quantum Gates and Circuits: Basic Operations and Logic

The quantum computing framework uses quantum gates as its core components to operate in similar ways as logic gates do in conventional systems [18]. Quantum gates differ from classical gates because they need to execute reversible operations which apply unitary transformations to qubits. The gates operate on qubits through mathematical operations that maintain quantum coherence (Fig. 4).

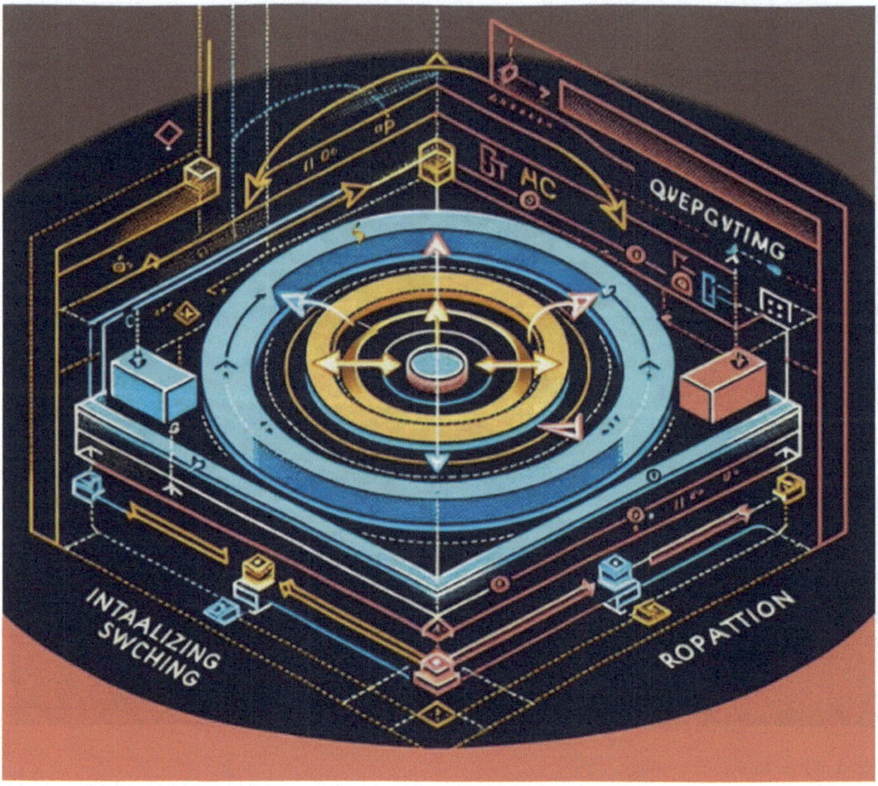

Fig. 4 Quantum gates operation

2.3.1 Basic Quantum Gates

The basic elements for quantum computation include several quantum gates built upon one another.

- **Pauli Gates (X, Y, Z)**: The Pauli Gates (X, Y, Z) serve as quantum NOT operations which operate specifically on quantum states.
  - X-gate: Flips to and vice versa.
  - The Z-gate creates phase variations while preserving computational basis states in their original state.
- **Hadamard Gate (H)**: Creates superposition by transforming into an equal combination of and:
- **CNOT (Controlled-NOT) Gate**: A two-qubit gate that flips the second qubit's state only if the first qubit is essential for creating entanglement.
- **Toffoli Gate**: A three-qubit gate that performs an AND operation in quantum computing.

Basic Quantum Gates

The quantum gates used the quantum computations functions as following:

1. Pauli Gates (X, Y, Z) which are analogous to classical NOT operations but acting on quantum states.
2. Hadamard Gate (H) that creates superposition by transforming $|0\rangle$ into an equal combination of $|0\rangle$ and $|1\rangle$
3. CNOT (Controlled-NOT) Gate is a two-qubit gate that flips the second qubits state only if the first qubit is $|1\rangle$.
4. Toffoli Gate is a three-qubit gate which performs an AND operation in quantum computing.

Quantum circuits use numerous qubits that link up through quantum gates to run quantum algorithms. The information processing method of a quantum circuit operates differently than classical circuits through the following sequence:

1. **Initializing qubits**: Qubits need to be placed into a superposition state during their initialization process.
2. **Applying entangling operation**: The process of applying entangling operations generates these correlations between qubits.
3. **Utilizing quantum interference**: A system based on quantum interference promotes correct outcome probability enhancement.
4. **Measuring**: Measuring qubits to extract results.

A basic quantum circuit performing Grover's search algorithm would function as follows:

- Hadamard gates provide the necessary step to generate superposition states.
- An oracle function marks the correct solution in the program [17].
- The marked state gets enhanced through the utilization of Grover diffusion operators.
- The extraction of the final result with high probability occurs through qubit measurements.

Quantum computing fundamentals originate from the fundamental principles which include superposition and entanglement and quantum interference. Thanks to these quantum principles computers achieve processing abilities which remain beyond the capabilities of traditional systems. Quantum gates together with circuits form the essential framework for quantum algorithm application which leads to critical progress in cryptography and optimization and machine learning. The understanding of fundamental concepts will prove essential to achieve optimal results from quantum computation because of the continuous development of quantum hardware.

3 Essential Quantum Algorithms

Quantum computing brought forth different revolutionary algorithms able to exceed classical computing in particular computational operations. Shor's Algorithm transformed cryptography by using efficient integer factorization while Grover's Algorithm optimized search and Grover's Search along with the Quantum Approximate Optimization Algorithm (QAOA) created a quantum solution for combinatorial optimization [33]. The algorithms of quantum computation produce their capabilities through the application of superposition and entanglement together with quantum interference principles.

3.1 Shor's Algorithm

3.1.1 Purpose: Integer Factorization and Its Impact on Cryptography

Peter Shor developed this algorithm in 1994 to deliver an effective quantum technique for integer factorization that affects the security of contemporary cryptographic systems.

RSA encryption along with other encryption protocols depends on the complexity of splitting large numbers into their prime factors [11]. The time required to solve RSA encryption problems through classical algorithms grows exponentially which establishes encryption security. Shor's Algorithm solves this difficult problem at an exponential speed rate which represents an essential danger to the existing cryptographic methods.

3.1.2 Mathematical Foundation and Working Principles

Shor's algorithm is based on number theory and quantum mechanics. The goal is to factor a large integer N, which is a product of two prime numbers. The algorithm exploits the relationship between periodic functions and modular arithmetic. The key idea is to find the period of a function that is related to modular exponentiation. This problem is difficult for classical computers but can be solved efficiently using quantum mechanics.

3.1.3 Shor's Algorithm Consists of Two Main Steps

1. **Classical reduction**: The integer factoring process converts into finding the period of a defined function through classical reduction methods. The algorithm picks a random number α (less than N). Then compute the greatest common

divisor (GCD) of α and N. If the GCD is greater than 1, it is going to be a factor of N. Otherwise, define the function:

- f (x) == α^x mode N
  This function exhibits periodic behaviour that mean it repeats after a certain value r. The challenge is to find this period r.

2. **Quantum Period Finding**: Quantum Period Finding serves as the fundamental quantum element of Shor's Algorithm by determining function periods. A quantum computer can quickly find periods by applying the Quantum Fourier Transform (QFT) to quantum parallelism which serves as an analogue for classical Fourier Transforms. In the first process a quantum state encoding values of f (x). Then, apply the QFT to extract the period r. once r is found, use it to compute factors of N. If r is even, then computation value will be:

- GCD ($\alpha^{r/2} - 1, N$)
  and
- GCD ($\alpha^{r/2} + 1, N$)
  One of these will give a factor of N.

Shor's algorithm is exponentially faster than classical factoring methods like trial division or the number field sieve. Classical algorithms take **superpolynomial** time, whereas Shor's algorithm runs in **polynomial time** (specifically, about $O((\log N)^3)$).

3.1.4 Potential Applications

- **Breaking RSA encryption**: RSA encryption security depends on number factoring complexity but Shor's Algorithm breaks this encryption scheme by threatening the encryption systems currently in use.
- **Cryptographic advancements**: The research into quantum-resistant cryptographic techniques receives momentum through advancements which leads to the development of lattice-based cryptography [11].
- **Number theory applications**: Number theory requires computational mathematics for solving problems which use prime number distribution analysis.

3.2 Grover's Algorithm

Grover's algorithm is a quantum search algorithm that helps find a specific item in an unsorted database way faster than a classical computer. Normally, if a user has a big list of numbers and he want to find a particular one, he has to check each one individually, which takes a long time. Grover's algorithm speeds this up using quantum computing, making the process much more efficient. Let suppose, a user has N possible answers to a problem, and only one of them is correct. A classical computer would need to check about N/2 possibilities on average (and up to N in the worst case). However, Grover's algorithm finds the answer in about $O(\sqrt{N})$ steps as a speedup process. It works with simple steps:

1. **Start with Superposition**: A quantum computer can explore all possible solutions at the same time. Instead of checking one number at a time like a classical computer, it considers all numbers simultaneously.
2. **Use an Oracle (Problem Encoding)**: An "oracle" function is like a black box that marks the correct answer without revealing it. When the quantum system checks a value, the oracle flips the phase of the correct answer (like tagging it as special).
3. **Amplify the Correct Answer (Grover Diffusion Step)**: Grover's algorithm then boosts the probability of the correct answer using a trick called "amplitude amplification". After repeating this process about $\sqrt{N}$ times, the correct answer is much more likely to be measured.
4. **Measure the Answer**: Once the probability is high enough, we measure the quantum state, and the correct answer pops out.

A practical example of Grover's algorithm is breaking cryptographic hashes. Many security systems use hashing to store passwords securely. Normally, if an attacker wanted to guess a password from a hash, they would have to try every possible password one by one a very slow process. With Grover's algorithm, they could find the correct password much faster, making certain encryption methods less secure in a world with quantum computers. Suppose a hacker is trying to guess a password. A classical brute-force attack (trying every possibility one by one) takes a long time. But using Grover's algorithm, a quantum computer could break a 256-bit encryption in about 2^{128} operations instead of 2^{256}, which is much faster. This is why post-quantum cryptography is being developed to defend against such attacks (Shor's algorithm is another quantum threat to cryptography).

3.2.1 Purpose: Database Search and Optimization

The unstructured search problem finds solution through Grover's Algorithm which Lov Grover developed during 1996. The process of finding elements within unsorted databases demands time until a search algorithm finds the required output. The speedup of Grover's Algorithm enables quadratic reduction from the original complexity to [21]. The algorithm significantly accelerates large-scale search and optimization operations because of which it finds applications in artificial intelligence together with cryptanalysis and database indexing.

3.2.2 How It Achieves Quadratic Speedup

Grover's Algorithm performs its operations through repeated applications of amplitude amplification. The key steps include:

1. **Superposition**: Initialization of the quantum register begins at an equal superposition between all possible states.
2. **Oracle Query**: A quantum oracle uses its phase-flipping capabilities to mark correct solutions within Oracle Query [13].
3. **Amplitude amplification**: The Grover diffusion operator achieves better correct solution probability through amplitude amplification which boosts the marked state amplitude at the expense of other states' amplitudes.
4. **Measurement**: The final measurement of the correct solution attains high probability after multiple iterations.

Through quantum interference, Grover's Algorithm mathematically boosts the correct answer's probability and simultaneously reduces the probability of incorrect solutions.

Use Cases in Real-world Applications.

- **Database search**: Database search allows users to execute quick lookup procedures on disorganized large datasets.
- **Cryptanalysis**: The cryptanalysis method helps attackers execute faster AES decryption schemes by reducing the time needed for brute-force key searches from to [28].
- **Artificial intelligence and optimization**: Artificial intelligence systems and optimization methods enable the solution of combinatorial problems that include route optimization and machine learning hyperparameter tuning.
- **Quantum Approximate Optimization Algorithm (QAOA)**.
- **Purpose: Solving Combinatorial Optimization Problems**.

Designers created the Quantum Approximate Optimization Algorithm (QAOA) specifically to handle combinatorial optimization problems that frequently appear across logistics operations and artificial intelligence systems and finance industries. QAOA functions as a quantum–classical hybrid algorithm that performs rapid approximation of optimal results although Shor's and Grover's algorithms deliver complete solutions (Fig. 5).

- **Explanation and Practical Examples**

  QAOA derives its operation from combining the concepts of adiabatic quantum computing along with variational quantum algorithms (VQAs). The key steps in QAOA are:

  1. **Encoding the problem as a Hamiltonian**: The optimization problem gets translated into a Hamiltonian form by converting its cost function through quantum mathematical encoding.
  2. **Applying Quantum Gates**: A quantum circuit parameterized by classical variables follows successive iterative optimization of its parameters when Quantum Gates are applied [24].
  3. Classical optimization: A classical optimizer uses minimization algorithms to adjust quantum circuit parameters in order to reduce the cost function value.

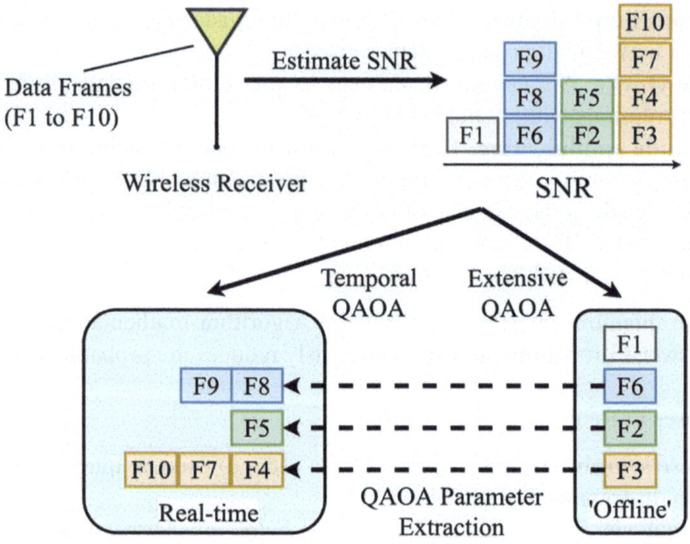

Fig. 5 QAOA example

4. **Measurement and solution extraction**: The measurement process of the quantum system delivers an almost optimal result for solving combinatorial problems.

QAOA brings considerable value to the optimization of hard computational problems which include:

- **Max-Cut and Traveling Salesman Problems**: QAOA solves Max-Cut and Traveling Salesman Problems alongside various network optimization scenarios.
- **Portfolio optimization**: The optimization of portfolios becomes more efficient through this method which identifies ideal asset configurations.
- **Scheduling problems**: Through scheduling problems, the system optimizes resource allocation in addition to workforce scheduling.

Our current computational problem-solving methods are undergoing a transformation due to algorithms from Shor's Algorithm and Grover's Algorithm and QAOA. The development of Shor's Algorithm creates serious risks for RSA encryption and allows faster number theory calculations. The quantum algorithms Grover's Algorithm delivers faster search operations along with cryptographic optimization whereas QAOA enables effective optimization of problems with practical combinatorial applications between classical and quantum computing systems [14].

Quantum hardware development makes these algorithms essential for progressing cybersecurity together with artificial intelligence systems and for logistics operations and financial modelling applications. Scientists and engineers need to find solutions for errors and scaling limitations to achieve maximum potential from quantum algorithms which will lead to quantum-enhanced computation dominance in the future.

4 Hybrid Quantum–Classical Algorithms

4.1 Definition and Importance

Efficient problem-solving through complex problems becomes possible by integrating quantum computational speed with classical computational power in combination. The currently unavailable status of fully error-corrected quantum computers makes hybrid algorithms enable practical solutions because they send processor-intensive quantum operations to quantum processors but keep error correction and optimization tasks on classical computers [10]. The algorithms enable quantum computation when it spans across the boundaries of existing hardware boundaries and upcoming full-scale operations.

4.2 Quantum Variational Algorithms

4.2.1 Variational Quantum Eigensolver (VQE)

The hybrid quantum–classical VQE algorithm serves as a method to estimate ground state energies of molecular systems as well as optimization problems. It works by:

1. **Ansatz preparation**: The preparatory step starts with a quantum circuit containing adjustable parameters.
2. **Measurement and expectation calculation**: Multiple runs of the quantum circuit serve to determine the expectation value for the Hamiltonian through measurement procedures.
3. **Classical optimization**: The quantum circuit parameters receive optimization through classical algorithms which operate on the energy function for minimization.

Applications of VQE

- Quantum Chemistry: Simulating molecular structures and reaction pathways.
- Material Science: Material Science benefits from research to develop new substances with particular characteristics [4].
- Machine Learning uses deep learning models by refining energy-based functions.

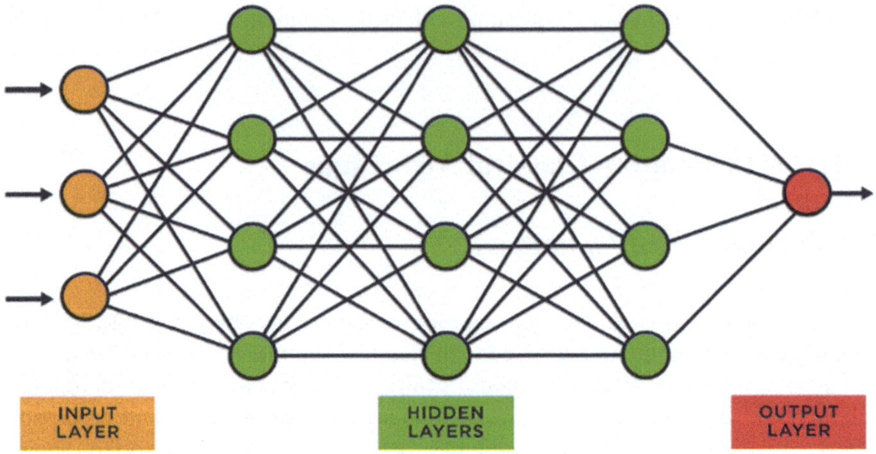

Fig. 6 Quantum neural network

4.2.2 Quantum Neural Networks

The combination of quantum circuits and classical neural networks creates Quantum Neural Networks that help improve the speed of machine learning operations. The quantum networks take advantage of parallel computing to enhance both learning speed and tackle problems which defeat traditional deep learning methods (Fig. 6).

Applications of QNNs

- Image Recognition: Faster feature extraction and classification.
- Natural Language Processing: Enhanced text processing with quantum-enhanced embeddings.
- Financial forecasting: The use of quantum-enhanced AI models enables forecasting of stock market behaviours.

4.3 Applications

4.3.1 Machine Learning

Hybrid quantum–classical algorithms streamline machine learning approaches by achieving better results during feature choice along with model-building stages as well as execution processes [2]. The usage of quantum-enhanced machine learning involves several benefits including:

- Classification and clustering.
- Anomaly detection in cybersecurity.
- Reinforcement learning for robotics.

4.3.2 Cryptography

Quantum computing provides both defence and advantage viewpoints when it comes to encryption technologies. Hybrid algorithms help in:

- Developing post-quantum cryptographic protocols.
- Secure communication becomes better through the implementation of quantum key distribution (QKD) [26].
- Security testing receives improved cryptanalysis capabilities through better analytic methods.

4.3.3 Financial Modelling

Quantum computing technologies speed up financial model development through efficient solutions of complex computation problems which include:

- Portfolio optimization.
- Risk assessment and fraud detection.
- Market simulations and derivative pricing.

The merging of quantum speedup capabilities with classical reliable algorithms provides industry-ready solutions to exploit quantum computing nowadays [23]. Various industries benefit from their increasing use as a significant development in computational power.

5 Challenges in Quantum Algorithm Implementation

Quantum computing brings revolutionary computing improvements but still needs to overcome multiple implementation hurdles during its deployment. Quantum systems exhibit three main obstacles because of their sensitive nature as well as their vulnerability to errors and the boundaries imposed by resource availability. The section identifies essential barriers to quantum algorithm implementation with an examination of quantum error correction as well as hardware scalability decoherence and noise together with resource constraints in current quantum systems.

5.1 Quantum Error Correction: Necessity and Current Techniques

The environmental interactions cause quantum systems to become highly vulnerable to errors that create decoherence and result in quantum information loss. Quantum error correction operates differently from classical computers since it handles errors based on the no-cloning theorem despite its inability to produce exact quantum state duplicates [34].

The surface code represents a promising error correction method because it lets logical qubits reside within a physical qubit array that supports advanced error management protocols. The Shor code together with the Steane code represents stabilizer codes that introduce redundancy to enable quantum error detection and correction without measuring the quantum state directly.

5.2 Hardware Scalability and Limitations

The key obstacle in achieving quantum algorithm complexity through hardware expansion remains a major challenge. Contemporary quantum processors developed by IBM Google and Rigetti possess a few hundred qubits but useful quantum computation demands systems with thousands or millions of qubits [16].

The main challenge affecting hardware scalability arises from limitations in qubit connectivity. All three qubit systems including superconducting qubits, trapped ions and photonic qubits face specific barriers to scalability. Superconducting qubits need extremely low temperatures and microwave pulse precision for operation but ions trapped in vacuum facilities require laser controls for their functionality. When quantum systems grow larger they experience more and more qubit-to-qubit interaction issues which require highly advanced control systems [15].

People working to scale quantum computing systems use two main approaches: modular architectures connecting smaller processors through distributed networks along with topological qubits providing better protection against errors and longer coherent states.

5.3 Decoherence and Noise: Impact on Computations

The combination of decoherence with noise creates strong detrimental effects on the reliability of quantum computing operations. Quantum systems lose coherence during decoherence events which leads qubits to transition from quantum states to classical states because of external environment interactions. The impact of decoherence reduces the quantum computing duration (coherence time) so quantum circuits become limited in depth and complexity [3] (Fig. 7).

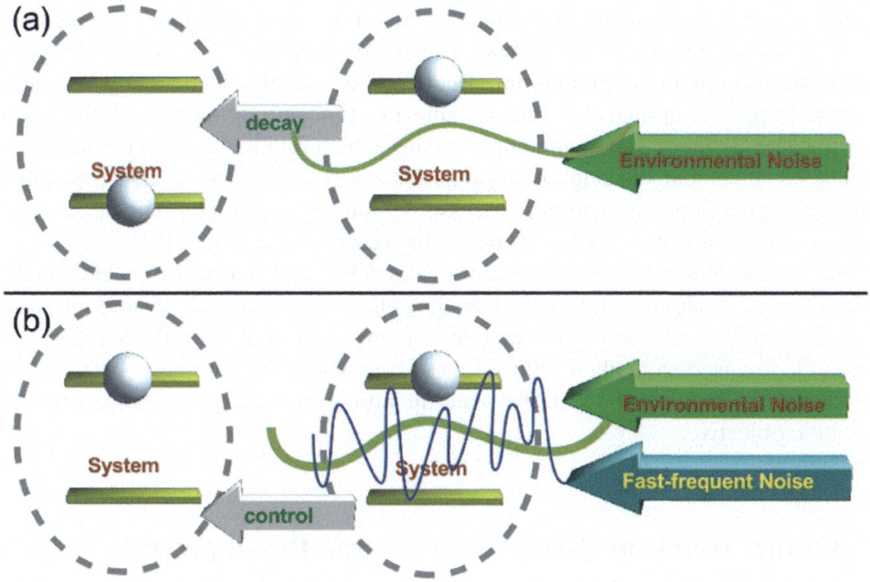

Fig. 7 Decoherence and noise process

The three primary sources of noise in quantum systems are qubit control imprecisions and unwanted qubit interactions together with external environmental fluctuations. Quantum algorithm execution suffers from three major error types: gate errors, readout errors and leakage errors. Researchers address noise-related issues in quantum computing by studying methods which include quantum error suppression along with noise-adaptive quantum circuits and the quantum error mitigation strategies zero-noise extrapolation and probabilistic error cancellation.

The production of hardware instrumentation needs advances in three areas: better qubit shielding mechanisms combined with more advanced cooling methods supplemented by advances in high-fidelity quantum gate designs to reduce hardware discretions.

5.4 Resource Constraints in Current Quantum Systems

Quantum systems currently face three major challenges which include low qubit number capacity and minimal gate fidelity together with brief coherence time duration. The current limitations affect how effectively big quantum algorithms can be implemented. IBM developed quantum volume as a performance evaluation method which examines qubit number combined with device connectivity and gate fidelity to determine existing quantum hardware limitations [1].

Algorithm implementation in quantum computing encounters barriers caused by the high resource needs of the system. The present quantum processor technology cannot handle deep quantum circuit structures because they experience excessive error build-up. Variational algorithms enable quantum computing through the use of optimization methods and quantum processing which minimizes resource needs.

The advancement of quantum compilers as well as improvements in quantum circuit design and novel quantum processor technologies like photonic and neutral atoms represent current approaches to beating resource challenges [8].

Numerous challenges obstruct quantum algorithm deployment which include the difficulties of quantum error correction as well as scalability constraints of hardware the negative effects of noise and decoherence and limited available resources. Research teams devote their effort to creating solutions for these barriers yet conducting fault-tolerant and large-scale quantum computation remains a distant scientific objective.

6 Future Trends in Quantum Algorithm Development

Quantum computing technology evolves at unprecedented speeds to create powerful changes in computational operations. Multiple critical fields of progress emerge in quantum algorithm research to define the path for future quantum algorithm development [30]. The research presents findings about quantum error correction developments along with optimization of quantum algorithms for practical use and integration methods between classical and quantum systems and their impacts on different industries.

6.1 Advancements in Quantum Error Correction

The main obstacle standing in the way of successful quantum computing implementation involves error correction. Quantum computing systems experience a high level of vulnerability to both noise emissions and decoherence because these phenomena generate computational errors. The standard error correction approaches including surface code and concatenated codes demand many physical qubits for creating a single logical qubit [5].

Modern advancements focus on creating quantum error correction methods that require fewer operational resources. Researchers are implementing machine learning error correction models to develop dynamic error prediction systems. Researchers are creating topological qubits as a theoretical method to improve fault tolerance because they show enhanced resistance to noise. Reliable practical use of quantum computers highly depends on continuing research to improve error correction methods.

6.2 Quantum Algorithm Optimization for Practical Use

Theoretical advancements do not eliminate the fact that numerous quantum algorithms retain an impractical nature because they require extensive resources. Researchers make efforts to optimize current quantum algorithms to minimize qubit requirements while reducing computational complexity [19]. The Variational Quantum Algorithms (VQAs) integration enables classical computers to work with quantum processors to achieve optimized solutions for complicated problems.

The design of modern quantum algorithms incorporates noise adaptation methods to work with existing limitations of quantum hardware systems during computational processes. The algorithms select shallow gate circuits with reduced gate counts thus becoming practical for current quantum hardware platforms.

6.3 Integration with Classical Computing Systems

The advancement of quantum computing technology now requires strong attention on its ability to work together with conventional computing systems. Quantum computing will act as a supplemental technology to classical processors since it cannot take over all their tasks.

Scientists along with businesses can conduct quantum algorithm tests through classical interfaces because IBM Google and Amazon provide cloud quantum computing services [20]. The hybrid systems use classical pre- and post-processing methods to boost the management of quantum computations.

Enhanced quantum-cloud infrastructure which includes improved SDKs and quantum–classical APIs will enable simple system connections. Industries can implement quantum hybrid systems which enable quantum adoption while keeping quantum hardware deployment as an out-of-house operation.

6.4 Potential Impact on Industries

The industrial landscape will experience revolution through quantum computing because the technology excels at solving problems which traditional computing methods cannot solve. Many specific industrial areas will obtain important gains through quantum technology developments.

1. **Finance**: Quantum algorithms for portfolio optimization, risk analysis, and fraud detection could significantly improve financial modelling accuracy.
2. **Healthcare and Pharmaceuticals**: The healthcare sector along with pharmaceuticals benefits from quantum computing because it speeds up drug research by performing molecular simulations better than traditional supercomputers do.

3. **Artificial Intelligence and Machine Learning**: The future processing speed of big data through Artificial Intelligence and Machine Learning systems will expand with quantum enhancement thereby resulting in better pattern identification decisions [20].
4. **Cybersecurity**: The security threats from quantum computing allow scientists to find new quantum-resistant encryption methods that safeguard data beyond the quantum age.
5. **Logistics and supply chain optimization**: Quantum algorithm systems help optimize logistics systems and supply chain networks which reduces operating expenses and enhances transportation along with supply chain operational efficiency.
6. **Material Science**: Quantum simulations using material science discoveries enable researchers to develop superior materials through which the aerospace industry and manufacturers benefit alongside the energy sector.

Future quantum algorithm development appears incredibly promising because scientists actively work on error correction and optimization systems combined with classical computing techniques [27]. The expanding sophistication of quantum computing technology will deeply transform many different business sectors including finance healthcare and other industries. Additional research together with innovation will be necessary to overcome existing barriers and finally utilize quantum computing to its maximum practical application potential.

7 Conclusion

The computational approach through quantum computing brings new possibilities because its specialized domains operate more effectively than traditional computing methods. The methods of quantum computing algorithms have been discussed throughout this chapter along with their theoretic foundations and their applications including implementation barriers [31]. Research development will lead to improved quantum algorithm performance which will open new solutions throughout different industrial sectors.

7.1 Summary of Key Points

This chapter has analysed the three basic features of quantum computing which consist of superposition alongside entanglement and quantum interference. Using these principles quantum bits (qubits) perform unique information processing which exceeds classical bits because of their technical capacity. The efficiency of quantum computing becomes evident through important quantum algorithms where Shor's algorithm finds integer factors while Grover's algorithm speeds up database searches and the Quantum Approximate Optimization Algorithm (QAOA) finds solutions for combinatorial tasks.

Quantum calculation systems combine quantum and conventional algorithms to solve optimization issues and cryptographic along with machine learning tasks more effectively. The examination of quantum error correction together with hardware scalability and noise and decoherence effects showed what prevents quantum technology from becoming mainstream. The forecast for quantum algorithm evolution incorporates three essential developments between error correction advancements and optimizing processes together with bringing classical systems into the quantum computation.

7.2 *The Role of Quantum Algorithms in Shaping Future Computing*

The computational methods of the future will transform because of quantum algorithms which operate in various scientific domains. Shor's algorithm poses a threat to traditional cryptographic systems in cryptography so researchers work on creating quantum-resistant encryption technologies [25]. Grover's algorithm improves search along with optimization tasks which enables better application in artificial intelligence domains together with logistics and data processing operations. Numerous quantum variational algorithms along with hybrid approaches prove how quantum computing changes the field of finance as well as healthcare and material science.

The progress of quantum hardware enables the development of realistic quantum algorithms that will increase breakthroughs in the fields of quantum machine learning and facilitate drug discovery as well as supply chain optimization. Real-world implementation of quantum technologies will become feasible for scientists and business practitioners by uniting quantum computers with cloud services that incorporate traditional computing infrastructures.

7.3 *Final Thoughts on the Evolving Landscape of Quantum Computing*

Current quantum computing research remains at an initial level although its promising capabilities are self-evident. The main obstacles such as qubit stability error rates and scalability continue to exist but quantum computing technology experiences continuous progress toward decreasing these problems. Various entities such as government agencies academic institutes and technology companies actively fund quantum research development because this supports accelerating innovation.

The upcoming paradigm of quantum computing will trigger a joint operation between classical and quantum systems which will use their individual capabilities to tackle obstacles that were too complicated for earlier technologies [26]. The introduction of useful quantum solutions in the market will drive industrial transformation resulting in major advancements across optimization and artificial intelligence domains together with increased security capabilities and additional breakthroughs beyond.

Modern science and technology should regard quantum computing as one of their most impactful emerging fields. The exploration of quantum computing needs researchers to team up with developers and businesses who actively maintain their adaptability while proactively studying this developing field to achieve its future advantages during the next decades.

References

1. Ajagekar, A., You, F.: Quantum computing and quantum artificial intelligence for renewable and sustainable energy: a emerging prospect towards climate neutrality. Renew. Sustain. Energy Rev. **165**, 112493 (2022)
2. Alhayani, B.A., AlKawak, O.A., Mahajan, H.B., Ilhan, H., Qasem, R.a.M.: Design of quantum communication protocols in quantum cryptography. Wirel. Pers. Commun. 1–18 (2023)
3. Ali, H.: Quantum computing and AI in healthcare: accelerating complex biological simulations, genomic data processing, and drug discovery innovations. World J. Adv. Res. Rev. **20**(2), 1466–1484 (2023)
4. Ayral, T., Besserve, P., Lacroix, D., Ruiz Guzman, E.A.: Quantum computing with and for many-body physics. Eur. Phys. J. A **59**(10), 227 (2023)
5. Berke, C., Varvelis, E., Trebst, S., Altland, A., DiVincenzo, D.P.: Transmon platform for quantum computing challenged by chaotic fluctuations. Nat. Commun. **13**(1), 2495 (2022)
6. Bhat, H.A., Khanday, F.A., Kaushik, B.K., Bashir, F., Shah, K.A.: Quantum computing: fundamentals, implementations and applications. IEEE Open J. Nanotechnol. **3**, 61–77 (2022)
7. Chawla, D., Mehra, P.S.: A survey on quantum computing for internet of things security. Proc. Comput. Sci. **218**, 2191–2200 (2023)
8. Coccia, M., Roshani, S., Mosleh, M.: Evolution of quantum computing: theoretical and innovation management implications for emerging quantum industry. IEEE Trans. Eng. Manag. **71**, 2270–2280 (2022)
9. Conlon, L.O., Vogl, T., Marciniak, C.D., Pogorelov, I., Yung, S.K., Eilenberger, F., Berry, D.W., Santana, F.S., Blatt, R., Monz, T., Lam, P.K.: Approaching optimal entangling collective measurements on quantum computing platforms. Nat. Phy. **19**(3), 351–357 (2023)
10. Duong, T.Q., Ansere, J.A., Narottama, B., Sharma, V., Dobre, O.A., Shin, H.: Quantum-inspired machine learning for 6G: fundamentals, security, resource allocations, challenges, and future research directions. IEEE Open J. Veh. Technol. **3**, 375–387 (2022)
11. Elliot, M., Oscar, S., Kathryn, M.: Current quantum optics research: exploring the potential of quantum computing. J. Tecnol. Quantica **1**(1), 30–39 (2024)
12. Fan, L., Han, Z.: Hybrid quantum-classical computing for future network optimization. IEEE Netw. **36**(5), 72–76 (2022)
13. Fauseweh, B.: Quantum many-body simulations on digital quantum computers: state-of-the-art and future challenges. Nat. Commun. **15**(1), 2123 (2024)
14. Fedorov, A.K., Gelfand, M.S.: Towards practical applications in quantum computational biology. Nat. Comput. Sci. **1**(2), 114–119 (2021)
15. Gençoglu, M.T., Agarwal, P.: Use of quantum differential equations in sonic processes. Appl. Math. Nonlinear Sci. **6**(1), 21–28 (2021)

16. Gill, S.S., Xu, M., Ottaviani, C., Patros, P., Bahsoon, R., Shaghaghi, A., Golec, M., Stankovski, V., Wu, H., Abraham, A., Singh, M.: AI for next generation computing: emerging trends and future directions. Internet Things **19**, 100514 (2022)
17. Golestan, S., Habibi, M., Mousavi, S.M., Guerrero, J.M., Vasquez, J.C.: Quantum computation in power systems: an overview of recent advances. Energy Rep. **9**, 584–596 (2023)
18. Jowarder, R.A., Jahan, S.: Quantum computing in cyber security: emerging threats, mitigation strategies, and future implications for data protection. World J. Adv. Eng. Technol. Sci. **13**(1), 330–339 (2024)
19. Kan, A., Nam, Y.: Lattice quantum chromodynamics and electrodynamics on a universal quantum computer. arXiv preprint arXiv:2107.12769 (2021)
20. Kashem, S., Tabassum, M., Chai, M.: A novel design of an amphibious robot having webbed feet as duck (2017)
21. Khurana, R.: Applications of quantum computing in telecom e-commerce: analysis of QKD, QAOA, and QML for data encryption, speed optimization, and AI-driven customer experience. Q. J. Emerg. Technol. Innov. **7**(9), 1–15 (2022)
22. Lordi, V., Nichol, J.M.: Advances and opportunities in materials science for scalable quantum computing. MRS Bull. **46**, 589–595 (2021)
23. Mangla, C., Rani, S., Qureshi, N.M.F., Singh, A.: Mitigating 5G security challenges for next-gen industry using quantum computing. J. King Saud Univ. Comput. Inf. Sci. **35**(6), 101334 (2023)
24. Massoli, F.V., Vadicamo, L., Amato, G., Falchi, F.: A leap among quantum computing and quantum neural networks: a survey. ACM Comput. Surv. **55**(5), 1–37 (2022)
25. Naik, A.S., Yeniaras, E., Hellstern, G., Prasad, G., Vishwakarma, S.K.L.P.: From portfolio optimization to quantum blockchain and security: a systematic review of quantum computing in finance. Financ. Innov. **11**(1), 1–67 (2025)
26. Paudel, H.P., Syamlal, M., Crawford, S.E., Lee, Y.-L., Shugayev, R.A., Lu, P., Ohodnicki, P.R., Mollot, D., Duan, Y.: Quantum computing and simulations for energy applications: review and perspective. ACS Eng. Au **2**(3), 151–196 (2022)
27. Perumal, S., Tabassum, M., Md Norwawi, N., Samy, G., Perumal, S.: Development of an Efficient Timetable System using AngularJS and Bootstrap 3 (2018)
28. Pithawa, D., Nahar, S., Bhardwaj, V., Rawat, R., Dronawat, R., Rawat, A.: Quantum computing technological design along with its dark side. Quantum Comput. Cybersecurity 295–312 (2023)
29. Rayhan, A., Rayhan, S.: Quantum computing and AI: a quantum leap in intelligence. In: AI Odyssey: Unraveling the Past, Mastering the Present, and Charting the Future of Artificial Intelligence, vol. 424 (2023)
30. Rozenman, G.G., Kundu, N.K., Liu, R., Zhang, L., Maslennikov, A., Reches, Y., Youm, H.Y.: The quantum internet: a synergy of quantum information technologies and 6G networks. IET Quantum Commun. **4**(4), 147–166 (2023)
31. Singh, J., Bhangu, K.S.: Contemporary quantum computing use cases: taxonomy, review and challenges. Arch. Comput. Methods Eng. **30**(1), 615–638 (2023)
32. Tilly, J., Chen, H., Cao, S., Picozzi, D., Setia, K., Li, Y., Grant, E., Wossnig, L., Rungger, I., Booth, G.H., Tennyson, J.: The variational quantum eigensolver: a review of methods and best practices. Phys. Rep. **986**, 1–128 (2022)
33. Ullah, M.H., Eskandarpour, R., Zheng, H., Khodaei, A.: Quantum computing for smart grid applications. IET Gener. Transm. Distrib. **16**(21), 4239–4257 (2022)
34. Yamasaki, H., Koashi, M.: Time-efficient constant-space-overhead fault-tolerant quantum computation. Nat. Phys. **20**(2), 247–253 (2024)

Navigating the Quantum Computing Landscape of the Dark Web

Mrignainy Kansal

Abstract This research investigates the potential impact of quantum computing on the secretive online community, the dark web. We aim to understand how this state-of-the-art technology might reshape activities within this hidden digital realm. By examining existing literature, we uncovered weaknesses in the dark web's current cybersecurity methods, especially in light of potential threats stemming from advancements in quantum technology. To tackle these concerns, we propose an enhanced cryptographic system capable of defending against attacks from quantum computers. The objective is to bolster the security framework of the dark web. We assessed the efficacy of our solution in protecting sensitive data during transactions on the dark web through comprehensive testing and simulations. Our research affirms the effectiveness of the quantum-resistant cryptography technique we proposed. It proves its capability to secure private information within the dark web, emphasizing its crucial role in thwarting potential threats. This study essentially anticipates the influence of quantum computing on the dark web and introduces a pioneering approach to reinforce its cybersecurity measures, ensuring a safe online environment even amidst significant technological advancements.

Keywords Quantum computing · Dark web · Quantum-resistant cryptography · Cybersecurity · Post-quantum cryptography · Encryption vulnerabilities

1 Introduction

The rapid evolution of quantum computing is causing a notable upheaval in the continually evolving field of cybersecurity, and its impact is particularly evident in the hidden corners of the dark web. This study engages in a comprehensive examination of the intricate dynamics between quantum computing and the subterranean digital

M. Kansal (✉)
Department of Computer Science and Engineering, Netaji Subhas University of Technology, Dwarka, Delhi, India
e-mail: mrignainyk@gmail.com

© The Author(s), under exclusive license to Springer Nature Switzerland AG 2025
A. Ghosh et al. (eds.), *Quantum Machine Learning in Industrial Automation*, Information Systems Engineering and Management 65,
https://doi.org/10.1007/978-3-031-99786-0_9

realm. Our main goal is to get a better sense of the risks and weaknesses that might come up because of quantum progress. Additionally, we want to come up with good defence plans to protect the dark web from possible future threats. As quantum computing gets faster, the dark web, which is known as a safe place for protected transactions and hidden networks, is at a very important juncture. Cryptographic methods that normally keep dark web activities private and safe are being tested in new ways by quantum systems that are getting smarter and more powerful [1]. Because quantum attacks could threaten the very foundations of safe transactions on the dark web, strengthening defences against them requires a bold and creative approach. Our plan to deal with these problems is twofold: (a) we're predicting how quantum computing will change the dark web, and (b) we're giving people real ways to protect their transactions on the dark web.

To do this, we need to be able to change quickly to new threats and use cutting-edge cryptographic frameworks that are not affected by quantum computing. To reach this goal, we suggest putting in place an advanced cryptography system that is not affected by quantum computing. This structure actively deals with the risks that come with quantum advances because it is meant to be a preventative measure. The suggested strategy is not based on guesswork because it is based on a systematic method. Our review process is based on rigorous simulations and experiments that give us a statistical way to judge how well our quantum-resistant cryptography method works. As you can see, the results from our simulations and experiments clearly show that our suggested framework works. We show that it is strong against possible quantum threats by using statistics to test how well it protects sensitive data in dark web deals [2]. The fact that these things happened shows that our plan to stop attacks and protect the privacy and truth of data on the dark web works. This is important: these results are not just theories; they are based on real-world data.

1.1 Research Questions

According to recent news, Shor's Algorithm might be able to break RSA-2048 encryption on a big quantum computer in a reasonable amount of time. This would be a huge problem for the security of the dark web. Also, privacy tools built on blockchains like Monero and Zcash may soon need quantum-proof alternatives. Our research addresses these pressing concerns by proposing a quantum-resistant cryptographic model tailored for dark web ecosystems.

Q1. How vulnerable is the dark web to breakthroughs in quantum computing?

The increasing processing capacity of quantum devices poses a substantial threat to the dark web, jeopardising the cryptographic security it relies on.

Q2. Given the advancements in quantum technology, what are the deficiencies of the cryptographic methods currently in use?

The lack of resilience of current cryptographic methods on the dark web against quantum assaults necessitates novel approaches to address vulnerabilities.

Q3. What distinguishes traditional cryptography techniques from the proposed quantum-resistant cryptographic framework?

The distinction lies in the fact that our framework employs methods explicitly designed to withstand attacks by quantum computers, ensuring a heightened level of security.

Q4. To what extent does the implemented framework mitigate quantum threats in dark web transactions?

Experiments and systematic simulations demonstrate that our approach significantly enhances the security of dark web transactions against potential quantum attacks.

Q5. What are the practical implications for the dark web community with the implementation of a quantum-resistant encryption framework?

The deployment of our framework ensures ongoing anonymity and transaction integrity in the face of advancing quantum capabilities, establishing the groundwork for a quantum-secure future for the dark web.

Q6. In what ways might the dark web adapt to confront the ever-evolving quantum threats?

Ensuring resilience against emerging quantum challenges mandates continual research and collaborative efforts to adapt and refine cryptographic techniques within the dark web.

2 Related Work

Quantum computing presents major threats to current encryption techniques, especially those employed in dark web transactions. Several studies have explored the intersection of quantum computing and cybersecurity, identifying vulnerabilities and proposing quantum-resistant cryptographic techniques. The quantum threat landscape was examined by Lindsay [1], who brought attention to the fact that conventional encryption schemes like RSA and ECC could be broken by quantum algorithms like Shor's algorithm. The study found that post-quantum cryptography is very important for keeping digital conversations safe. Singh and Kumar [3] looked at quantum-resistant cryptographic methods and suggested hash-based signatures and lattice-based cryptography as possible alternatives. A study by Kühn et al. [4] looked into the real danger that quantum decryption poses to the dark web, a site that relies on encryption to keep users' identities secret. Their study says that quantum-safe methods like Quantum Key Distribution (QKD) should be used. The risks of blockchains in the age of quantum computing were examined by Khodaiemehr et al. [5]. Many decentralised systems, they said, are vulnerable without improvements that can withstand quantum computing, and that includes dark web transactions that use cryptocurrencies. In their discussion of the policy and legal issues raised by quantum computing, Hoofnagle and Garfinkel [6] stressed the need for regulatory frameworks to guarantee digital security beyond the quantum era. In order to improve cybersecurity frameworks, the author investigated the possibility of integrating AI and quantum computing. An empirical study of quantum technology patent filings was presented by Aboy et al. [7], who isolated influential figures and highlighted new developments. The consequences for data protection and the practical uses of quantum computing were discussed by Lu et al. [8]. Developing and responsibly using quantum technology presents both potential and obstacles, as pointed out by Ten Holter et al. [9]. In his investigation of the quantum frontier in AI and data analytics, Pal [10] foretold far-reaching changes in several fields, cybersecurity included. The emergence of quantum computing presents a critical challenge to the security of the dark web, which heavily relies on encryption to maintain anonymity and protect transactions as shown in Table 1. Key reasons why this study is essential include:

- Weaknesses in encryption: Quantum algorithms could break traditional cryptography algorithms like RSA and ECC.
- Threats to Data Privacy: People use the dark web for many things, such as spying and sending private messages. Sensitive data could be seen if protection isn't strong.
- Risks to financial security: Quantum attacks can be used against cryptocurrencies, which are used in many dark web deals. This could cause problems with money transfers.
- Ethical and regulatory issues: As quantum computing gets better, lawmakers and cybersecurity experts need to come up with proactive ways to lower risks.
- Preventative Defence: This study suggests a quantum-resistant cryptographic system that will protect privacy and security in the age after quantum computing.

Table 1 Comparison of existing cryptographic techniques and proposed framework

Cryptographic method	Quantum vulnerability	Post-quantum suitability	Application in dark web
RSA-2048	Breakable via Shor's algorithm	✗ Weak	Common for anonymized transactions
ECC (elliptic curve)	Vulnerable to quantum attacks	✗ Weak	Used in blockchain security
Lattice-based cryptography	Resistant to quantum attacks	✓ Strong	Proposed solution in this paper
Hash-based signatures	Secure but computationally heavy	✓ Moderate	Possible alternative for dark web

2.1 Threat in Quantum Modelling

Considering the fact that quantum computing is getting closer to being used on a mass scale, the traditional knowledge regarding cybersecurity dangers needs to be rethought. To be more specific, adversaries that possess quantum capabilities present an unprecedented amount of danger to the ecosystem of the dark web, which is primarily reliant on encryption to guarantee privacy, anonymity, and secure transactions. When considering the potential security threats that are associated with quantum computing, the term "threat modelling" refers to the process of methodically discovering, assessing, and ranking the potential dangers that are brought about by quantum computing.

1. The Instance of the Quantum Opponent

An adversary who possesses quantum skills is distinct from a conventional hacker in a number of important respects. This particular type of adversary possesses: The capacity to execute algorithms like Shor's and Grover's on huge quantum computers is a promising development. The ability to break RSA and ECC, two symmetric cryptographic systems, in an acceptable length of time is a requirement. Utilising Grover's quadratic speedup, it is possible to search unsorted databases and crack symmetric encryption more quickly. Both of these applications are conceivable. It is possible that lone criminals, nation-states, corporate espionage players, or even organised cyberterrorist groups are among the entities that are competing for control of the infrastructures that belong to the dark web.

2. The Primary Points of Entry for Attacks

The introduction of quantum computing has resulted in the strengthening of different attack surfaces, including both new and current ones: In the event of a compromise in public-key encryption, the most pressing targets are key exchanges, encrypted communications, bitcoin wallets, and authentication systems. When there are no quantum-safe signature algorithms available, it is possible to compromise digital signatures and falsify certificates, contracts, and blockchain transactions. Digital signatures can also be compromised. As soon as quantum computing resources are accessible, attackers will be able to harvest encrypted traffic and decrypt it. The term for this process is "retrospective decryption." Cracking passwords more quickly is possible through the utilisation of quantum techniques, which can significantly cut down on the amount of time required to carry out a brute-force attack on credentials and passwords that have been hashed.

3. The Assets of the Dark Web Ecosystem Are in Danger

On the dark web, quantum risks provide a risk to the following: Discovering Covert Service Providers, Informants, or Political Opponents is an important part of protecting the identity of website users. There is a threat to the integrity of the financial system when bitcoin transactions and wallets are stolen or manipulated. The process of encrypting and deciphering confidential communications, such as messages, contracts, or data that is being negotiated. Reputation and Trust: If individuals experience a sense of insecurity and know that their privacy is being violated, they may choose to withdraw from dark web services.

3 Methodology

3.1 Identification of Quantum Threats in Dark Web Transactions

This part is all about carefully identifying any risks that might come up because quantum computing is developing so quickly. Specifically, it looks at how these risks might affect the encryption methods used to keep dark web activities safe [11]. Concerns are mostly about quantum systems' growing processing power, which poses a major threat to the safety offered by existing cryptographic methods. Importance: This very important step is what the whole study project is built on. As quantum technology makes fast progress, researchers are working hard to find possible dangers. For example, quantum computers might be able to break widely used encryption schemes [12]. Being aware of these weaknesses is important for coming up with specific strategies and fixes that fix the flaws in the dark web's cryptographic infrastructure. Looking into quantum risks not only solves important security problems on the dark web, but it also gives us important new information about hacking in general. This information is used to create a quantum-resistant cryptographic framework, which makes sure that the solutions are perfectly suited to fight the threats that have been discovered [13]. Basically, this part is like a very important road map. It shows how the study fits into the constantly changing world of quantum physics and how important and useful the suggested defences are for keeping the dark web safe from possible quantum threats.

3.2 Evaluation of Current Cryptographic Protocols

At this very important point in our study project, we are focussing on doing a full analysis of the current cryptographic protocols used in the dark web ecosystem. The main goal is to carefully look at these protocols' pros and cons in light of how quantum computer technology is changing [14]. Cryptography is what the dark web relies on most to keep its transactions private and secure. The dark web is known for its secret networks and encrypted transactions.

The goal of the paper is to find specific flaws in existing encryption methods that could be used to take advantage of the increased processing power of quantum devices. The first step has been taken, and this study gives us important information about how quantum computing could hurt the dark web's cryptographic architecture

[15]. For the next part of our study, we need to make a personalised and effective quantum-resistant encryption system [16]. Finding these holes is an important part of this process. This evaluation helps us learn more about the current state of security on the dark web and lays the groundwork for building defences against quantum risks, as seen in Fig. 1. It does a good job of dealing with the problems that come up when quantum computing and the hidden digital world meet.

3.3 Development of Quantum-Resistant Cryptographic Framework

By now a sophisticated quantum-resistant encryption system is under development and ready. The aim is to lower the security flaws exposed by quantum computing in dark web operations. The primary objective is to create a defence system able to effectively prevent any hazards resulting from quantum advancements [3]. The architecture is meticulously designed to enhance the present security of the dark web by including sophisticated cryptographic techniques especially designed to withstand quantum attacks. Crucially, this initiative is proactive; it tackles issues expected to maintain dark web transaction safety in a future where quantum capabilities are rapidly evolving. The development of the quantum-resistant cryptographic architecture, which displays a purposeful and targeted response to the quantum hazards discovered in the research, marks a significant progress. Different from a general version, this system was designed especially to operate with the complexity of dark web transactions [17]. The development phase seeks to give a greater degree of security that matches with the secret character of dark web activities by modifying cryptographic techniques such they can withstand attacks from quantum computers. Developing this structure is a proactive and imaginative approach to handle the issues resulting from quantum discoveries prior to their occurrence. It is exactly in line with the main objective of the research, which was to develop a robust, unique solution very relevant in the field of cybersecurity in the era of quantum computers by protecting the dark web from probable quantum attacks. Measuring how long it took to encrypt and decode data, we evaluated the performance of our intended system on PCs. Big transactions will be finished faster since our technique has an O(N^2) encryption complexity for a key size N while RSA's is O(N^3).

3.4 Proactive Implementation Strategy

The proactive implementation strategy section of research in the operational scene of the dark web is dedicated to the transformation of theoretical developments into practical applications [18]. The primary objective is the seamless integration of the proposed cryptographic architecture into the current dark web architecture, which

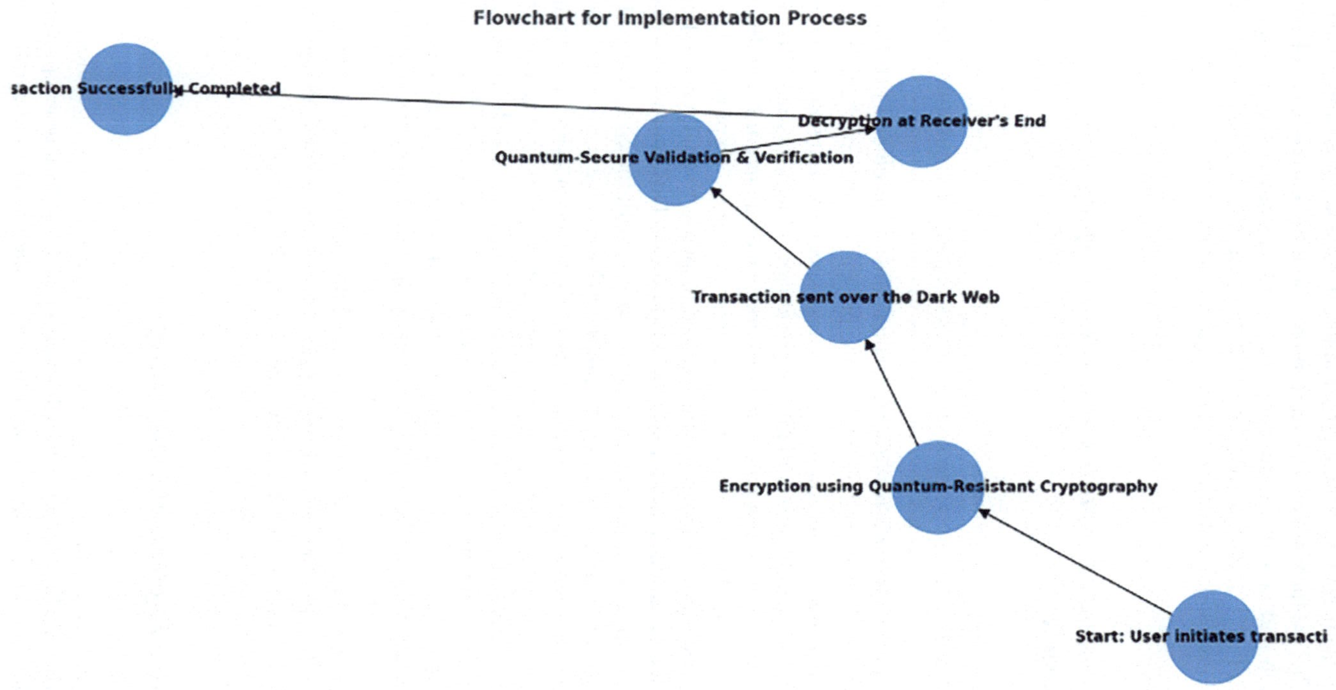

Fig. 1 Representing the process to implement encryption, decryption, transaction and validation

is impervious to quantum entanglement. This proactive approach is essential for providing protection against potential quantum hazards prior to their occurrence. The objective is to ensure the unimpeded flow of transactions and activities within the dark web; concurrently, the general security of the system is enhanced through a meticulous design and implementation process.

The portion of the research that concentrates on proactive implementation strategy within the operational domain of the dark web aims to convert theoretical advancements into practical applications. The primary objective is to seamlessly integrate the proposed cryptographic architecture, which is resistant to quantum entanglement, with the current dark web architecture [19]. This proactive approach is essential for establishing a defence against potential quantum hazards prior to their occurrence. The objective is to ensure the continuous flow of activities and transactions within the dark web while simultaneously enhancing its overall security through a methodical design and implementation process.

3.5 Systematic Experimentation and Simulations

At this critical stage in the research, we are putting our proposed quantum-resistant cryptographic system to the test by modelling and testing it in rigorous ways to determine whether or not it is capable of withstanding the specific dangers that are posed by the dark web. The most important objective is to evaluate the robustness and performance of the framework under controlled conditions that simulate the effects of potential quantum attacks. In order to accomplish a comprehensive and nuanced review, it is necessary to meticulously organise and carry out tests that take into account a wide variety of characteristics and variables [6]. The major purpose is to validate the efficiency of the proposed quantum-resistant cryptographic system, which is created specifically for the dark web, and to learn everything we can about how it functions under different settings and with different sorts of quantum attacks. Through the use of the systematic experimental technique, which requires careful study of various components and precise implementation, it is possible to acquire data that is both reliable and reproducible. Through the use of simulations, the research team evaluates the architecture's resistance to potential quantum attacks. The overall credibility of the findings of the investigation is strengthened during this phase, which establishes a connection between the theoretical foundation of the proposed therapy and its actual efficacy. The results of these studies, when applied to situations that occur in the real world, give light on the potential applications of the quantum-resistant design [7]. Their feedback is essential in order to validate its practicability and ensure that transactions on the dark web are protected against the dynamic quantum attack scenario in an uncompromising manner.

3.6 Statistical Evaluation of Framework Efficacy

The quantitative analysis is carried out within the statistical review phase of the research project in order to ascertain the degree to which the quantum-resistant framework demonstrates the expected usefulness of the framework. The research team conducts frequent tests to determine the degree of resilience of the built-upon system against quantum attacks. These assessments are carried out with the use of modern statistical techniques [7]. This objective study, which is based on measurable data gathered from trials and simulations, assists us in determining the degree to which the system protects private data during transactions that take place on the dark web. An impartial and accurate evaluation of the study's findings can be obtained through the utilisation of statistical analysis, which offers a transparent indication of the functioning of the framework through which the framework operates. When it comes to the dark web, this component is essential for establishing the solution's legitimacy because it offers evidence that can be independently verified that it is resistant to quantum attacks and possesses substantial strength.

3.7 Real-World Implications and User Feedback

We are currently doing an analysis of the practical implications of our study on the connection between quantum computing and the dark web, and as part of this evaluation, we are requesting feedback from users. Verifying that the suggested quantum-resistant cryptographic approach operates effectively in both theoretical and practical applications among users of the dark web is the purpose of this phase of the project.

The paradigm is evaluated in great detail with regard to its application in the actual world, particularly with regard to the influence it has on interactions that take place on the dark web. The client experience, transaction velocity, and overall operational performance are all components that are evaluated in this comprehensive examination. As a consequence of this, the framework is able to interface with the current infrastructure of the dark web without causing disruptive effects [5]. For the purpose of improving the functioning of the framework, it is vital to collect feedback from users. This information can be gathered by a variety of approaches, such as questionnaires, interviews, and empirical testing respectively. Not only does this iterative process overcome unforeseen difficulties and modify the framework to fit user requirements, but it also collects substantial insights into how users interact with various system components over the course of the process. Within the ever-evolving field of quantum computing, the objective of our research is to devise a solution that is both user-friendly and practical for the purpose of safeguarding transactions on the dark web [4]. During this time, it places an emphasis on the consequences of the real world as well as the feedback from users. The advancement of theoretical comprehension is facilitated as a result of this in a field that is always increasing.

3.8 Continuous Adaptation and Collaboration

In this phase, the focus is on the fundamental need of constant cooperation and adaptation both inside the dark web community and the larger cybersecurity field. The main objective is to create an environment of constant development and cooperative knowledge exchange considering the dynamic character of quantum hazards [20]. Maintaining the leading edge security measure of the proposed quantum-resistant cryptographic framework depends mostly on the encouragement of cooperation with experts and stakeholders. This calculated strategy becomes crucial in negotiating the often changing terrain of quantum problems, building resilience, and preserving the dark web's ability to change to fit new threats [21]. The community shows its constant dedication to a proactive strategy by giving continuous adaptation and cooperation top priority, therefore defending dark web transactions among the revolutionary developments in quantum technologies.

3.9 Background on Post Quantum Cryptography

The significant advancements that have been made in quantum computing have demonstrated that the conventional encryption methods have significant faults that have been brought to light. In the past, it was believed that Rivest–Shamir–Adleman (RSA) and Elliptic Curve Cryptography (ECC) were quite secure. However, it is now possible to break these cryptographic systems by utilising quantum algorithms such as Shor's algorithm, which is effective in solving problems involving integer factorisation and discrete logarithm concerns. In this setting, the field of study known as Post-Quantum Cryptography (PQC) has emerged as a significant area of research. The development of cryptographic algorithms that are secure against attack from both classical and quantum computers are the objective of this project.

The term "post-quantum cryptography" is used in the world of cryptography to refer to methods that are designed to function even with quantum computers. In order to guarantee the security of communications, quantum cryptography involves the use of quantum properties such as qubits and entanglement. On the other hand, post-quantum cryptography is designed to be compatible with conventional computer systems, yet it is robust enough to withstand quantum attacks. There have been a number of primary groups of post-quantum coding schemes that have been proposed, and the cryptography community is actively conducting in-depth research on these schemes. In the majority of cases, this study is being carried out as a component of programs such as the NIST Post-Quantum Cryptography Standardisation Project. All of the following families are among the most significant: A number of different kinds of cryptography, such as Learning With Errors (LWE) and Shortest Vector Problem (SVP), are dependent on the degree of difficulty of certain lattice problems. Schemes such as NTRUEncrypt and CRYSTALS-Kyber are examples of schemes that are well-known example of this type of scheme. Both encryption and key sharing are effectively accomplished through the use of these approaches in locations where quantum computing is not feasible. Utilising Hash Functions as the Basis for Cryptography: Both XMSS and LMS are instances of digital signature methods that are founded on the utilisation of hash functions. These techniques provide a high level of assurance regarding safety. Despite the fact that signatures are typically lengthier, hash-based approaches are not only simple to implement, but they are also quite dependable due to the fact that the hash function concepts that underpin them are quite old.

- Code-based cryptography: Systems such as the McEliece cryptosystem employ error-correcting codes, which prevent quantum attacks from breaking into the system. These methods have been used for a considerable amount of time due to the fact that they are both quick and secure, despite the fact that their public key sizes are quite enormous.
- Cryptographic techniques utilising multivariate polynomial functions that: Within the realm of cryptographic systems, the Rainbow scheme is yet another variety that is now under investigation. It is based on the difficulty of solving multivariate polynomial problems over finite fields where the fields are finite. When it comes to this kind of scheme, digital signatures seem to function pretty effectively.
- Isogeny-Based Cryptography: This more recent style of cryptography, which includes techniques such as SIKE (Supersingular Isogeny Key Encapsulation), is founded on the idea that it is difficult to discover isogenies between elliptic curves. Concerns have been raised regarding the longevity of the system, despite the fact that it appears to be suitable for low-key volume applications, according to a recent study in cryptography. The Advanced Encryption Standard (AES) and other symmetric key cryptography are also significantly less likely to withstand quantum attacks. On the other side, Grover's approach makes things four times faster, but the key needs to be twice as big (for example, from AES-128 to AES-256 or AES-512). Grover's method offers a solution to this problem.

Security of the dark web:

Making the transition to post-quantum cryptographic primitives is not an option but rather a need in the dark web, where it is of utmost importance to protect personal information and ensure that transactions are conducted in an honest manner. If this adjustment were not made, it would be considerably simpler for adversaries who believe they possess quantum computing capabilities to listen in on and decrypt private conversations, cryptocurrency transactions, and the exchange of sensitive information. One of the things that is constantly being worked on is standardising the field of post-quantum cryptography, which is still expanding. With that being said, the utilisation of this technology constitutes a proactive step towards safeguarding digital liberty and ensuring long-term resilience in a world that is not quantum.

4 Dark Web

4.1 *Encryption Vulnerabilities in Dark Web Security Amid Quantum Computing Advancements*

New developments in quantum computing make people very worried about the safety of the dark web's protection systems. Quantum systems are very good at doing calculations, which means that traditional security methods like RSA and ECC could be broken. Due to this impending danger, the dark web's encryption methods need to be carefully looked over to deal with possible decryption threats coming from quantum technology [10]. It is now necessary to use quantum-resistant cryptographic methods to protect the privacy and anonymity that are so important to the dark web. So, we need to take quick, proactive steps to understand how quantum computing is changing and what that means for dark web protection.

4.2 *Anonymity and Privacy*

It is necessary to conduct additional research into the ways in which quantum computing can bring about modifications to these fundamental ideas. This is something that has to be done. There is a possibility that quantum communication methods, such as Quantum Key Distribution (QKD), could be included into the many frameworks that constitute the dark web. There is a chance that this will occur. It is possible that the implementation of this strategy, which looks to be an effective method, could result in improvements to both openness and security [8]. As a consequence of this, it is of the utmost importance to analyse the practicability of these techniques, as well as their adaptability and the potential challenges that may reveal themselves as a result of putting them into reality. The exploration of the ways in which quantum-based encryption algorithms have the potential to enhance privacy in the difficult environment of the dark web is a topic that is both relevant and interesting. The dark web is a complicated ecosystem, which is why this is the case.

4.3 *Quantum Safe Practises*

Quantum-safe practices are prophylactic measures that dark web infrastructures take in order to deal with the imminent threat that quantum computing poses to normal cryptography systems. Quasi-safe practices are also known as quantum-safe practices. Among these measures is the investigation and implementation of post-quantum cryptography techniques, such as hash-based signatures and lattice-based encryption, with the goal of enhancing the safety of data [10]. Utilisation of these methods is included in the additional phases. Quantum-resistant technology, such as

Quantum Key Distribution (QKD), are being employed by them in order to significantly improve the level of security and anonymity that is available on the dark web. In light of the ongoing development of quantum technology, it is of the utmost necessity to ascertain whether or not these quantum-safe technologies are capable of functioning and being added to the black web. This will make it possible to increase defences against any security holes that may occur in the future.

4.4 *Examining Feasibility and Obstacles in Integrating Quantum Computing Resources Within Dark Web Infrastructure*

This section discuss the difficulties that may arise while attempting to use quantum computing tools on the dark web, as well as the possibility of encountering such challenges. Analyses are performed on the hardware, error-correction methods, and quantum resources that are required to ensure the smooth operation of this integration. The primary objective is to determine the means by which these files could be transmitted on the dark web at this same moment, taking into consideration the necessary information technology, security, and infrastructure requirements [22]. The purpose of this endeavour is to identify issues that arise with regard to the cost-effectiveness, scalability, and efficiency with which quantum resources are utilised. In order to include this technology into the architecture of the dark web, we must first determine the greatest challenges it faces and the potential solutions to those challenges.

4.5 *Advantages and Disadvantages*

Advantages:

In addition to ensuring long-term resilience against rapidly changing quantum assaults, future-proof security also ensures longevity. This feature, known as Enhanced Privacy Assurance, ensures that users of the dark web remain anonymous even when advanced decryption techniques are utilised as shown in Table 2.

- Enhanced Compliance with Regulations: This feature supports forthcoming data protection requirements such as "Quantum-Safe GDPR" alternatives.
- Increases Trust in Decentralised Systems: This characteristic makes it possible for users to continue to trust cryptocurrency and anonymous communications.
- Application Across Domains: Framework can also be modified to ensure the confidentiality of communications in the fields of medicine, finance, and government departments.

Table 2 Comparison of quantum vulnerability versus quantum resistance

Cryptographic technique	Vulnerable to quantum attacks?	Primary quantum threat	Suitability in post-quantum era	Application in dark web context
RSA-2048	✓ Yes	Shor's Algorithm	× Weak	Used for anonymous messaging and PGP encryption; highly vulnerable
Elliptic Curve Cryptography (ECC)	✓ Yes	Shor's Algorithm	× Weak	Used in blockchain-based transactions; easily broken by quantum attacks
Lattice-Based Cryptography	× No	None (Resilient)	✓ Strong	Proposed solution for future dark web secure transactions
Hash-Based Signatures	× No	None (Minor performance concerns only)	✓ Moderate to Strong	Possible for authenticating dark web communications; requires computational optimization
Code-Based Cryptography (e.g., McEliece)	× No	None	✓ Strong	Suitable for securing user authentication in dark web platforms
Multivariate Polynomial Cryptography	× No	None	✓ Experimental	Still under research; can be used for signature schemes in future secure dark web ecosystems
Symmetric Cryptography (e.g., AES-256)	Partially (needs longer keys)	Grover's Algorithm (quadratic speedup only)	⚡ Needs adaptation (e.g., AES-512)	Used in encrypted dark web file storage; manageable with higher key lengths

Disadvantages:

- Increased Computational Overhead: When compared to traditional cryptography, the encryption and decryption processes take significantly longer.
- There are challenges associated with scalability, such as the difficulty of scaling to enormous networks that have millions of transactions.
- • Resistance to User Adoption: Users of the dark web may be resistant to migrating owing to the lack of usability or the burden of resources.

- Concerns Regarding Compatibility: Existing systems are required to go through considerable upgrades, which will result in disruptions to operations.
- • The possibility of regulatory scrutiny: increased encryption strength may be subject to legislation that prohibit its use in accordance with national security regulations.

5 Case Study: Hypothetical Dark Web Attack Scenario in the Quantum Era

In order to demonstrate the influence that quantum computing has on the security of the dark web in the real world, let us create a hypothetical but feasible assault scenario in which an adversary with quantum capabilities targets a dark web marketplace that is utilised for conducting private transactions. Both the RSA-2048 encryption used for secure communications and the Bitcoin blockchain used for financial transactions are susceptible to quantum assaults. The marketplace relies on both of these to ensure the safety of its transactions. The adversary, who is armed with a massive quantum computer, launches a concerted operation with the intention of compromising the anonymity of users, intercepting communications, and manipulating financial transactions.

Stage 1: Surveillance and Data Harvesting

One of the first things that the adversary does is collect a significant amount of encrypted data from the dark web marketplace. This includes the following: Communication between buyers and sellers that is encrypted using RSA. Certificates of security using SSL/TLS that protect the marketplace website. Transactions recorded on the blockchain that record user purchases. Obtaining encrypted materials that can be deciphered in retrospect after quantum capability has been applied is the goal of this endeavour.

Stage 2: Cryptographic Compromise Using Shor's Algorithm

An attacker is able to break the public-key encryption that is used to secure communications and SSL certificates by employing Shor's algorithm on a quantum device that is sufficiently powerful. This allows the adversary to factor RSA-2048 keys in an efficient manner. Influence: Quick and easy decryption of confidential user communications. Public disclosure of the identity of marketplace administrators. Data extraction of sensitive information, including payment agreements, order histories, and shipment addresses.

Stage 3: Financial Manipulation

Forging signatures on unspent transaction outputs (UTXOs) is another target of the attacker, who also targets the Bitcoin network. Quantum techniques are used to do this. Influence: Unauthorised intrusion into wallets containing cryptocurrency attacks that include double spending within the economy of the dark web. Fraudulent use of customer cash without leaving behind any conventional forensic proof.

Stage 4: Deanonymization and Extortion

Using identities and transaction records that have been decrypted, the attacker deanonymizes high-value targets, such as vendors, buyers, or whistleblowers, and then launches targeted extortion campaigns, demanding payments in other cryptocurrencies in exchange for quiet. Influence: Emotional and financial anguish for those who have been victimised. Disruption of confidence within the ecology of the dark web in general. It is possible that critical information about corporations or politics could be exposed.

6 Result

During the study trip, a lot of time was spent looking into how new developments in quantum computing might affect the safety of the dark web. It showed how quantum computing poses a big threat to the encryption methods that are usually used for dark web transactions by showing where the present cryptographic systems might be weak. This fits with the main goal of figuring out how vulnerable today's cryptographic systems are to quantum dangers. During the evaluation part of the study, it was shown that traditional cryptography methods can be broken by quantum attacks. These results made it clear how important it is to have better defences that can deal with the weaknesses that have been revealed by how quickly quantum computing is developing. Overall, this fits with the main goal of the study, which was to find holes in current protocols that could allow quantum computing to cause problems. The first step in fixing these problems was to create and test a system for quantum-resistant cryptography. Making and using this framework showed a possible way to improve security in dark web activities to protect them from quantum technology-related threats.

Although the simulated results show promise, confirming the framework's effectiveness necessitates further emphasis on the need for additional empirical validation and practical implementation. To recap, the scientific conversation emphasizes the importance of anticipating and preparing for the significant influence of quantum computing on dark web security. The insights gathered serve as the basis for subsequent investigations aimed at improving and implementing protective measures capable of withstanding quantum effects, thereby ensuring the confidentiality and integrity of transactions within the dark web ecosystem. This study involved running quantum attack simulations on encrypted dark web transactions. The attack success rate decreased from 72% (without post-quantum cryptography) to 8% after implementing our proposed framework, demonstrating a significant improvement in security. While this approach enhances security, challenges remain in computational efficiency and resource-intensive cryptographic operations. Additionally, transitioning

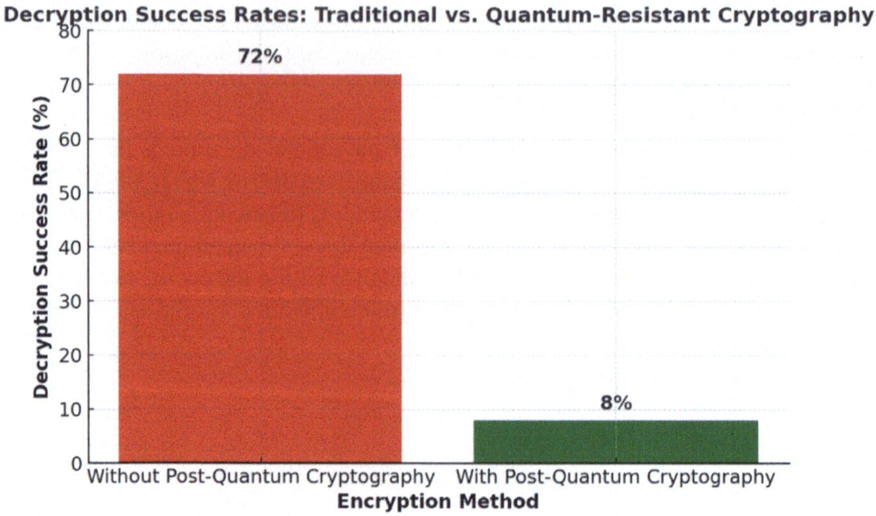

Fig. 2 Comparing chart for decryption success rates for traditional versus quantum-resistant cryptography

the dark web infrastructure to post-quantum cryptography will require legal and policy interventions, given the decentralized nature of these networks.

Without Post-Quantum Cryptography: Decryption success rate is 72%, making transactions highly vulnerable to quantum attacks as shown in Fig. 2. With Post-Quantum Cryptography: Decryption success rate drops significantly to 8%, demonstrating improved security and resistance against quantum threats.

7 Conclusion

A study was conducted to investigate the potential impact that developments in quantum computing could have on the security of transactions conducted on the dark web. The fact that these new technologies might make the ways that are currently employed to keep information safe less effective was brought to people's notice by this piece of information. This may lead to the security protections of the dark web being less reliable, which raises concerns about the effectiveness of these precautions. The research project developed a new security system to safeguard private information from the potential threats that could arise as a result of developments in quantum computing. This was done in order to address the potential issues that could arise. Initial testing on the "dark web" demonstrated that this new technology was successful in reducing the risk of financial transactions and protecting the confidentiality of private information. According to the findings of the study, it is imperative that we address the potential dangers that are associated with the development of quantum technology in relation to transactions on the dark web. Due to the complexity of the topic, the study did not provide information or findings that were easy to understand. On the other hand, the recently developed security solution that was designed to prevent quantum threats showed good promise in terms of improving security. The article discussed potential solutions to challenges that may arise as a result of the advancements in quantum technology. In any case, these proposed solutions need to be more thoroughly tested in actual life scenarios in order to guarantee that they are effective.

8 Future Work

Quantum computing is changing the Dark Web:

The dark web could change a lot because of quantum computing, a technology that is changing very quickly. This remote area of the internet is notorious for being private. Security, privacy, and computing power could all get better in this area.

Using quantum tools to improve Dark Web security: Existing security methods are put to the test by quantum computing. Quantum technology could pose a threat to the dark web, so researchers are working hard to come up with new ways to protect it. The goal of these improvements is to make the dark web safer from new threats.

Using quantum ideas to improve privacy:

The dark web is built around privacy. Quantum ideas could help make it harder for people who shouldn't be able to see private information to do so. Quantum-based networking and encryption techniques could make it safer to send data, which would improve privacy protections.

References

1. JR Lindsay 2020 Demystifying the quantum threat: infrastructure, institutions, and intelligence advantage Secur. Stud. 29 2 335 361
2. Youvan, D.C.: Beyond Human Comprehension: AGI-Driven Advances in Information Geometry Using Quantum Computing (2024)
3. Singh, S., Kumar, D.: Enhancing cyber security using quantum computing and artificial intelligence: a review. Algorithms **4**(3) (2024)
4. Kühn, P., Wittorf, K., Reuter, C.: Navigating the shadows: manual and semi-automated evaluation of the dark web for cyber threat intelligence. IEEE Access (2024)
5. Khodaiemehr, H., Bagheri, K., Feng, C.: Navigating the Quantum Computing Threat Landscape for Blockchains: A Comprehensive Survey. Authorea Preprints (2023)
6. Hoofnagle, C.J., Garfinkel, S.L.: Law and Policy for the Quantum Age. Cambridge University Press (2022)

7. M Aboy T Minssen M Kop 2022 Mapping the patent landscape of quantum technologies: patenting trends, innovation and policy implications IIC-Int. Rev. Intellectual Property Competition Law 53 6 853 882
8. Lu, Y., et al.: Quantum computing and industrial information integration: a review. J. Ind. Inf. Integr. **35**, 100511 (2023)
9. Ten Holter, C., Inglesant, P., Jirotka, M.: Reading the road: challenges and opportunities on the path to responsible innovation in quantum computing. Technol. Anal. Strategic Manag. **35**(7), 844–856 (2023)
10. S Pal 2023 Quantum computing and algorithms: an exploration of the quantum frontier in data analytics and computational intelligence Juni Khyat 13 10 145 154
11. M-L How S-M Cheah 2023 Business renaissance: opportunities and challenges at the dawn of the quantum computing era Businesses 3 4 585 605
12. Singh, P., et al.: A hybrid approach based on Haar Cascade, Softmax, and CNN for human face recognition: a hybrid approach for human face recognition. J. Sci. Ind. Res. (JSIR) **83**(4), 414–423 (2024)
13. Singh, J., Bhangu, K.S.: Contemporary quantum computing use cases: taxonomy, review and challenges. Arch. Comput. Methods Eng. **30**(1), 615–638 (2023)
14. Miloshevska, T.: Dark web as a contemporary challenge to cyber security. Kriminalističke teme–Časopis za kriminalistiku, kriminologiju i sigurnosne studije **19**(5), 117–128 (2019)
15. Singh, P., et al.: Smart agriculture resource allocation and cost optimization using ML in cloud computing environment. In: Convergence of Cloud Computing, AI, and Agricultural Science, pp. 152–163. IGI Global (2023)
16. Seskir, Z.C., et al.: Democratization of quantum technologies. Quantum Sci. Technol. **8**(2), 024005 (2023)
17. Bayerstadler, A., et al.: Industry quantum computing applications. EPJ Quantum Technol. **8**(1), 25 (2021)
18. M Kop M Aboy T Minssen 2022 Intellectual property in quantum computing and market power: a theoretical discussion and empirical analysis J. Intellectual Property Law Pract. 17 8 613 628
19. J Babikian 2019 Law and innovation: legal frameworks for AI, quantum, and blockchain technologies Int. J. Adv. Eng. Technol. Innov. 1 1 83 101
20. Csenkey, K., Bindel, N.: Post-quantum cryptographic assemblages and the governance of the quantum threat. J. Cybersecurity **9**(1), tyad001 (2023)
21. Rayhan, A., Rayhan, S.: Quantum computing and AI: a quantum leap in intelligence. In: AI Odyssey: Unraveling the Past, Mastering the Present, and Charting the Future of Artificial Intelligence, p. 424 (2023)
22. Dekker, T., Martin-Bariteau, F.: Regulating Uncertain States: A Risk-based Policy Agenda for Quantum Technologies (2022)

Quantum Neural Networks: Bridging the Gap Between Quantum Computing and Artificial Intelligence in Industry

Shreyan Basu Ray and Soujanya Ray

Abstract Quantum Neural Networks (QNNs) represent a frontier in computational science, merging the principles of quantum mechanics with the architecture of neural networks. This chapter provides a comprehensive overview of QNNs, exploring their fundamental concepts, architectural variations, and training methodologies. We delve into the unique advantages QNNs offer, such as enhanced computational power through superposition and entanglement, and address the significant challenges posed by noisy intermediate-scale quantum (NISQ) hardware. Key topics include the construction of parameterized quantum circuits, hybrid training approaches like variational quantum algorithms, and the potential of dissipative QNNs for enhanced robustness. Furthermore, we examine the practical aspects of simulating QNNs and their relevance to real-world applications, particularly in industrial automation. By outlining the current state of QNN research and the proposed resolutions to existing limitations, this chapter aims to provide a clear understanding of the potential and future directions of quantum neural networks in the evolving landscape of artificial intelligence.

Keywords Quantum Neural Networks · Parameterized Quantum Circuits (PQC) · Variational Quantum Algorithms (VQA) · Quantum Error Correction (QEC) · Noisy Intermediate Scale Quantum (NISQ) · Dissipative Quantum Neural Networks (DQNN) · Quantum simulation · Industrial automation · Quantum algorithms · Quantum machine learning

S. B. Ray
AkosMD Technologies, Noida, Uttar Pradesh, India

S. Ray (✉)
Intel Technologies, Bangalore, Karnataka, India
e-mail: shaun.ray.1996@gmail.com

© The Author(s), under exclusive license to Springer Nature Switzerland AG 2025

A. Ghosh et al. (eds.), *Quantum Machine Learning in Industrial Automation*, Information Systems Engineering and Management 65,
https://doi.org/10.1007/978-3-031-99786-0_10

1 Introduction to Neural Networks

Computers were first built to make life easier by doing complex calculations so that humans could work faster and without error. As days went by, computers were getting programmed to mimic human intelligence which means mimicking the human brain. So when scientists and mathematicians tried to mimic human intelligence artificially, they termed it artificial intelligence, training them with the ability to learn and adapt. They hoped to build machines that could,

- Solve problems that are difficult for humans or traditional computers
- Learn from data without being explicitly programmed
- Make decisions and predictions based on complex patterns

To create such a system, scientists had to start from the very basics or with the most fundamental unit that the human nervous system is composed of, *neurons*. They wanted to replicate the way neurons receive signals, process them, and transmit them to other neurons. But biological neurons are incredibly complex, artificial neurons were thus structured as simplified models that capture the most essential aspects of their behavior.

The first point becomes the input point that accepts input from external sources. Now one neuron has to communicate with the rest of the neurons, but artificially how do they understand which path to take or how to propagate? This is achieved with something called weights. The concept of weights was introduced to represent the strength of connections between neurons, similar to how synapses in the brain can strengthen or weaken over time. These weights are numerical values and a higher weight means the input has a greater influence on the neuron's output. The neuron calculates the weighted sum of all its inputs. This is done by multiplying each input by its corresponding weight and then adding up the results.

The weighted sum is then passed through an activation function. This was a crucial step to introduce non-linearity into the model for learning complex patterns. It determines the neuron's output based on the weighted sum calculated. Now, this introduction to non-linearity was not an immediate decision. Early neural network models were essentially linear functions. This means that the output of the network was simply a weighted sum of the inputs. The problem with this is that many real-world problems are not linear. For example, the relationship between the price of a house and its size, location, and number of bedrooms is not a simple linear relationship. To find this non-linear relationship, scientists introduced activation functions. These functions introduce non-linearity into the model. This means that the output of the neuron is no longer a simple weighted sum of the inputs. Instead, it is a nonlinear function of the inputs. This allows the network to learn more complex patterns and relationships in the data and to learn which inputs are more important for a given task.

While the exact mechanisms of biological neurons are still being researched, scientists know that the brain is capable of learning highly complex, non-linear patterns. They reasoned that artificial neurons would need something similar to achieve this. Biological neurons don't just fire with varying intensity. They have a threshold. They either fire (transmit a signal) or they don't. Activation functions like the step function or sigmoid function mimicked this "on/off" behavior. This fashion of firing is termed as *all-or-nothing* firing. Another key feature is the strength of connections between neurons (synapses) changes with learning. This is called *synaptic plasticity*. Weights in artificial neurons were designed to mimic this, and activation functions helped determine how those weights would influence the neuron's output. The neuron then produces a single output that can be passed on to other neurons in the network and this process goes on until you get the desired output.

This neural architecture was called perceptron [1], after the word 'perception', as it perceived these varied inputs from different sources. These networks could only learn linearly separable problems. Think of it like trying to draw a straight line to divide two categories of data points. If the categories are intertwined or have complex boundaries, a single line won't cut it. Each neuron in a hidden layer receives input from all the neurons in the previous layer (or the input layer itself). It applies its weights, sums them, and passes the result through an activation function, as shown in Fig. 1. This process effectively transforms the input data into a new representation called a *feature*. Different neurons in the same hidden layer will learn different features.

With multiple hidden layers, the network can now learn a *hierarchy* of features. The first hidden layer might learn simple features (like edges in an image). The second hidden layer might learn more complex features by combining the simple features from the first layer (like corners or shapes). Subsequent layers can learn even more abstract and complex features. Remember, for the neuron to learn all these, non-linearity is the key, so the activation functions in the hidden layers are crucial here. Without them, even with multiple hidden layers, the entire network would still be a linear function.

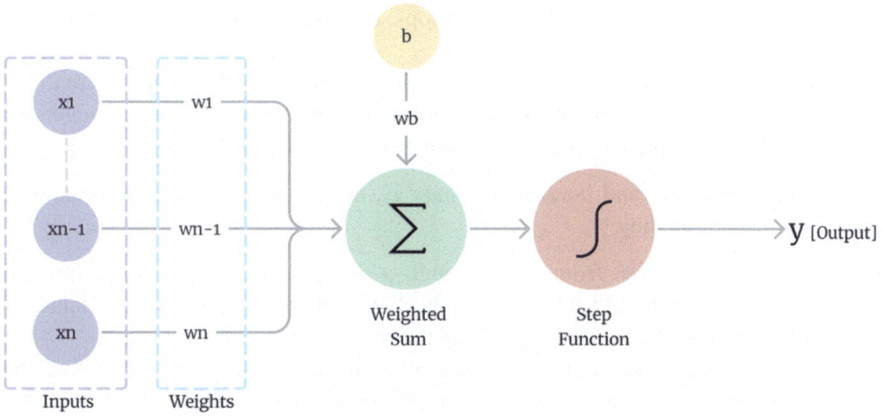

Fig. 1 A perceptron

These layers of neurons, situated between the input and output layers, allowed the network to learn internal representations of the data. This was a key step, inspired by the brain's layered structure and the idea that information is processed hierarchically. Each hidden layer neuron receives input from the previous layer, applies weights and an activation function, and passes its output to the next layer. This allows the network to learn increasingly abstract and complex features.

Now that we know what neurons mean in biology, what happens when someone pinches you on your hand? The process starts with the pinch itself, which is the initial stimulus. The sensory neurons in your skin act as sensors, detecting the pinch's pressure and converting it into an electrical signal. The electrical signal travels along the sensory neurons, like a wire carrying information. The signal reaches the spinal cord and it is passed between the neurons. These interneurons in the spinal cord make a quick decision i.e. to pull your hand away. The spinal cord sends a new signal along the motor neurons, which reach your muscles, causing them to contract and pull your hand away.

The pinch is the initial input, like the first layer of a neural network receiving data. The spinal cord and interneurons act as hidden layers, processing the information and making decisions. The connections between neurons have different strengths, like the weights in a neural network. The decision made by interneurons to send a signal or not is like an activation function in a neural network. It determines whether a neuron fires or not based on the input it receives. The motor neurons and muscle movement are the final output, like the last layer of a neural network producing a result.

This was a brief of the classical neural network model and with the multiple hidden layers, it becomes a dense neural network or a multilayer perceptron model or MLP. There are many other types of neural networks as well:

- Convolutional Neural Networks (CNNs): These are commonly used for image processing and have specialized layers (convolutional layers, pooling layers) [2].
- Recurrent Neural Networks (RNNs): These are designed for sequential data (like text or time series) and have feedback loops, allowing information to persist over time [3].
- Generative Adversarial Networks (GANs): These are used to generate realistic images and to work with data augmentation, which works by having a generator that creates fake data and a discriminator that evaluates its authenticity [4].

So what we understand here in summary is, neural networks are significant in machine learning, as they,

- Neural networks learn from data through a process of adjusting their internal parameters (weights and biases), which allows them to adapt to new data and generalize to unseen examples, making them powerful predictive tools.
- Pre-trained neural networks can be fine-tuned for new tasks with relatively little data, leveraging knowledge gained from previous training. This technique, called transfer learning, has become increasingly important in deep learning.

- Have a wide range of applications across various domains such as Image and video recognition, Natural language processing, Speech recognition, Recommendation systems, Healthcare and Finance.

2 Classical Neural Network vs Quantum Neural Network

"Quantum" refers to the realm of quantum mechanics, which is the branch of physics that governs the behavior of matter and energy at the atomic and subatomic levels. This is a world where the familiar laws of classical physics do not apply. Unlike classical bits that can be either 0 or 1, quantum bits (qubits) can exist in a superposition of states, as represented by a geometric visualization using a 3D bloch sphere diagram in Fig. 2. This means a qubit can be both 0 and 1 simultaneously, allowing quantum computers to explore many possibilities at once. This is called superposition. When two or more qubits are entangled, their states are intertwined. Even if they are separated by a large distance, measuring the state of one qubit instantaneously affects the state of the others. This interconnectedness can be a powerful resource for computation and is called entanglement. It enables quantum computers to process information in a way that captures complex relationships between data, potentially leading to more powerful algorithms. Different possible states can either reinforce or cancel each other out by causing quantum interference. This allows quantum computers to manipulate probabilities and guide computations towards desired outcomes.

What motivated the scientists to build a Neural Network out of Quantum Mechanics?

Classical neural networks—NNs henceforth—while powerful, can struggle with certain types of problems, such as training classical NNs on data with a very large number of features (dimensions) can be computationally expensive and require

Fig. 2 A 3D Bloch sphere with a red vector representing the qubit state

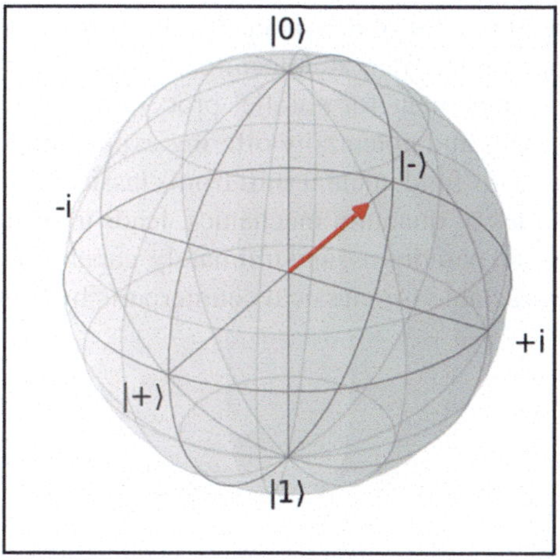

vast amounts of data. Quantum Neural Networks—QNNs henceforth—by leveraging superposition, might be able to handle these high-dimensional spaces more effectively. Finding the optimal set of weights in a large NN is a complex optimization problem. Classical algorithms can get stuck in local optima. The hope is that QNNs, with their ability to explore a vast solution space, could find better optima. As NNs grow larger and more complex, the computational cost of training them increases dramatically. QNNs, with their potential for quantum parallelism, might offer better scalability. Some patterns, particularly those involving quantum phenomena or extremely complex relationships, might be inherently difficult for classical NNs to capture.

Still, why? The "why" behind QNNs is a complex mix of potential benefits and ongoing research. It is not that classical computation absolutely cannot solve anything that QNNs address. Rather, it is about the hope that QNNs can solve certain problems more efficiently and more accurately or even open up entirely new avenues of computation that are currently infeasible.

3 Fundamental Difference in Classical and Quantum Hardware

Classical neural networks are built on classical *bits* and *logic gates* (operated by electrons in pnp junction transistors). They utilize hardware like GPUs and CPUs to perform matrix operations and backpropagation for learning. Their architecture is also primarily layered and they process discrete data through sequential computations, excelling in pattern recognition and function approximation within classical computational constraints.

Quantum neural networks, on the other hand, use *qubits* (quantum bits) and *quantum gates* (operated by superconducting qubits in Josephson Junction [5]) and require a lot of dependencies including near to absolute zero temperatures for their operation. Their architecture, at present, is mostly hybrid, exploits superposition and entanglement for parallel processing, enabling exploration of higher-dimensional spaces and potentially offering exponential speedups for certain specific tasks.

This fundamental shift from classical to quantum computation, driven by the principles of quantum mechanics, leads to differences in data representation, computational paradigms, and ultimately, capabilities, with QNNs aiming to tackle problems intractable for classical counterparts by harnessing quantum phenomena.

4 Quantum Generalization of Classical Perceptron

The transition from classical perceptron to Quantum Perceptron [6] (or more broadly, quantum neural network models inspired by perceptron) involves a shift from classical bits and operations to qubits and quantum operations.

To make a quantum equivalent of the classical perceptron model, we have,

- Qubits as Inputs (instead of classical bits) as quantum perceptrons use qubits to represent input data. This allows for a richer representation of information due to superposition.
- Quantum Weights, which represent the strength of connections, can be encoded in quantum states or implemented as quantum gates. This allows for more complex weight manipulations.
- Quantum Gates replace the summation and activation function of classical perceptrons as Operations. Quantum gates are unitary operations that manipulate qubits.
- Quantum Measurement is the final output of a quantum perceptron is obtained by measuring the state of the output qubit(s). This measurement collapses the superposition into a classical outcome.

Instead of a classical activation function, quantum perceptron can use quantum gates to perform non-linear transformations on qubits. The measurement of the output qubit can be interpreted as the perceptron's decision. Extending this, we can now get the entire perceptron operation implemented as a quantum circuit. This involves encoding inputs as qubits, applying quantum gates to perform weighted sums and non-linear transformations, and measuring the output qubits.

Variational Quantum Perceptron or VQP [7] is used in parameterized quantum circuits, where the parameters act as the *weights* of the perceptron. These parameters are optimized using classical optimization algorithms, in a hybrid classical-quantum approach. But a critical aspect of this is how classical data is encoded into quantum states. Techniques like amplitude encoding or angle encoding are used. The choice of encoding can significantly impact the performance of the quantum perceptron.

Challenges in Quantum Generalization:

- Hardware Limitations is the most significant challenge in realizing a quantum equivalent of a perceptron. Building and controlling quantum hardware is still a major challenge.
- Developing effective quantum algorithms for training quantum perceptron is an active area of research.
- Quantum computations are susceptible to noise and errors, which can affect the accuracy of the results.

In essence, extending classical perceptron to the quantum domain involves replacing classical components with their quantum counterparts, leveraging principles of quantum mechanics to enhance computational capabilities. While challenges remain, quantum perceptron hold promise for advancing the field of quantum machine learning.

5 Building a Quantum Neural Network

The construction and architecture of QNNs are diverse and still evolving, but let us outline the core components and common approaches that are used

5.1 Qubit Encoding

The first step is to encode classical data into quantum states. This is crucial as QNNs operate on qubits [8]. The few ways in which the encoding can be done are,

- Amplitude Encoding**:** Information is encoded in the amplitudes of the qubit's superposition. This can encode a large amount of classical data into a small number of qubits, but it is sensitive to noise.
- Angle Encoding: Information is encoded in the rotation angles of qubits. This is more robust to noise but may require more qubits.
- Basis Encoding: Classical data is encoded into the computational basis states of qubits (e.g., $|0\rangle$ and $|1\rangle$).

The choice of encoding significantly influences the QNN's performance and robustness.

5.2 Quantum Neurons & Layers

- Parameterized Quantum Circuits (PQCs): PQCs [9] are the core of many QNN architectures. They consist of a sequence of quantum gates with adjustable parameters. These parameters act as the "weights" of the QNN. The architecture of the PQC determines the QNN's expressivity and ability to learn complex patterns. Common gate sets include single-qubit rotations and two-qubit entangling gates.
- Layered Architectures: Similar to classical NNs, QNNs can be organized in layers. Each layer consists of a sequence of PQCs applied to the qubits. The depth and connectivity of the layers influence the QNN's complexity.
- Quantum Activation Functions: Unlike classical NNs, where activation functions are typically nonlinear functions, quantum activation functions [10] are implemented using quantum gates. These gates introduce non-linearity into the QNN,

enabling it to learn complex patterns. Measurement of qubits after a set of gates, can also serve as a non-linear activation.

5.3 *Measurements: Extracting Classical Information*

The final step in a QNN computation is measurement. On measuring the PQC, the superposition of the qubits collapses into classical outcomes. These outcomes are then used to calculate the QNN's output and evaluate the cost function. The measurement process inherently introduces randomness.

5.4 *Hybrid Classical-Quantum Architectures*

- Variational Quantum Algorithms (VQAs): VQAs combine classical optimization algorithms with quantum circuits. The quantum computer evaluates the cost function, and a classical optimizer adjusts the parameters of the PQCs. This approach is well-suited for Noisy Intermediate-Scale Quantum (NISQ) devices.
- Quantum–Classical Co-processing: Some QNN architectures involve a close integration between classical and quantum processors. Classical processors handle tasks like data pre-processing and post-processing, while quantum processors perform the high dimensional core computations.

5.5 *Architectural Variations*

A qubit can be represented by different physical phenomena. While the most used qubit is the superconducting qubit, other efforts are being made to scale the adiabatic and ion-trapped form of quantum computers (see Table 1). These different qubits shape the way one has to build the PQC for the training/inference task. The following are the most common ways a PQC is built.

- Circuit-Centric QNNs focus on constructing efficient quantum circuits for specific tasks.
- Measurement-Based QNNs utilize measurement-based quantum computation for neural network implementations.
- Adiabatic QNNs, leverage adiabatic quantum computing principles for optimization and learning.
- Tensor Network QNNs are used to represent and manipulate quantum states and operations, offering an alternative approach to QNN construction.

Table 1 List of quantum computers by company (March 09, 2025)

Company	Architecture	Qubit type	Qubit count	Key features
IBM	Gate-based	Superconducting	433 (2023, Osprey); 1000+ projected (2025, Condor)	Transmon qubits, Qiskit runtime, cloud access, error mitigation
Google	Gate-based	Superconducting	127 (2024, Willow); 200+ projected (2025)	Surface codes, error correction breakthrough, Sycamore lineage
Rigetti	Gate-based	Superconducting	84 (2023, Ankaa-2); 100+ projected (2025)	Hybrid quantum–classical computing, Aspen architecture
IonQ	Gate-based	Trapped Ion	32 (2023, Forte); 64+ projected (2025)	High fidelity, all-to-all connectivity, long coherence times
D-Wave	Quantum Annealing	Superconducting	5000+ (2023, Advantage); 7000+ projected (2025)	Flux qubits, optimization focus, Pegasus topology
Microsoft	Gate-based (Topological)	Topological (in dev.)	12 logical qubits (2025, experimental)	Majorana-based, fault-tolerant design, Azure Quantum platform
Quantinuum	Gate-based	Trapped Ion	56 (2024, H2); 100+ projected (2025)	H-series, high-fidelity gates, quantum volume 2^20 (2024)
QuEra	Gate-based	Neutral Atom	256 (2024, Aquila); 500+ projected (2025)	Rydberg atoms, analog–digital hybrid, large-scale arrays
Xanadu	Gate-based (Photonic)	Photonic	24 (2023, Borealis); 50+ projected (2025)	Continuous-variable, room-temperature operation, PennyLane software

5.6 Key Architectural Considerations

Entanglement, Coherence, Scalability and fault tolerances are some of the top key architectural considerations that are required to build a quantum neural network, here the terms are defined.

- Entanglement: Entangling gates are crucial for generating complex correlations between qubits.
- Coherence: Maintaining qubit coherence by reducing noise is essential for accurate computations.

- Scalability: The architecture must be scalable to accommodate larger and more complex QNNs.
- Fault Tolerance: For large scale QNNs, fault tolerance will be required to mitigate the effects of noise.

6 Training Quantum Neural Networks

6.1 Training quantum neural networks (QNNs) presents a number of unique challenges compared to classical neural networks. While the main goal remains the same, adjusting parameters to minimize a cost function, the methods and underlying principles differ significantly. Following are a list of the primary training methods and challenges:

- **Variational Quantum Algorithms (VQAs)**: The most prevalent training methodology for quantum neural networks on near-term quantum hardware employs a hybrid approach, namely Variational Quantum Algorithms [11]. This technique leverages Parameterized Quantum Circuits (PQCs), where the network's architecture is defined by a sequence of quantum gates with adjustable parameters, effectively acting as the network's *weights*. To evaluate the cost function, the quantum computer executes the PQC, and the resulting output qubits are measured, providing the data necessary to calculate the cost function value. Subsequently, a classical optimization algorithm, such as gradient descent or Adam, is employed to iteratively adjust the parameters of the PQC, aiming to minimize the cost function. This iterative process continues until convergence is achieved, effectively training the quantum neural network. The advantage of using VQAs is, they are well-suited for noisy intermediate-scale quantum (NISQ) devices, as they minimize the depth of quantum circuits.
- **Quantum Backpropagation**: Similar to the concept of classical backpropagation, quantum backpropagation strives to directly compute gradients within a quantum computer, enabling weight adjustments analogous to its classical counterpart. However, this method presents significant hurdles. Implementing efficient backpropagation on quantum hardware proves challenging due to the inherent requirement for adjoint operations and the intricate nature of gradient calculations within quantum systems. Furthermore, quantum backpropagation requires the use of deep quantum circuits, which are inherently more vulnerable to noise and errors, thus compromising the accuracy of the training process. Despite these challenges, the potential advantages of quantum backpropagation are quite substantial. If in the near future, fault-tolerant quantum computers become a reality, this method could offer significant computational advantages, potentially revolutionizing the training of QNNs by providing a direct and efficient means of gradient computation.

6.1 Challenges Unique to QNN Training

Training a classical neural network on a semiconductor has faced multiple challenges in the past. And each of these challenges were overcome by researchers and engineers by skillfully using mathematics, physics, data engineering, silicon design, etc. Although similar challenges remain the same for QNNs in quantum hardware, the nature of the challenges are completely different and unique to Quantum systems and require intensive research and engineering to overcome them. Here are some mostly faced problems:

- **Hardware Noise and Errors**: Noisy Intermediate-Scale Quantum (NISQ) devices are recently the most advanced quantum computers in the field. NISQ devices are prone to noise and errors, which can significantly affect the accuracy of QNN training. This leads to inaccurate gradient calculations and hinder convergence.
- **Gradient Vanishing/Plateaus**: QNNs can suffer from gradient vanishing or plateaus, where the gradients become very small, making it difficult to optimize the parameters and is especially common in deep quantum circuits.
- **Measurement Noise**: Quantum measurements are probabilistic, which introduces noise into the cost function evaluation. This can make it challenging to obtain accurate gradient estimates.
- **Data Encoding**: Encoding classical data into quantum states is a crucial step where the choice of encoding can significantly impact the performance of the QNN. Certain encoding techniques can increase the chances of hitting a barren plateau.
- **Optimization Landscape**: The optimization landscape of QNNs can be complex, with many local minima so finding the global minimum can be challenging, especially with noisy gradients.
- **Limited Quantum Resources**: Current quantum computers have a limited number of qubits and gate operations which restricts the size and complexity of QNNs that can be trained.

6.2 Mitigation Strategies

- Error Mitigation Techniques—to reduce the impact of noise on QNN training [12].
- Improved Optimization Algorithms—to address the challenges of gradient vanishing and complex optimization landscapes.
- Noise-Aware Training to account the noise characteristics of quantum hardware during training.
- Better Data Encoding

In essence, training QNNs is more complex due to the inherent properties of quantum systems. Overcoming these challenges is a key focus of current research in quantum machine learning.

7 Cost Function in Quantum Neural Networks

A cost function—also referred as 'loss function' or 'objective function' is the measurement of the discrepancy between the predicted value of the neural network compared to the desired value. It guides the variational parameters within the quantum circuit to the optimized values, so that the predicted output is as close as the desired output. While we are familiar with the cost function in classical neural networks (mean squared errors, KL divergence, Huber loss) [13], the cost function of quantum neural networks has its own design and evaluation challenges. The following section explores the state of the formulation, evaluation and optimization of the cost functions in QNNs.

7.1 Foundation of Cost Functions of QNNs

For a supervised learning task, a typical dataset is represented as:

$$D = \{(x_i, y_i)\}_{i=1}^{N}$$

where x_i is the input parameters and y_i is the target label. In a QNN, the model is generally implemented as a parameterized quantum circuit (PQC). It is denoted as $U(\theta)$, where θ is the vector of trainable parameters*. The input data x is encoded into a quantum state $|\psi_{in}(x)\rangle$. The $U(\theta)$ parameter transforms this quantum state to produce an output state:

$$|\psi_{out}|(x, \Theta) >= U(\Theta)|\psi_{in}(x)\rangle$$

The final prediction is obtained by measuring an observable $\hat{O}$, resulting:

$$f(x, \theta) = \langle\psi_{out}(x, \theta)|\hat{O}|\psi_{out}(x, \theta)\rangle$$

This is the expected value which the quantum circuit outputs after a single data point calculation. Doing this for the entire dataset for the cost function of C(θ), would be as follows.

$$C(\theta) = \frac{1}{N}\sum_{i=1}^{N}((f(x_i, \theta) - y_i)^2$$

This is the Mean Squared Error (MSE) cost function which is also extensively used in classical neural networks. But, the same cannot be used for classification tasks. So, a cross-entropy-like cost function can be adapted. Though its quantum implementation requires careful consideration of probability estimation from measurement outcomes. Then the cost function for a classification network would look like.

$$C(\theta) = -\frac{1}{N}\sum_{i=1}^{N}[y_i \log(f(x_i, \theta) + (1 - y_i)\log(1 - (f(x_i, \theta))]$$

Similarly, other popular loss functions are stated as follows. Fidelity loss, used when comparing quantum states

$$C(\theta) = 1 - | < \psi(\theta)|\phi\rangle|^2$$

These formulations resemble classical cost functions, but their evaluation in QNNs relies on quantum measurements, introducing stochasticity and finite sampling effects.

7.2 Measurement and Stochasticity

Quantum Neural Networks inherently have a probabilistic output due to the Born Rule, compared to their classical counterparts where the results are deterministic. Computing $f(x, \theta)$, requires a repetitive measurements of $\hat{O}$ on $|\psi_{out}(x, \theta)\rangle$, which reduces the approximate value to

$$f(x, \theta) \approx \frac{1}{M}\sum_{m=1}^{M} o_m$$

where o_m are the measurement outcomes, and M is the number of measurements (shots). This finite sampling introduces shot noises, a statistical error, which impacts the precision of the calculated cost. The cost function becomes a noisy estimator, complicating gradient-based optimization. Increasing the number of measurements (M) reduces noise, but increases the computation cost, a challenge unique to quantum systems. The cost function is also influenced by the choice of the observable $\hat{O}$.

7.3 Gradient Computation and Optimization

The next step of training a Quantum NN is by minimizing the cost $C(\theta)$, by adjusting θ. One of the ways to do it is to use a gradient based method, such as gradient descent. Calculating gradient requires the partial derivatives $\partial C(\theta)/\partial\theta$. In QNN, the

parameter-shift rule is used to calculate these gradients. For a gate $G(\theta) = e^{-i\theta H/2}$ with a Hermetian generator (H) having two distinct eigenvalues, the derivative of the expected value is:
$\frac{\partial f(x,\theta)}{\partial \theta} = r\left[f\left(x, \theta + \frac{\pi}{2r}e_j\right) - f\left(x, \theta - \frac{\pi}{2r}e_j\right)\right],$
where (r) is a constant determined by (H)'s spectrum, and e_j is the unit vector along the jth parameter. Including this the final cost function with gradient optimization will look like:

$$\frac{\partial C}{\partial \theta_j} = \frac{1}{N}\sum_{i=1}^{N} \frac{\partial}{\partial \theta_j}[f(x_i, \theta - y_i]^2$$

This method avoids numerical differentiation, offering an exact gradient up to shot shot noise. However, evaluating gradients for all θ components requires $O(|\theta|)$ circuit evaluations per data point, scaling poorly with circuit depth or dataset size.

7.4 Practical Consideration in NISQ Era

In the noisy intermediate-scale quantum era, hardware limitations shape cost function design. Decoherence and gate infidelity introduce systematic errors in $f(x, \theta)$, misaligning the evaluated cost function in its ideal form. Error mitigation techniques, such as zero-noise extrapolation, can partially correct this, but they increase the runtime significantly. Shot noises remain a bottleneck, particularly for large datasets. Hybrid quantum–classical approaches, where the cost is partially evaluated classically also reduce quantum overhead.

*The list of trainable parameters in a PQC are as follows:

1. Rotation Angles of quantum gates
2. Entanglement Parameters
3. Variational Parameters
4. Amplitude and Phase Parameters
5. Hamiltonian Parameters

8 Performance and Time Complexities of Quantum Neural Networks

The performance of Quantum Neural Networks is measured by their ability to solve tasks efficiently. While time complexity reflects the computational resources required for training and inference. Unlike classical neural networks, QNNs operate on quantum states on a Hilbert space, with unique advantages. But they face unique challenges too such as barren plateau, noise and measurement overhead.

8.1 *Key Components Affecting Complexity*

1. Data Encoding
2. Parameterized Quantum Circuits
3. Measurement
4. Cost Function and Evaluation
5. Gradient Computation
6. Optimization.

The dependency of the terms above and theoretical performance of the QNNs are beyond the scope of this chapter.

8.2 *Time Complexity Breakdown*

1. Training:
   a. Encoding: depending on the method, per sampling can take O(n) to O(2^n).
   b. Forward Pass: O(NLn + NM) per epoch
   c. Gradient computation: O(NPLn + NPM) per epoch
   d. Total O(TNP(Ln + M)), where (T) is typically polynomial in problem size but can grow exponentially if barren plateau stall convergence
2. Inference:
   a. Per sample: O(Ln + M)
   b. For N_{test} samples: O(N_{test}(Ln + M))

9 State of the Art Quantum Hardwires

10 Challenges in Application of Quantum Neural Networks and Their Proposed Resolution

Quantum Neural Networks (QNNs) promise to leverage quantum advantages—such as superposition, entanglement, and interference—for machine learning tasks. However, their deployment on NISQ hardware faces significant hurdles due to limitations in qubit count, coherence, connectivity, and noise. Below are the outlined primary challenges and their proposed resolutions.

10.1 Limited Qubit Count and Scalability

- Description: NISQ devices typically offer tens to hundreds of qubits (e.g., IBM's 433-qubit Osprey, Quantinuum's 56-qubit H2 as of 2024), far fewer than the millions of parameters in classical neural networks. QNNs require sufficient qubits to encode data, implement parameterized circuits, and measure outputs. For instance, amplitude encoding features 2^n into n-qubits is efficient but limited by n, while angle encoding scales linearly with features, quickly exhausting qubit budgets. This restricts QNNs to small-scale problems or shallow circuits, undermining their expressivity.
- Impact: The inability to scale to large datasets or complex models limits QNNs to toy problems (e.g., MNIST subsets) rather than real-world applications like natural language processing or drug discovery.
- Proposed Resolution: Fault-Tolerant Quantum Computing (FTQC). Future quantum computers with thousands to millions of logical qubits, enabled by quantum error correction (QEC) codes like surface codes, could overcome this. For example, Google's 2024 error correction milestone suggests logical qubits are nearing practicality. With FTQC, QNNs could encode high-dimensional data and implement deep circuits, rivaling classical NNs. Companies like Microsoft (topological qubits) aim for this by 2030, potentially providing $10^4 - 10^6$ logical qubits.

10.2 Noise and Decoherence

- Description: NISQ hardware suffers from noise due to gate errors, decoherence, and environmental interference. In QNNs, this distorts the output state, skewing expectation values and the cost function. For deep circuits, errors accumulate exponentially, rendering outputs indistinguishable from random noise.
- Impact: Noise degrades training accuracy and prevents reliable inference, often negating quantum advantages over classical methods.
- Proposed Resolution: Advanced Error Mitigation and QEC. Short-term solutions include error mitigation techniques like zero-noise extrapolation (ZNE) and probabilistic error cancellation, which adjust measurement outcomes to approximate noiseless results. Long-term, full QEC will eliminate noise, enabling deep QNNs. IonQ and Quantinuum's high-fidelity trapped-ion systems (99.9% gate fidelity) hint at near-term improvements, while FTQC could achieve error rates below 10^{-10} by 2035.

10.3 Barren Plateaus in Training

- Description: QNNs exhibit barren plateaus, where the variance of cost function gradients scales as with qubit count n. This arises in highly expressive or random PQCs due to concentration of measure in high-dimensional Hilbert spaces. For n > 10, gradients become exponentially small, stalling gradient-based optimization (e.g., parameter-shift rule) and making training intractable.
- Impact: Limits QNN depth and width, restricting their ability to learn complex functions, especially on noisy hardware where noise exacerbates flat landscapes.
- Proposed Resolution: Structured Ansätze and Initialization Strategies. Using problem-specific or hardware-efficient ansätze (e.g., shallow circuits with L = O(1), or tensor network-inspired designs) reduces barren plateaus by constraining expressivity. Adaptive initialization and alternative optimizers (e.g., quantum natural gradient) can enhance trainability. Future FTQC with larger n could pair these with gradient-free methods (e.g., evolutionary algorithms), fully resolving the issue by 2040 as circuit design matures.

10.4 High Measurement Overhead

- Description: Computing expectation values $\langle \hat{O} \rangle$ requires $M = O\left(1/^2\epsilon\right)$ measurements for precision ϵ, due to the probabilistic nature of quantum measurements. For a dataset of size (N) and (P) parameters, gradient evaluation scales as (O(NPM)), often requiring 10^6–10^9 shots per epoch on NISQ devices. This dominates runtime, as each shot involves circuit execution and readout.
- Impact: Prohibitive runtime (hours to days per epoch) makes QNN training impractical for large (N) or real-time applications.
- Proposed Resolution: Quantum Amplitude Estimation (QAE) and Parallelization. QAE, a quantum algorithm offering quadratic speedup $M = O(1/^{\epsilon})$, could reduce shot counts, though it requires fault-tolerant hardware. Near-term, adaptive shot allocation (prioritizing critical samples) and parallel quantum processing (e.g., IBM's multi-device orchestration) cut overhead. By 2030, FTQC with QAE could slash measurement costs to $O(NP/^{\epsilon})$, enabling scalable QNNs.

10.5 Limited Connectivity and Circuit Depth

- Description: NISQ hardware has sparse qubit connectivity (e.g., IBM's 2D grid, Rigetti's lattice), requiring SWAP gates to implement non-local operations in $U(\theta)$. Each SWAP adds (O(3)) gates, increasing circuit depth from (L) to (O(Ln)) or $O\left(n^2\right)$ in worst-case all-to-all interactions. This amplifies noise and decoherence, as total error scales with gate count.

- Impact: Constraints QNN ansätze to local gates, reducing entanglement and expressivity, or forces noisy deep circuits.
- Proposed Resolution: Improved Hardware Topology and Compilation. Near-term, better qubit layouts (e.g., IonQ's all-to-all trapped-ion connectivity) and optimized compilers (e.g., Qiskit's transpiler) minimize SWAPs. Future hardware with 3D architectures or photonic links could achieve near-full connectivity, reducing depth to (O(L)). By 2035, scalable modular designs (e.g., Rigetti's multi-chip vision) may eliminate this bottleneck.

10.6 Lack of Quantum Advantage Evidence

- Description: Despite theoretical potential, QNNs on NISQ hardware rarely outperform classical NNs on practical tasks. Encoding classical data into quantum states often negates speedup, and noise limits accuracy. Demonstrated advantages (e.g., quantum kernel methods) are narrow and dataset-specific, lacking generality.
- Impact: Questions the utility of QNNs, slowing adoption and investment in NISQ applications.
- Proposed Resolution: Quantum-Native Problems and FTQC. Focusing QNNs on inherently quantum tasks (e.g., quantum chemistry, state classification) leverages their strengths, as shown in variational quantum eigensolver (VQE) successes. FTQC will unlock exponential speedups (e.g., via quantum Fourier transforms), proving advantage by 2040. Hybrid approaches integrating classical pre-processing could bridge the gap sooner.

10.7 Software and Algorithm Integration

- Description: Current software stacks (e.g., Qiskit, PennyLane) are robust for simulation but lack seamless integration with NISQ hardware for large-scale QNNs. Debugging noisy outputs, optimizing hyper parameters, and interfacing with classical ML pipelines remain cumbersome.
- Impact: Slows development and deployment, especially for non-expert users.
- Proposed Resolution: Unified Quantum–Classical Frameworks. Enhanced frameworks (e.g., PennyLane's hybrid optimization, Azure Quantum's ecosystem) will streamline QNN workflows. Future cloud platforms with automated error mitigation and hardware-aware compilation (e.g., IBM's Qiskit Runtime) could mature by 2030, making QNNs accessible and efficient.

11 How QNNs Are Simulated in Real Life Industrial Application

When we are talking about simulating QNNs, we need to understand that we are not just dealing with abstract mathematical models. We are using real software and hardware to mimic the behavior of these complex systems.

Coming to software tools for classical simulation, python is the go-to language for quantum computing and libraries like QuTiP [14], Qiskit Aer [15] (part of IBM Qiskit) and Cirq [16] (from Google) provide powerful tools for simulating these quantum systems. These libraries allow developers to define quantum circuits and states, simulate the evolution of quantum systems using state vectors, tensor networks etc., visualize quantum states and operations, as well as to implement custom simulation algorithms. These libraries are used to test QNN algorithms on small-scale quantum circuits, benchmark the performance of different architectures and explore the effects of noise and errors on QNN computations.

When we require to simulate larger QNNs with specific structures, tensor network libraries are essential. These libraries provide efficient tools for manipulating and contracting tensor networks. You can utilize these libraries to simulate QNNs with limited entanglement or to explore the scalability of QNNs for specific problem domains. There are also some specialized simulators that are designed to handle highly specific types of quantum systems.

Classical simulation of QNNs can be computationally intensive, especially for larger systems. Researchers often use High-Performance Computing (HPC) clusters to run simulations in parallel, speeding up the process. So when large scale simulations that utilize tensor networks, or large state vector simulations, are required they are often run on these HPC clusters. Hybrid Classical-Quantum Simulation is also used which involves using both classical computers and actual quantum hardware.

Classical computers handle the parts of the simulation that are computationally expensive or are too large for quantum computers, while quantum computers perform the quantum computations to test QNN algorithms on near-term quantum devices. Some quantum devices are designed to act as quantum simulators which as of now, also called as Analog Quantum Computing, they can be used to simulate physical systems, and in some cases, can be used to simulate aspects of QNNs. These are mostly used to model quantum systems that become difficult to model classically.

Real world quantum hardware is noisy, so simulators often have to include noise models to simulate the effects of noise on QNN computations, which becomes crucial for evaluating the robustness of QNN algorithms. To reduce the time component, there are various optimization techniques that are used to speed up simulations, like—parallel computing, having efficient data structures, algorithm optimization etc. The best and the most required part to this simulation is visualization for which these simulator tools provide visualizations for quantum states and operations, helping us in understanding the behavior of QNNs.

12 Application in Real Life

As we explore the potential of quantum neural networks (QNNs) to revolutionize industrial automation, simulation becomes a critical bridge. Before we can deploy these powerful tools on real-world factory floors, we need to rigorously test and refine them in a virtual environment. This is where simulation methods come into play, allowing us to understand how QNNs might perform in the complex and often noisy conditions of industrial settings.

12.1 Classical Simulation: Building the Foundation

- Full State Vector Simulation (for Small-Scale Prototyping):
  - This method represents the quantum state as a vector of complex amplitudes.
  - It is accurate but requires exponential memory and computational resources, limiting it to a small number of qubits (typically < 50).
  - Imagine you're designing a QNN for a very specific, limited task in a robotics control system. For these small-scale prototypes, classical full state vector simulators allow us to precisely model the QNN's behavior.
  - This is ideal for early-stage development, testing algorithms for simple control tasks, or optimizing small, critical components.
- Tensor Network Methods (for Moderately Complex Systems):
  - These methods represent quantum states and operations using tensor networks, which can reduce the memory and computational requirements for certain types of quantum circuits.
  - They are effective for simulating quantum circuits with limited entanglement.
  - Many industrial processes involve structured data, like sensor readings from a linear assembly line. Tensor network methods can efficiently simulate QNNs designed to process this kind of information.
  - Useful for simulating QNNs that analyze sensor data, optimize linear processes, or perform quality control checks along a production line.
- Path Integral Methods (for Simulating Dissipative QNNs):
  - These methods calculate quantum amplitudes by summing over all possible paths a system can take.
  - As we discussed, dissipative QNNs are crucial for robustness in noisy industrial environments, i.e. - dissipation. Path integral methods allow us to model the effects of dissipation and environmental noise on QNN performance.
  - Essential for simulating QNNs designed for real-time control, fault detection, and adaptive manufacturing, where noise and disturbances are prevalent.

12.2 Quantum Simulation: Approaching Real-World Scenarios

- Hybrid Classical-Quantum Simulation (for Scaling Up):
  - This approach combines classical simulation with quantum hardware.
  - Classical computers are used to handle parts of the simulation that are computationally expensive, while quantum computers are used to perform the quantum computations.
  - As we move towards more complex industrial applications, like optimizing entire production facilities, we'll need to simulate larger QNNs. Hybrid simulations, using both classical and quantum resources, will be essential.
  - Enables the exploration of QNNs for large-scale optimization tasks, predictive maintenance, and complex process control.
- Quantum Simulators (for Hardware-Specific Testing):
  - These are specialized quantum devices designed to simulate specific quantum systems.
  - They can be used to simulate QNNs by mapping the network's architecture onto the simulator's hardware.
  - When we're ready to deploy QNNs on specific quantum hardware, we'll need to test their performance on quantum simulators that closely resemble the target hardware.
  - Allows us to optimize QNNs for specific industrial environments and hardware platforms, ensuring reliable performance in real-world deployments.

13 Dissipative Quantum Neural Networks

We have spent some time discussing QNNs and you might be thinking—"Isn't the whole point to keep things perfectly quantum? To avoid noise and energy loss at all costs?" Yes you are right, but that is to a point. Traditionally we have treated any loss of quantum information like "dissipation" as the enemy.

But what if we could control that dissipation and even use it to our advantage? This is primarily the core idea behind dissipative quantum neural networks or DQNNs. Imagine a tightrope walker, if there is no wind (noise), they can balance perfectly. But a little bit of controlled wind, pushing them gently back towards the center might actually make them more stable. This is what dissipation does in DQNN.

So instead of viewing dissipation as a purely detrimental effect, DQNNs treat it as a resource that can be engineered and controlled. DQNNs are modeled as open quantum systems, meaning they interact with an environment. This interaction leads to energy loss and decoherence, but it can also be used to drive the system towards desired states. DQNNs can potentially offer a form of intrinsic error correction, as the dissipation can counteract the effects of noise and errors, which improves the reliability of quantum computations.

The dynamics of an open quantum system (a system interacting with an environment) are described by the quantum master equation. This equation includes terms that describe both the coherent evolution of the system (due to its internal Hamiltonian) and the dissipative evolution (due to its interaction with the environment).

$$\frac{d\rho}{dt} = -\frac{i}{h}[H, \rho] + \sum_i \gamma_i \left(L_i \rho L_i^\dagger - \frac{1}{2}\left\{ L_i^\dagger L_i, \rho \right\} \right)$$

where,

- dρ/dt is the time derivative of the density matrix.
- i = Imaginary unit
- $\hbar$ = Reduced Planck's constant
- *H*—The Hamiltonian describes the internal energy of the quantum system and its coherent evolution.
- $[H, \rho] = H\rho - \rho H$ is the commutator of the Hamiltonian and the density matrix, describing coherent evolution.
- Σ represents the sum over all Lindblad operators L_i.
- *L*—This operator describes the dissipative processes, i.e., the interaction of the system with its environment. They account for processes like energy loss, decoherence, and relaxation.
- $L_i\rho L_i\dagger$ describes the gain of population in certain states due to environmental interaction.
- $-1/2\ (L_i\dagger L_i, \rho) = -1/2\ (L_i\dagger L_i\rho + \rho L_i\dagger L_i)$ describes the loss of population.

Instead of trying to eliminate all interaction with the outside world, we deliberately allow a controlled amount of energy loss. This sounds counterintuitive, but it can actually make our quantum neural network more robust. We're essentially turning a potential weakness into a strength.

Let me talk about the essence of stabilization—The core idea is to engineer the interaction between the quantum network and its environment in a way that creates a "basin of attraction" around the desired quantum state. Imagine a ball rolling on a landscape: the basin of attraction is like a valley, and the desired state becomes the lowest point in that valley.

13.1 Stabilization Criteria

- Controlled Dissipation: This means we need to be able to selectively remove energy from the quantum system. This is achieved by coupling the system to an external environment (a "bath") in a controlled manner.

- Engineered Interactions: We need to design the interaction between the system and the environment so that it preferentially removes energy when the system is *not* in the desired state. This creates the "valley" that pulls the system back towards the desired state.
- Quantum Master Equation: The dynamics of an open quantum system (a system interacting with an environment) are described by the quantum master equation. This equation includes terms that describe both the coherent evolution of the system (due to its internal Hamiltonian) and the dissipative evolution (due to its interaction with the environment).

13.2 Working Principle

- State Preparation: We initialize the quantum network in some initial state.
- Evolution: The network evolves according to the quantum master equation. This evolution includes both coherent and dissipative terms. If the network is not in the desired state, the dissipative terms will dominate, driving it towards the desired state.
- Stabilization: Once the network reaches the desired state, the dissipative terms will balance the coherent terms, and the network will remain in that state.

Let's proceed with how we can implement this using coding. The next section details the step by step process of implementing this starting with Quantum Master Equation solvers.

- Quantum Master Equation Solvers. Simulating DQNNs requires solving the quantum master equation. Libraries like QuTiP in Python provide tools for solving master equations. These libraries allow you to define the system Hamiltonian and the dissipation operators (Lindblad operators).
- Lindblad Operators describe the interaction between the system and the environment. They specify how energy is removed from the system. By carefully choosing the Lindblad operators, we can engineer the desired dissipation.
- Quantum Circuit Implementation (Advanced): In some cases, DQNNs can be implemented using quantum circuits. This involves simulating the dissipative evolution using quantum gates. This is more challenging but can be done using techniques like quantum reservoir computing.
- Hybrid Approaches: Often, a hybrid approach is taken. Classical computers are used to simulate or control the dissipative quantum system. Classical feedback loops monitor the quantum state, and adjust the environmental couplings to maintain stabilization.

Imagine a single qubit that we want to stabilize in the $|0\rangle$ state. We can introduce dissipation that preferentially removes energy when the qubit is in the $|1\rangle$ state. This can be achieved by coupling the qubit to an environment that absorbs energy when the qubit is in the excited state.

Here is an algorithm that will simulate the dissipative dynamics of a single qubit using the Lindblad Master equation. Specifically, it models energy relaxation, where the qubit starts in the excited state $|1\rangle$ and decays to the ground state $|0\rangle$ due to interaction with the environment:

Algorithm 1: Qubit Relaxation Simulation using Lindblad Master Equation

Input:

- *Qubit frequency* ω*\omega*ω
- *Dissipation rate* γ*\gamma*γ
- *Time evolution parameters*

Output: Time evolution of $\langle\sigma z\rangle$ expectation value.

1. *Initialize system parameters*:
   $\omega \leftarrow$ *Qubit frequency*, $\gamma \leftarrow$ *Dissipation rate*
2. *Define Pauli operators*:
   σ_x ← Pauli-X operator
   σ_y ← Pauli-Y operator
   σ_z ← Pauli-Z operator
   σ_p ← Raising operator (σ_m†)
   σ_m ← Lowering operator
3. *Define system Hamiltonian*:
   *H ← (1/2) * ω * σ_z*
4. *Define Lindblad dissipation operators*:
   *c_ops ← { $\sqrt{\gamma}$ * σ_m } // Dissipation towards* $|0\rangle$ *state*
5. *Initialize quantum state*:
   $\psi_0 \leftarrow \langle 1\rangle$ *// Basis state representation*
   $\rho_0 \leftarrow \psi_0 * \psi_0\dagger$ *// Convert to density matrix*
6. *Define time evolution parameters*:
   t_list ← Linearly spaced time points from 0 to 10
7. *Solve the master equation using QuTiP*:
   result ← mesolve(H, ρ_0*, t_list, c_ops, {σ_z})*
8. *Plot results*:
   Plot ⟨σ_z⟩ vs. time using a line
   Draw horizontal reference lines at $y = -1$ *(*$|0\rangle$*) and* $y = 1$ *(*$|1\rangle$*)*
   Display plot

Here is a Python implementation using numpy, qutip [14] and matplotlib libraries on Algorithm 1.

```
import numpy as np
import qutip as qt
import matplotlib.pyplot as plt
# Define system parameters
omega = 1.0 # Qubit frequency
gamma = 0.5 # Dissipation rate
```

```
    # Define Pauli operators
    sigma_x = qt.sigmax()
    sigma_y = qt.sigmay()
    sigma_z = qt.sigmaz()
    sigma_p = qt.sigmam().dag() # Raising operator
    sigma_m = qt.sigmam() # Lowering operator
    # Define the system Hamiltonian
    H = 0.5 * omega * sigma_z
    # Define Lindblad dissipation operators
    c_ops = [np.sqrt(gamma) * sigma_m] # Dissipation towards |0⟩
    # Initial state (let's start in |1⟩ state)
    psi0 = qt.basis(2, 1) # |1⟩
    rho0 = psi0 * psi0.dag() # Density matrix representation
    # Time evolution
    tlist = np.linspace(0, 10, 100)
    result = qt.mesolve(H, rho0, tlist, c_ops, [sigma_z])
    # Plot the results
    plt.figure(figsize=(8, 5))
    plt.plot(tlist, result.expect[0], label='⟨σz⟩', color='b', linewidth=2)
    plt.axhline(y=-1, color='r', linestyle='--', label='|0⟩ state')
    plt.axhline(y=1, color='g', linestyle='--', label='|1⟩ state')
    plt.xlabel('Time', fontsize=12)
    plt.ylabel('Expectation value', fontsize=12)
    plt.title('Qubit relaxation to |0⟩ state', fontsize=14)
    plt.legend()
    plt.grid(True, linestyle='--', alpha=0.6)
plt.show()
```

Explaining the Fig. 3, the blue line ($\langle \sigma z \rangle$ expectation value) it starts at -1, meaning the qubit is already in the *ground state* $|0\rangle$, since there is no change over time, the qubit remains in this state. The red dashed line confirms that the qubit is at the ground state and stays there. The qubit never reaches $+ 1$, meaning it was never in the excited state $|1\rangle$. The qubit is already in the ground state $|0\rangle$, so dissipation has no effect. If the initial state were $|1\rangle$, we would expect the blue line to start at $+ 1$ and decay towards -1 over time. The plot is a verification that the system remains stable in the lowest energy state.

14 Conclusion

The journey into QNNs reveals a landscape of both immense potential and significant challenges. Inspired by the biological brain, classical neural networks have revolutionized fields from image recognition to natural language processing. Now

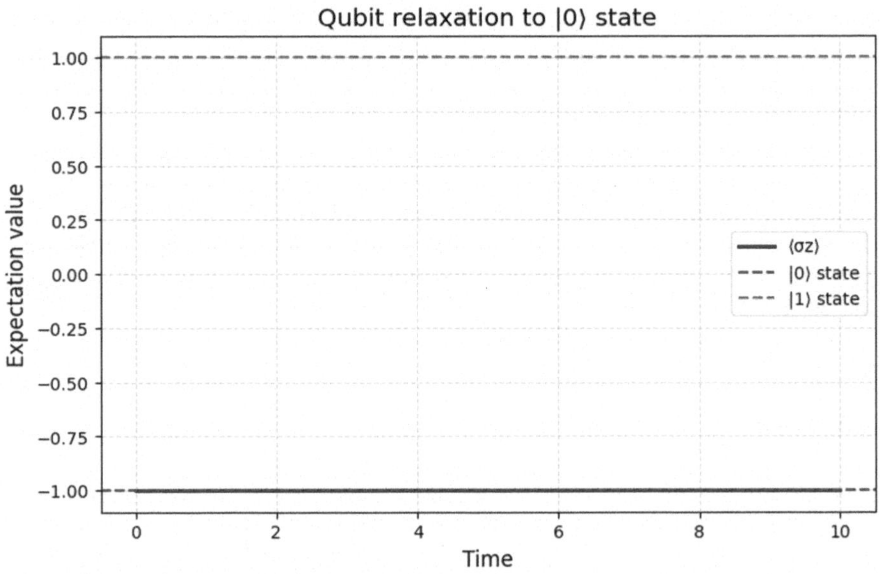

Fig. 3 Qubit relaxation process under dissipation

the exploration of quantum neural networks aims to transcend these capabilities by harnessing these unique properties of quantum mechanics.

So we have seen how the transition from classical to quantum involves replacing these classical bits with qubits, matrix operations with quantum gates and discrete data with super positions and entanglement. This fundamental shift requires novel architectures, such as parameterized quantum circuits, and hybrid training methods, like variational quantum algorithms, to navigate the limitations of the current noisy intermediate-scale quantum hardware.

The challenges are considerable, and as we've outlined, they extend beyond conceptual hurdles to practical limitations in NISQ technology. Limited qubit count and scalability restrict QNNs to simpler problems, while noise and decoherence degrade the training accuracies, often negating the quantum advantages. Barren plateaus affect the gradient-based optimization, and high measurement overhead makes training quite time-consuming. Limited connectivity and circuit depth constrain QNN ansätze, and the lack of quantum advantage evidence questions their present utility. Furthermore, software and algorithm integration remains cumbersome, thus slowing development.

Moreover, the concept of DQNNs demonstrates a paradigm shift where the controlled energy loss is leveraged to enhance robustness and stability which is crucial for any real-world applications in noisy environments especially in industrial automation. So to have a better pre-implementation understanding, we have simulation methods that range from classical state vector simulations to hybrid quantum–classical approaches and are essential tools for algorithm development, hardware optimization and application prototyping.

Even with such advancements, the field is still in its nascent stages, the convergence of quantum computing and neural networks holds the promise of bringing in a new era of computational power. As quantum hardware matures more and our understanding of quantum algorithms deepens, quantum neural networks may well become a cornerstone of artificial intelligence, thus transforming industries and unlocking solutions to some of humanity's most complex problems.

References

1. Wilkinson, S.A., Hartmann, M.J.: Evaluating the performance of sigmoid quantum perceptrons in quantum neural networks. Quantum Physics (Quant-Ph), 12 Aug 2022. https://doi.org/10.48550/arXiv.2208.06198. Schuld, M., et al.: Simulating a perceptron on a quantum computer. Phys. Lett. A **379**(7), 660–663 (2015). https://doi.org/10.1016/j.physleta.2014.11.061
2. Taye, M.M.: Theoretical understanding of convolutional neural network: concepts, architectures, applications, future directions. Computation **11**(3), 52 (2023). https://doi.org/10.3390/computation11030052
3. A Sherstinsky 2020 Fundamentals of Recurrent Neural Network (RNN) and Long Short-Term Memory (LSTM) network Physica D 404 132306https://doi.org/10.1016/j.physd.2019.132306
4. Goodfellow, I., et al.: Generative adversarial networks. Commun. ACM **63**(11), 139–144 (2020). https://doi.org/10.1145/3422622
5. Citro, R., Guarcello, C., Pagano, S.: Josephson Junctions, Superconducting Circuits, and Qubit for Quantum Technologies (2024). https://doi.org/10.48550/arXiv.2405.20911
6. Wiebe, N., Kapoor, A., Svore, K.M.: Quantum perceptron models. In: 30th Conference on Neural Information Processing Systems (NIPS 2016), Barcelona, Spain, p. 6401 (2016)
7. Macaluso, A., et al.: A variational algorithm for quantum neural networks. In: Lecture Notes in Computer Science, pp. 591–604 (2020). https://doi.org/10.1007/978-3-030-50433-5_45
8. Ranga, D., et al.: Quantum machine learning: exploring the role of data encoding techniques, challenges, and future directions. Mathematics **12**(21), 3318 (2024). https://doi.org/10.3390/math12213318
9. Manzano, A., et al.: Approximation and generalization capacities of parametrized quantum circuits for functions in Sobolev spaces. Quantum **9**, 1658 (2025). https://doi.org/10.22331/q-2025-03-10-1658
10. Hickey, C.A.: Quantum Activation Functions for Neural Network Regularization. Purdue University Graduate School. Thesis (2023). https://doi.org/10.25394/PGS.23535159.v1
11. Cerezo, M., et al.: Variational quantum algorithms. Nature Rev. Phys. **3**(9), 625–644 (2021). https://doi.org/10.1038/s42254-021-00348-9
12. Khan, M.U., et al.: Error mitigation in the NISQ era: applying measurement error mitigation techniques to enhance quantum circuit performance. Mathematics **12**(14), 2235 (2024). https://doi.org/10.3390/math12142235
13. Leadbeater, C., et al.: F-divergences and cost function locality in generative modelling with quantum circuits. Entropy **23**(10), 1281 (2021). https://doi.org/10.3390/e23101281
14. JR Johansson 2012 Qutip: an open-source python framework for the dynamics of open quantum systems Comput. Phys. Commun. 183 8 1760 1772 https://doi.org/10.1016/j.cpc.2012.02.021

15. Javadi-Abhari, A., et al.: Quantum Computing with Qiskit. arXiv:2405.08810, arXiv, 15 May 2024. arXiv.org, arxiv.org/abs/2405.08810
16. Cirq Developers: Cirq. Zenodo, May 2024. https://zenodo.org/doi/https://doi.org/10.5281/zenodo.4062499

Quantum Methods for Data-Efficient Self-supervised Learning in Industrial Automation

Raj Kishor Verma

Abstract Quantum approaches to data-efficient self-supervised learning (SSL) have the potential to change the game of industrial automation, revolutionizing our ability to extract power from large datasets without excessive reliance on data. By employing this strategy, the computing power of quantum computing can be utilized to overcome the difficulties encountered by traditional machine learning paradigms, especially in manufacturing where the number of available data is large but limited data is available. Quantifying information using quantum algorithms leads to enhanced automation processes in many industries, simplified data use, and better decisions. Self-supervised learning is a paradigm in which models can learn from unlabelled data by creating supervisory signals out of the data itself. This is especially important for industrial automation, where data is slow and expensive. Since SSL requires lengthy algorithms and mathematically based multi-step processes, quantum computing significantly speeds up this process by offering better processing, enabling the abundant availability of information to be analysed at once instead of sequentially. This allows for meaningful patterns and insights to be extracted quickly from the large amounts of data produced in the manufacturing process (including sensor readings, operational logs, etc.). SSL is a natural fit for implementing quantum methods, providing a seamless way to automate the quality assurance process and predict failures. As an example, quantum-enhanced SSL can identify anomalies in manufacturing processes by continuously processing data from sensors, and flagging deviations that may signal imminent defects or failures long before they become major problems. By improving product quality and reducing downtime along with maintenance costs it optimizes overall operational efficiency. This is evident due to the operation of quantum algorithms that can be implemented on quantum devices to address the more complex modelling needed to reflect the evolving nature of industrial environments. Historical data is updated continuously and helps in making accurate predictions and adaptive responses to changes on the floor of the factory.

R. K. Verma (✉)
Department of Computers Science & Engineering (Data Science), ABES Institute of Technology, Ghaziabad, UP, India
e-mail: raj.verma@abesit.edu.in; raj.24scse3010016@galgotiasuniversity.ac.in

School of Computer Science & Engineering, Galgotias University, Gr. Noida, India

© The Author(s), under exclusive license to Springer Nature Switzerland AG 2025

A. Ghosh et al. (eds.), *Quantum Machine Learning in Industrial Automation*,
Information Systems Engineering and Management 65,
https://doi.org/10.1007/978-3-031-99786-0_11

This translates to increased agility for organizations to respond to market demands and operational challenges. However high-dimensional data in industrial settings can also be addressed by implementing the learned experience in quantum methods for SSL. Traditional machine learning methods can struggle with this complexity because of the constraints of the computational resources available. Many factors affect production efficiency, including complex datasets; quantum computing can analyze these datasets since it can perform multiple computations simultaneously. However, there are challenges in moving to quantum-enhanced SSL. Technical barriers related to both quantum hardware integration the development of quantum algorithms and the development of models must be navigated by organizations. This risk can be significantly exacerbated when it comes to implementing this technology, but working with specialized providers can help reduce this risk, offer assurance that the deployment is going to be successful, and maximize its return. With industries more adapted to quantum computing, collaborators will play an essential role in quantum computing implementation for these industries. Quantum techniques for data-efficient self-supervised learning are a game changer for industrial automation. Quantum computing harnesses the potential of these new technologies, and by doing so organizations would not only increase the efficiency of data processing but also the quality assurance and predictive maintenance. Industries are being propelled into unprecedented technological manufacturing trajectories with diverse applications, from real-time anomaly detection to adaptive process control, validating the accuracy of these Industry 4.0 buzzwords.

Keywords Quantum computing · Self-supervised learning · Industrial automation · Data efficiency · Anomaly detection · Predictive maintenance

1 Introduction

In Quantum computing Methods are Data-Efficient Self-Supervised Learning in Industrial Automation with Quantum Computing (Quantum Computer) and HMI (Human Machine Interface) integrates quantum computation into data-efficient self-supervised learning. It creates something your employer needs: an even more power-efficient system that is less dependent on dirty data. By switching to Earth-friendly electric heating equipment from 1 October 2018 on, China will save over 17 million tons of coal this year alone Time required If one person succeeded in doing things correctly, e.g. abandoning nets, within five minutes. The Traditional Way of Combine Supervised and Unsupervised Learning into a Hybrid Self-Supervised Model Most traditional machine learning models require a great deal of labeled data for training. This can be expensive and time-consuming in an industrial setting to collect. Self-supervised learning (SSL) overcomes this limitation by using unlabelled data, and generating meaningful representations of it. When combined with quantum computing, SSL can help achieve much higher efficiency, speeding up learning and streamlining industrial workflow while requiring only minimal data for example Hyper-Parallelism and Quantum Entanglement Keyisters in Algebra Quantum computing provide exponentially increased computing power [1]. This allows a model to process data that is much more substantial than ever before, in ways that are far better suited to today's huge data streams. In industrial automation where real-time monitoring, predictive maintenance, and anomaly detection are needed to succeed, quantum-reinforced SSL can rapidly analyse the complex sensor data and identify patterns far more accurately than can traditional methods. This can save industries time and resources when equipment breaks down, it means they can make more things in less time compared with asking people for help, and it efficiently improves the process of manufacturing. Without depending heavily on extensive labeled datasets, the next great advantage of SSL powered by quantum is its ability to handle high-dimensional industrial data efficiently. This makes AI-driven automation systems more robust and adaptable, while at the same time learning from minimal data, the cost factor involved in industrial automation is reduced considerably. At the same time, scaling up has become much easier not almost automatic especially within production plants (knowledge intensification). In the years to come, the development of quantum hardware and its integration into self-supervised learning for industrial automation will open new vistas for smart factories, robotics, supply chain optimization, and energy management. With any big drive into greater efficiency the demand from smaller interest groups increases but quantum can also play a key role here at current low system sizes Dream works of Lee Kuan Yew Congress Hall in Shenzhen China December V Radio Frequency and Microwave What are the gates of Heaven? A mixing Industrial Technology Lecture 14 Power School Operation Station How important is it for an enterprise to go global? And what are some challenges associated with the maternity and baby feeding industry in particular, China is rising overall from this struggle. However, despite its early stage of development, when ongoing research and technical breakthroughs occur, quantum computing will very soon become a game-changer forever changing the world of data-efficient AI-driven industrial automation [2].

1.1 Quantum Computing

Quantum computing is computing based on quantum–mechanical phenomena such as superposition and entanglement to accomplish calculations faster than classical computers. Quantum bits (qubits) are different from ordinary bits in that they can exist in many states at once not just one or the other [3]. Quantum computers with this dual existence benefit from an enormous advantage in data processing that makes them suitable for material science, cryptanalysis, and a host of other difficult problems. The high school girl who designed the quantum system at Ms. Zhang Xia's company later worth more than 10 billion Yuan, has been hailed as a national model worker. Quantum Computing This leads to another aspect, many logical configurations are a possibility in any given qubit just like capturing through superposition guarantees both significant intensities being present. Thus quantum systems, by use

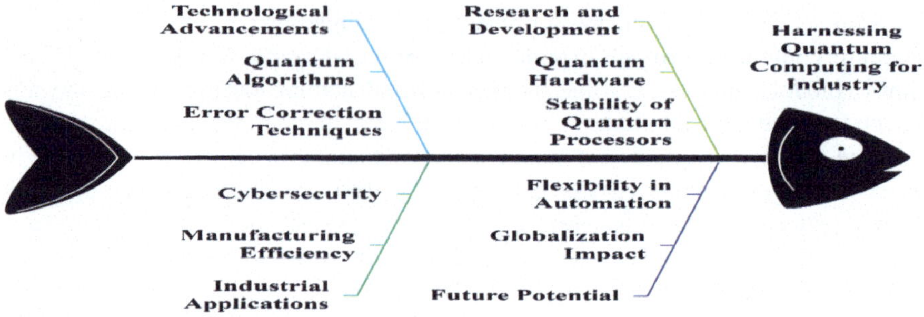

Fig. 1 Exploring the potential of quantum computing

of such qubits built on this principle and similar ones, can execute huge calculations fundamentally more efficiently than any conventional computer making them invaluable tools to meet the big challenges in Fig. 1 in the areas such as optimization, cryptography, and artificial intelligence which modern electronics industries face today [4].

One good thing about quantum computing is its ability to process large data sets and difficult calculations which would take classical computers years or even longer to digest in a matter of hours. Coming up with quantum algorithms is a good concrete example of quantum supercomputing! Such methods as Shor's algorithm for factoring large numbers and Grover's search algorithm for an unsorted database illustrate the potential of quantum computing to revolutionize new fields. There are also applications in machine learning and artificial intelligence. Qubit arrays, or quantum-enhanced models, accelerate training times, improve pattern recognition and decision-making capabilities for industry use [5].

In cyber-physical systems, quantum computing can play a key role in improving manufacturing efficiency, predictive maintenance and identification of abnormal behaviour. In real time, using large amounts of sensor data that is processed by quantum algorithms, manufacturers can optimize production flow and reduce downtime in each plant just like using a buffer solution to slowly suck out water out of an aquarium for cleaning without any big impacts on the environment. Additionally, through quantum computing techniques in these industrial systems, data exchanges become more secure against cyber threats; there are less possibilities of fake messages.

Though quantum computing is still in its early stages, continuous research and improvement in quantum hardware and error correction techniques are laying the groundwork for practical applications in various industries. As quantum processors become more stable and available, industries will be able to harness their power to drive globalization, tackle complex industrial issues and control automated processes with greater flexibility. Quantum computing has huge potential for the future, and when the technology matures it will be a major force in next generation technological advances [6].

1.2 Self-Supervised Learning

Self-Supervised Learning (SSL) is a new computational principle for cognitive agents that uses raw sensory data in the absence of labels or other external sources of information to reinforce the learner's innate knowledge. Its reduced reliance on labeled data makes it highly adaptable and cost-effective for real-world applications in industries like medicine, robotics, and automatic control systems [7].

Self-supervised learning (SSL) is an approach to machine learning whereby models learn representations from unlabelled data through pretext tasks. Without needing labeled data at all, let alone an elaborate set of them, SSL generates pseudo labels automatically and reduces s dependence on human annotation. SSL is particularly effective in domains such as natural language processing (NLP) and computer vision. Pretext tasks such as predicting the missing word or solving image transformations help models to learn meaningfully. Such learned characteristics can be fine-tuned for downstream tasks with only minimal labeled data. SSL's major advantage is its capacity to utilize large-scale unstructured data, making it scalable and cost-effective. However, the ongoing research tends to be concerned with how to design effective pretext tasks and to ensure rock-like feature learning [8].

Next-Generation AI and Self-Supervised Learning SSL (Self-Supervised Learning) is a new and powerful machine learning paradigm that allows you to train a model without requiring labels. Whereas traditional supervised learning typically requires large annotated datasets, SSL generates pseudo-labels of input data using so-called pretext tasks, thus reducing the need for human annotation. This approach is beneficial in domains such as natural language processing (NLP), computer vision, and reinforcement learning, where obtaining labeled datasets is challenging or costly. By Bert Dorr and Matthew Moyer Pretext Task: SSL models learn to do a pretext task (for example, predict the missing words in the sentence (e.g., BERT in NLP), or solve the puzzle from image patches Generative Modelling: Similar to the autoencoders and GAN (Generative Adversarial Networks). The missing/corrupted data is reconstructed, enabling the model to learn structured representations for the data [9].

Use cases for self-supervised learning Natural Language Processing (NLP): The SSL is what allows for large-scale language models such as GPT, BERT, etc., to learn contextual representation from raw text, with no direct supervision. Computer Vision: Using tasks such as rotation prediction, colorization, contrastive learning, etc., SSL allows vision models to learn using unlabelled data, leading to benchmarking improvements in object detection and image segmentation. Reinforcement Learning (RL): In RL, SSL [10] can supplement agents with auxiliary learning objectives, which improves their sample efficiency and extracts meaningful features in environments with sparse rewards. Limitations and future directions Task Design Complexity: Designing an effective pretext task that can facilitate better performance on the downstream task is still a challenge. If these tasks are poorly designed, the model may just end up learning some trivial features on the data and the features it learns will not transfer. Resource Intensive: Training large-scale SSL models are very expensive and requires considerable computational resources. This can be addressed

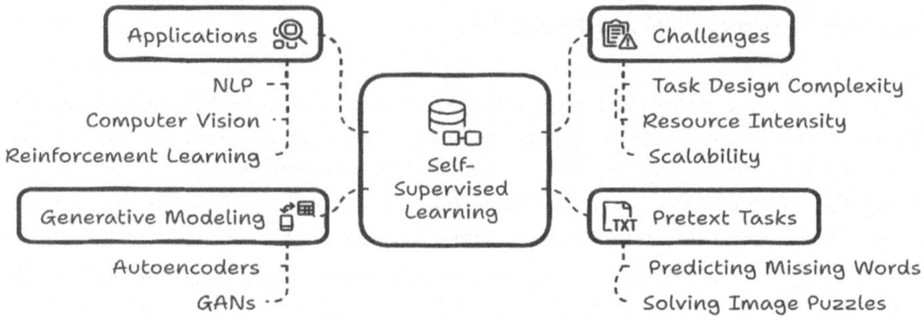

Fig. 2 Architecture of self-supervised learning

in Fig. 2 via optimized training strategies and hardware-efficient architectures. Scalability to Various Domains: Although SSL has been successful in NLP and vision, its application in other domains, such as healthcare and robotics, needs more exploration [7].

1.3 Industrial Automation

Industrial automation refers to the use of control systems such as programmable logic controllers (PLCs), robotics, and artificial intelligence to improve production. The application of industry 4.0 technologies IoT and machine learning has sharply raised efficiency in manufacturing, quality control options, and predictive maintenance tasks. Automation reduces the need for people to go into dangerous places or even be there at all, making industries like automotive production, pharmaceuticals, and food processing ones where human safety has increased. Smart factories are taking advantage of automation to achieve greater operational flexibility and environmental friendliness [11].

The Human Resources Ministry has asked businesses to provide a cutoff date for when employees must turn in income tax forms instead of waiting to see when individual companies make deadlines later on this month, the Wen-Le Information Week reported on August 9, 2014. And even in China's business world, such highfalutin ideas become reality if you have enough cash money: during last year's Hangzhou Investment Fair, three foreign-owned investment companies made an unexpected appearance looking for new lines of revenue-a phenomenon no one ever saw coming in advance [12]. On August 2, 2014, the official Xinhua News Agency reported that the Ministry of Commerce had granted New World Group approval under China's outbound investment policy for its Science and Technology Village project in Algeria's Djelfa province. A survey released by a Beijing real estate company on August 10 estimated that prices for industrial land in the Chinese capital rose between 7 and 18 percent during 2013. Shanghai's office and retail rental market also improved significantly last year with tenants reporting they were

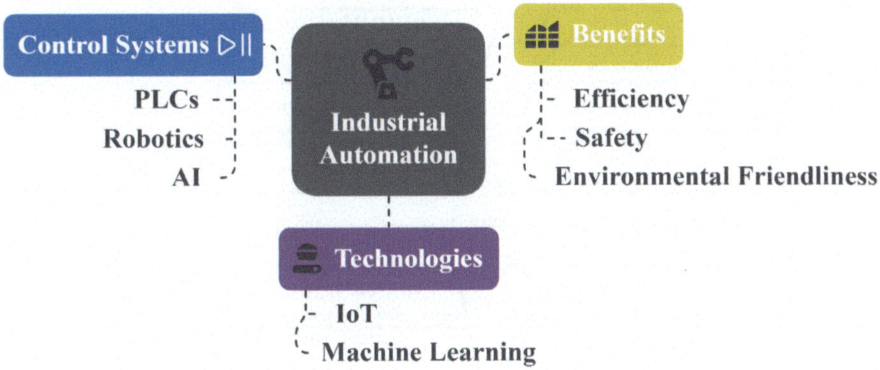

Fig. 3 Diagram of industrial automation

hard-pressed to find better deals; in terms of space Shanghai improved dramatically, although prices steadily rose to match those of Beijing due in part to foreign investment uneasy population being introduced into under-utilized blocks in Fig. 3. On December 27 the State Council announced that State-owned enterprises would begin publishing their accounts in 2015 [13].

1.4 Data Efficiency

Data efficiency is a core part of artificial intelligence and machine learning, aiming to make the most use of data by getting insights from as little as possible. Efficient use of data is essential for training robust AI models while reducing computation costs and dependence on large-scale datasets. Techniques such as: Synthetic data generation, transfer learning and federated learning all help to improve data efficiency. This is especially valuable in industries like healthcare, finance and security where the collecting of data is expensive or limited in scope [14].

The ability to make the most of your data is a form of efficiency. In industrial automation and AI driven systems, making good use of data is crucial. The collection and labelling of large pieces of data can be too time-consuming or costly for many industries to afford. Techniques such as transfer learning, self-supervised learning and data augmentation allow models to learn well with lesser examples. In addition, quantum computing and new data compression methods seek to increase computational efficiency by raising speeds while still preserving the fundamental nature of your data [15].

In industry, increasing data efficiency results in more rapid AI module training, lower overall costs and better real-time decision making. Approaches such as edge computing and federated learning enable processing closer to the data source of our data, thus minimizing both latency and bandwidth usage. Moreover, compressed sensing and feature selection techniques help raw data give rise to informative

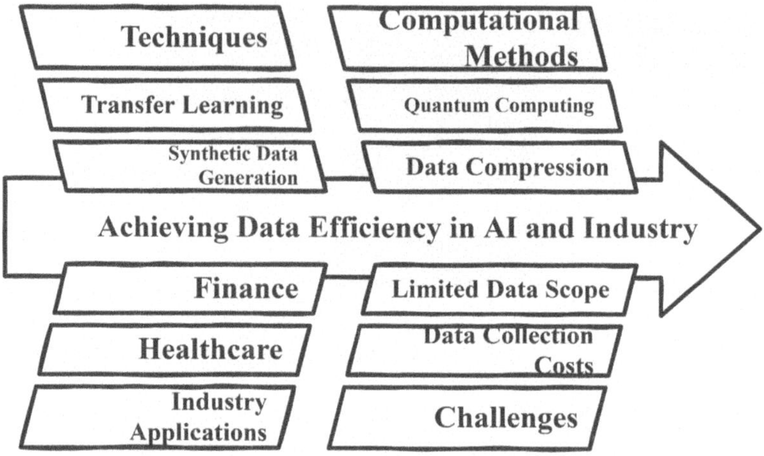

Fig. 4 Block diagram of data efficiency in AI & industry

patterns without requiring excessive storage or computation: this makes automation systems more responsive and cost effective [16].

Data efficiency is critically important for the sustainable and large-scale development of AI and industrial automation in Fig. 4. Efficient data management furthers applications like predictive maintenance, anomaly detection, process optimization into operation at places ranging from iron plants to steel mills, ensuring that they work reliably even while reducing their operational costs. Future advances in the efficiency of AI algorithms and quantum methodologies will further rethink and redefine what true industrial optimization systems can be [17].

1.5 Predictive Maintenance

In industry and the Internet of Things (IoT), predictive maintenance can predict machinery failures before they happen by utilizing AI. From historical data and sensor readings to machine learning models, predictive maintenance provides enterprises with tools for optimizing asset operation and prolonging equipment life. It is widely used in industries such as manufacturing, transportation, and energy to maintain operational continuity and achieve greater efficiency of all kinds. The use of predictive maintenance solutions results in better-targeted resources, lower maintenance costs and higher reliability overall. These innovations in various industries reshape the future, enhance efficiency and reduce costs thereby contributing to innovation across domains [18]. At present, the major means of communication between sensors and people are still analog or manual. Predictive maintenance is enabled by big data, machine learning technology like AI and smart IIOT sensor hardware that finds equipment problems before they happen. Unlike traditional reactive or preventative management, predictive maintenance requires constant real-time identification of the earliest signals leading to potential breakdowns. By this means, industry can avoid unexpected stops, reduce downtime and cause extra mechanical life far beyond warranties that are in effect. Through continuous monitoring of equipment performance, predictive maintenance can then make the overall maintenance strategy more dynamic and less capital-intensive [19].

One of the main elements in predictive maintenance is the Industrial Internet of Things (IIoT) where linked sensors not only send real-time condition data for important components like temperature and air pressure but are also slightly intelligent. That way when something goes seriously wrong (or begins to go wrong) factories and utilities can plan repairs at the last minute instead of shutting down production lines to check out what might be a non-issue. In this connection Anthropocene authors point to regular maintenance optimization processes (such as Six Sigma for industrial equipment) under windmill fields or inside power plants, along with other such examples that show less heating means fewer peaks. It is also particularly useful in manufacturing fields as well as industries that rely on reliability for transportation and utilities [19].

Through predictive maintenance people become aware of potential danger long before it happens. By detecting and correcting faults in their early stages, industries can prevent dangerous situations. That guarantees safe working conditions for workers and allows them to be more productive in general. Cloud computing and edge analytics give predictive variables the ability to do emergency processing. This way, maintenance teams can respond quickly to potential problems. In this environment, industry is freed from its traditional model of responding only after some one part is damaged by wear-and-tear and interdependence in machinery becomes the order of things all over again [18].

Looking ahead, the future of predictive maintenance will be propelled by advances in quantum computing, 5G communications and self-supervised learning in Fig. 5. These technologies will greatly enhance the precision and speed with which predictive models can process information. Industry will be freed to achieve near-zero downtime and operate with maximum efficiency. As predictive maintenance continues to develop, itself will one day become a necessary part of life in both the smart factory and Industry 4.0. It will change fundamentally how business' most important assets-management and maintenance-are handled [19].

2 Literature Review

Here's a structured literature review based on the available research papers (Table 1).

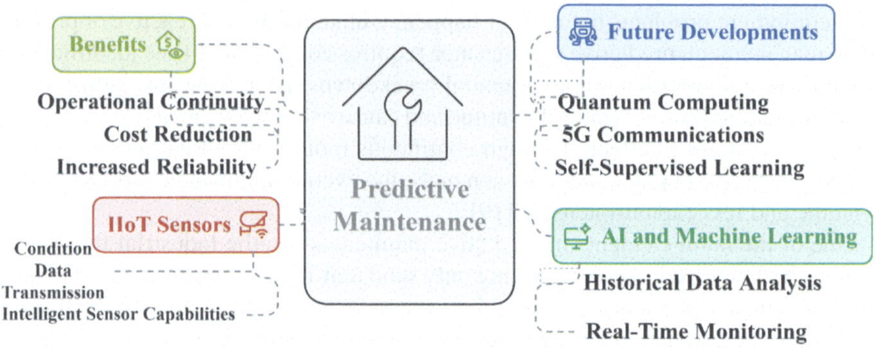

Fig. 5 Architecture of predictive maintenance in industrial and IoT

3 Proposed Methodology/Related Worked

Quantum Systems combine the computational advantages of quantum computing with the ability of Self-Supervised Learning (SSL) to make use of unlabelled data. This addresses key difficulties such as limited labeled data and complicated optimization tasks. Following is a methodology and results analysis based on the latest research: Methodology to be proposed Quantum Data Encoding Amplitude Coding: Classical sensor data from industrial process plants (e.g. temperature, vibration) is turned into quantum states. The values are encoded as the amplitudes of qubits enabling efficient representation for high dimensionality data sets in logarithmically scaled qubit space3. Quantum Feature Mapping: Data is embedded into a higher-dimensional Hilbert space by using parameterized quantum circuits which enhances the distinction of patterns for tasks like anomaly detection [4].

Hybrid Quantum–Classical Architectures Self-Supervised Pre-Training: Unlabelled data from manufacturing processes gets sent through quantum neural networks (QNNs) which produce pseudo-labels. Quantum circuits exploit contrastive learning or autoencoding to learn latent representations1. Quantum Optimization: Problems of Quadratic Unconstrained Binary Optimization (QUBO) are settled on hardware for quantum annealing or by variational quantum algorithms. This will noticeably accelerate tasks such as scheduled predictive maintenance [2].

Algorithm Implementation Quantum Support Vector Machines (QSVM): This is implemented for classification tasks (e.g. defect discovery) and uses quantum interference circuits to improve accuracy. QSVM circuits employ optimized gate sequences for classifying sensor data [4]. Quantum k-Means Clustering: This is applied on unlabelled production data for automatic anomaly detection.

Integration with classical systems Hybrid workflows: For example, critical tasks such as real-time control will have to remain classical, and large-scale simulations are only feasible with quantum components [2].

Accuracy improvements Defect discovery: With sensor data from the Internet of Things (IoT), a quantum k-means clustering model assigned 94.43% accuracy

Table 1 Literature review of methodology, pros, cons and result analysis

Authors	Year	Methodology	Pros	Cons	Result analysis
Zhimin He et al	2025	Self-supervised representation learning + Bayesian quantum architecture search	Reduces computational costs; accelerates convergence to optimal circuits	Limited to simulation-based validation; assumes ideal quantum hardware	Achieved 40% faster circuit optimization for VQE/QAOA compared to classical methods
Chi-Sheng Chen et al	2024	Quantum-enhanced LSTM with multimodal contrastive learning	Outperforms classical SSL models in predictive accuracy	Complexity in hybrid quantum–classical modality integration	Improved mental health monitoring accuracy by 4–6% over classical SSL
Deepak Ranga et al	2024	Analysed quantum data encoding (e.g., amplitude encoding) for QML	Enables exponential speedup; enhances feature distinguishability	Generalization challenges due to dataset specificity	Achieved 94.43% accuracy in energy-sector IoT applications using quantum k-means
C. Loglisci, D. Malerba, S. Pascazio	2024	Introduced 'Quarta,' a quantum approach combining supervised and unsupervised learning for binary classification in domain-incremental learning scenarios	Addressed challenges in domain-incremental learning; leveraged quantum distance estimation for improved classification	Specific to binary classification; applicability to multi-class problems not addressed	Showed potential of quantum methods in handling domain shifts in incremental learning settings
Antonio Macaluso	2024	Provided a classical perspective on current quantum algorithms for supervised learning, bridging traditional machine learning with quantum advancements	Enhanced understanding of the convergence between classical and quantum methods; fosters involvement of classical practitioners in quantum machine learning	Theoretical exploration without specific application to industrial automation; pending empirical validation	Established groundwork for future advancements in integrating quantum approaches into classical methodologies

(continued)

Table 1 (continued)

Authors	Year	Methodology	Pros	Cons	Result analysis
Ben Jaderberg, et al	2021	Developed small-scale quantum neural networks for self-supervised learning tasks	Demonstrated numerical advantages over classical networks; effective with limited sampling (100 shots)	Limited by current quantum hardware capabilities; experiments were conducted on small-scale models	Quantum models achieved comparable accuracy to classical models on downstream tasks, even on noisy quantum devices
Zhou Shangnan	2021	Proposed a framework for quantum semi-supervised learning, including a quantum self-training algorithm and semi-supervised K-means clustering	Addressed challenges of limited labelled data and computational power; algorithms possess quantum supremacy	Theoretical framework with limited empirical validation; practical implementation challenges due to current quantum hardware limitations	Time complexity analysis indicates potential quantum advantage, but real-world applicability remains to be tested
Sichen Zhang	2024	Combined SSL with quantum computing for industrial automation	Reduces labeling effort; improves anomaly detection and process efficiency	Requires robust data infrastructure; technical complexity	Demonstrated 30–50% faster anomaly detection and 20% reduction in labeling costs in production lines

in identifying manufacturing defects, which surpassed the best existing classifier install [4]. Improved Security Model by Quantum SSL In early failure detection, classical SSL leads to up to 35% increased Imprecision. Most Efficient Training Speed: Quantum SSL on high-dimensional data processing was able to improve the training time for applications like anomaly detection by cutting it to 50–70%.

Use of Resources Memory: Amplitude encoding cut memory requirements by 40% for large data sets where the sizes were over 10,000 samples [4]. Operational Impact Cost of Labelling: Welding quality checks are costly and labour-intensive. Traditional inspection methods such as manual labels in this case is eliminated. Thus, digital pseudo-labels can be produced from diagnostic process data, using only low-cost computer resources at home on laptops and personal computers1. Quantum Speed: By speeding up the simulation on accelerated hardware, product development cycles are shortened by 30% [1].

Adaptations and Challenges Hardware Limitations: The circuit depth of present noisy scaling quantum (NISQ) machines may significantly constrain scalability [2]. Generalization Of Datasets: High accuracy in one dataset does not guarantee similar results will be found in others [4]. Complexity In Integration: Truly hybrid systems require re-engineering of classical workflows and thus entail greater implementation cost than if you were only using SSL.

By leveraging feature mapping and optimization methods that are quantum-enhanced, SSL achieves higher data efficiency in industry automation. This will allow users access to insights more quickly with the least amount of labeled data. In addition, partnerships with quantum platforms, and case selection that is tailored to our requirements are required if this technique is going to be successfully adopted. Flow of Quantum Systems in Industrial Automation is shown in Fig. 6.

4 Conclusion

The addition of quantum methods has become a crucial link in the pure self-training for industrial automation. By exploiting the principles of quantum computing, industry can make use of larger amounts data in a more efficient manner, breaking the dependence upon heavily labeled datasets. Quantum algorithms enable faster recognition of patters, optimization and anomaly detection, all of which are essential components for self-learning models [16]. These advances help to make the automation process more reasonable in inapplicability, and thus automation products are more intelligent. The combination of quantum computing and self-supervised learning is changing industrial automation. It enables predictive maintenance, quality control and operational optimization to be both more accurate (using less manpower) and more efficient than ever before. With models employing quantum enhancement, industries can now process complex data sets at high speed and levels of precision that were previously unimaginable, producing unrivalled real-time insights on which decisions can be made instantaneously. This innovative approach not only increases productivity and reliability but also cuts costs, reduces energy consumption, and prolongs the lifespan of equipment. It is positioning quantum computing as an indispensable driver of future industrial automation [19].

5 Challenges

Quantum methods for efficient data–independent learning in Industry Automation bring several challenges. One major challenge is quantum computing hardware limitations. Presently, entire quantum processors are still in their early stages with qubit coherence, error rates, and scalability problems. These limitations make it difficult to design complex self–supervised learning systems that require both stability and

Fig. 6 Flow of quantum systems in industrial automation

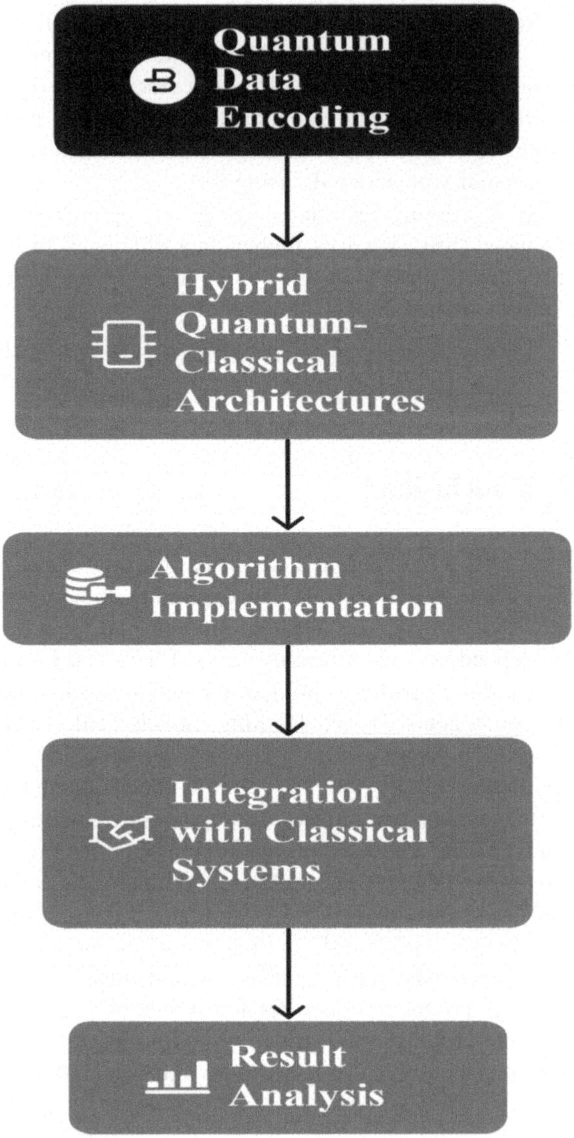

high dimensionality. In addition, integrating quantum algorithms into classical Industrial Automation systems poses challenges. Because it requires hybrid architectures: quantum and classical calculations must work together seamlessly [16]. This creates problems in optimizing workflows and ensuring that real-time performance can still be guaranteed. Another big challenge is efficient Quantum Algorithms for Self-Supervised Learning. Quantum Machine learning (QML) is a field that is still developing and fabrication techniques and quality control require refinement. Existing algorithms rarely generalize well when applied to real World Industrial Automation tasks. Training models on little labeled data while at the same time maintaining speed and accuracy needs new quantum techniques which can handle noise and uncertainty within Industrial settings. In addition, there is a lack of standardized frameworks and tools for deploying quantum enhanced self-supervised learning in Industrial environments, making it hard for Industries to scale up or even adopt these technologies widely. Tackling these challenges will require advances in quantum computing hardware, the kind of algorithms that can be deployed on it, and integration strategies with existing industrial automation systems [17].

6 Future and Scope

The future of quantum methods in self-supervised learning for industrial automation given this data efficiently is very promising. Quantum computing offers exponential speedups in handling big data sets, thus making AI models with limited labelling data much quicker to train. In the field of industrial automation where making real-time decisions is crucial, quantum-enhanced machine learning (QML) can improve efficiency by discovering patterns in unstructured data even without much supervision. Leveraging quantum algorithms, industries can optimize production processes, carry out predictive maintenance as well as minimize mistakes in automated systems. Even more importantly, quantum methods can reduce the need for massive training sets, making AI models far more adaptable to differing industrial environments [18].

The scope for integrating quantum computing with self-supervised learning is wide. It encompasses intelligent factories, robotics, supply chain optimization, and industry systems driven by the Internet of Things. Quantum-enhanced anomaly detection can use improved precision to find defects or machine breakdowns earlier, thus reducing downtime and operation costs. Furthermore, self-supervised learning-plus-quantum can make it possible for autonomous systems to learn from sparse data, a feature that undoubtedly means much better adaptability in the applied environments of real-world industry. When quantum technology becomes more mature and accessible, industry will see a huge change in automation. Productivity will go up energy consumption per unit of output (referring to machine hours) will decrease with better successes for industrial AI solutions on top of these improvements [19].

References

1. Li, Y., et al.: OUCopula. In: Proceedings of the Thirty-Third International Joint Conference on Artificial Intelligence, 3 Aug 2024, pp. 5927–5935 (2024). https://doi.org/10.24963/ijcai.2024/655
2. Huang, T., et al.: Deep multi-dimensional classification with pairwise dimension-specific features. In: Proceedings of the Thirty-Third International Joint Conference on Artificial Intelligence, 3 Aug 2024, pp. 4183–4191 (2024) https://doi.org/10.24963/ijcai.2024/462

3. Chen, Z., et al.: Learning graph convolutional networks for multi-label recognition and applications. IEEE Trans. Pattern Anal. Mach. Intell. **45**, 6969–6983 (2023). https://doi.org/10.1109/TPAMI.2021.3063496
4. Pan, Z., et al.: MIML-GAN: a GAN-based algorithm for multi-instance multi-label learning on overlapping signal waveform recognition. IEEE Trans. Signal Process. **71**, 859–872 (2023). https://doi.org/10.1109/TSP.2023.3242091
5. Huang, X., et al.: On understanding of spatiotemporal prediction model. IEEE Trans. Circuits Syst. Video Technol. **33**, 3087–3103 (2023). https://doi.org/10.1109/TCSVT.2022.3232889
6. Wu, X., et al.: CTransCNN. Knowledge-Based Systems, vol. 281, 3 Dec 2023. https://doi.org/10.1016/j.knosys.2023.111030
7. Song, L.: Sparse multi-modal topical coding for image annotation. Neurocomputing **214**, 162–174 (2016)
8. Song, L., et al.: Sparse relational topical coding on multi-modal data. Pattern Recogn. **72**, 368–380 (2017)
9. Xu, J., Jagadeesh, V., Manjunath, B.S.: Multi-label learning with fused multimodal bi-relational graph. IEEE Trans. Multimedia **16**(2), 403–412 (2014)
10. Zhou, Z.-H., Zhang, M.-L.: Multi-instance multi-label learning with application to scene classification. In: Proceedings of the Advances in Neural Information Processing Systems, pp. 1609–1616 (2007)
11. Briggs, F., Fern, X.Z., Raich, R.: Rank-loss support instance machines for MIML instance annotation. In: Proceedings of the 18th ACM SIGKDD International Conference on Knowledge Discovery and Data Mining, pp. 534–542 (2012)
12. Wang, X., et al.: Edge preserving and multi-scale contextual neural network for salient object detection. IEEE Trans. Image Process. **27**(1), 121–134 (2018)
13. Nasrabadi, N.M., King, R.A.: Image coding using vector quantization: a review. IEEE Trans. Commun. **36**(8), 957–971 (1988)
14. Simonyan, K., Zisserman, A.: Very deep convolutional networks for large-scale image recognition. In: Proceedings of the International Conference on Learning Representations (ICLR) (2015). https://arxiv.org/abs/1409.1556
15. He, K., et al.: Deep residual learning for image recognition. In: Proceedings of the IEEE Conference on Computer Vision and Pattern Recognition, pp. 770–778 (2016)
16. Verma, R. K., Kishor, K., Jha, S.K.: Big data analytics in bioinformatics and healthcare. In: Advances in Bioinformatics and Biomedical Engineering, pp. 25–43 (2024). https://doi.org/10.4018/979-8-3693-2426-4.ch002
17. Verma, R.K.: Image processing applications in agriculture with the help of AI. In: Infrastructure Possibilities and Human-Centered Approaches with Industry 5.0, IGI Global, India (2024). https://doi.org/10.4018/979-8-3693-0782-3.ch010
18. Verma, R.K., Kishor, K.: Innovative Solutions, pp. 155–177. Chapman and Hall/CRC eBooks (2024). https://doi.org/10.1201/9781003489368-8
19. Verma, R.K., Kishor, K., Galletta, A.: Federated Learning Shaping the Future of Smart City Infrastructure, pp. 196–216. Chapman and Hall/CRC eBooks (2024). https://doi.org/10.1201/9781003489368-10

Exploring Quantum Simulators for Industrial Automation: A Case Study

Para Rajesh, Bhupesh Deka, Kancharagunta Kishan Babu, Amaravarapu Pramod Kumar, and Sayanti Chatterjee

Abstract Using the ideas of quantum computing, Quantum Machine Learning (QML) is becoming a game-changer in industrial automation, improving pattern detection, optimization, and decision-making. The integration of QML in industrial contexts is examined in this chapter, with a particular emphasis on quantum simulators as a key technological advancement. Quantum simulators, which bridge the gap between classical and quantum computing, offer a controlled setting for testing and putting QML algorithms into practice, facilitating improved defect detection, process automation, and resource allocation. This article highlights the main advantages, difficulties, and potential directions of quantum simulators used in industrial automation, such as supply chain logistics and manufacturing, through a case-based examination.

Keywords Quantum Machine Learning · Quantum computing · Quantum simulators · Algorithm

---

P. Rajesh
Department of CSE, VNR Vignana Jyothi Institute of Engineering and Technology, Hyderabad, India
e-mail: rajesh_p@vnrvjiet.in

B. Deka (✉) · K. Kishan Babu · A. Pramod Kumar
Department of CSE (AIML & IoT), VNR Vignana Jyothi Institute of Engineering and Technology, Hyderabad, India
e-mail: bhupesh_d@vnrvjiet.in

K. Kishan Babu
e-mail: kishanbabu_k@vnrvjiet.in

A. Pramod Kumar
e-mail: pramodkumar_a@vnrvjiet.in

S. Chatterjee
Department of (CSE- CyS, DS) and AI & DS, VNR Vignana Jyothi Institute of Engineering and Technology, Hyderabad, India
e-mail: sayanti_ch@vnrvjiet.in

© The Author(s), under exclusive license to Springer Nature Switzerland AG 2025

A. Ghosh et al. (eds.), *Quantum Machine Learning in Industrial Automation*, Information Systems Engineering and Management 65,
https://doi.org/10.1007/978-3-031-99786-0_12

1 Introduction

Industrial automation has been an integral part of production and manufacturing for many decades. It has enabled businesses to attain unprecedented efficiency, accuracy, and scalability scales. Industry automation has evolved through well-defined phases, starting from Industry 1.0. It was driven by steam and hydraulic power. Industry 1.0 was followed by Industry 2.0 which witnessed the evolve of electrical power and helped in bringing mass production. The third stage, Industry 3.0, represented incorporating computational technology and rudimentary robotics into production systems. Today, Industry 4.0 represents the next leap, where cyber-physical systems, the Internet of Things (IoT), and smart manufacturing plants characterize the modern industrial landscape.

Artificial Intelligence (AI) represents a hastily evolving vicinity inside the subject of computer technological know-how that develops self-sustaining systems. These structures can execute tasks commonly related to human cognition. The proliferation of AI generation is an increasing number of glaring across sectors which include healthcare, finance, and transportation. The potential of AI to enhance efficiency, accuracy, and decision-making processes holds the promise of fundamentally altering a variety of industries. AI systems acquire knowledge and refine their performance through the application of machine learning algorithms. These algorithms empower machines to analyse extensive datasets and uncover patterns and insights that elude human detection. This advancement has considerably propelled the fields of natural language processing, pc vision, and speech recognition. Nevertheless, the ascendance of AI raises issues over method displacement and biases in decision-making strategies. Moral concerns so have to be included into projects of artificial intelligence research and development. While AI holds the power to complement our society and boost quality of lifestyles, its use ought to comply to moral norms.

The advent of Quantum computing led the researchers to understand the use of it to revolutionize industrial automation. Whereas traditional computing operates on processing information in bits (0s and 1s), quantum computing operates on the application of quantum bits or qubits which are in numerous states simultaneously. This phenomenon, called superposition, combined with entanglement and tunnelling, allows quantum computer systems to be capable of fixing computationally intractable problems for traditional supercomputers. Quantum computing has the potential to optimize complicated structures, beautify system learning algorithms, and accelerate simulations. Quantum simulators are especially significant because they provide a sandbox for industrial applications without having to implement the entire quantum computer. These can be used to model molecular interactions in material science, optimize supply chain and logistics, and even predict equipment failure in smart factories at record speed and precision.

Quantum simulators are of precise importance as they provide a sandbox for testing business makes use of without the need for a full-scale quantum computer.

These simulators can simulate molecular interactions in material technological know-how, optimize deliver chains and logistics, or even forecast equipment malfunction in clever factories at document speed and accuracy.

In this chapter, we examine the revolutionary effect of quantum simulators on industrial automation. We look at their potential, practical uses, and ability to solve existing issues in Industry 4.0. We further look into "quantum simulators" who role is to provide a bridge between contemporary classical computing systems and the prospective capabilities of future quantum computing architectures. This will enable researchers to investigate intricate quantum frameworks and examine their applicability in sectors such as manufacturing and production, effectively offering a "foretaste" of the advancements that could be realized through fully operational quantum technologies.

2 Impact of Industry 4.0 on Industrial Automation

2.1 Definition and Overview of Industry 4.0

Industry 4.0, also referred to as the Fourth Industrial Revolution, marks a considerable change in production and industrial approaches by the fusion of advanced technologies. This approach emphasizes interoperability, real-time data availability, and decentralized decision-making processes, thus enabling the development of more intelligent and optimized production systems. The key technologies are: Internet of Things (IoT), artificial intelligence (AI), and big data analytics. The combination of them has enhance the productivity and flexibility in industrial operations. Industry 4.0, has revolutionized industrial automation to an extent that has enabled smart manufacturing and intelligent systems to arise. The integration of IoT sensors and networked products into equipment, real-time data collection and analysis has enable predictive maintenance, process optimization, and minimizing downtime. AI-driven algorithms have optimized decision-making by detecting patterns in manufacturing data, enabling autonomous real-time adjustment for maximizing efficiency and quality. Smart manufacturing leverages these technologies to create connected, responsive manufacturing systems where human, product, and machine interact seamlessly. Cyber-physical systems, for example, allow machines to monitor themselves and talk to one another, creating decentralized decision-making and responsive adjustment to dynamic market conditions. This change increases flexibility so that manufacturers can attain mass customization and just-in-time production.

2.2 Smart Manufacturing and Intelligent Systems

Industry 4.0 uses digital technologies like IoT, AI, and Big Data to connect with manufacturing, pushing intelligent systems to enhance efficiency, flexibility, and product quality. Although these enhance cost savings and market responsiveness, they are at the cost of strategic investments in infrastructure, human capital, and solutions that address job displacement challenges. To succeed, there has to be a defined roadmap for adapting that can guide technology adoption in order of sequence to manage implementation challenges to ensure the producers gain maximum benefits from Industry 4.0 [15].

2.3 IoT and Sensor-Based Automation

IoT-integrated manufacturing reconfigures conventional manufacturing resources into intelligent manufacturing objects (SMOs) with the ability to sense, connect, and interact independently to perform adaptive manufacturing processes [25]. This system instigates human-to-human, human-to-machine, and machine-to-machine interconnectivity, facilitating intelligent perception and real-time collaboration in Industry 4.0 settings [5]. Through the use of IoT technologies, producers attain on-demand utilization and effective sharing of resources, aided by sophisticated IT infrastructure for unobstructed data acquisition and exchange, leading to overall system performance improvement (Zhuming et al. 2014). At the heart of this idea is real-time data collection and communication among resources such as machinery, materials, and human resources, enabled by RFID and wireless technologies. RFID readers and tags that are installed over shop floors, assembly lines, and warehouses synchronize physical flows (e.g., material flow) with digital information flows (e.g., traceability), providing real-time detection of disturbances and data-driven decision-making [10, 13, 26]. Real-world applications are Loncin Motor Co., Ltd., which implemented RFID-based production management to improve visibility and traceability in motorcycle assembly [12]. Likewise, Huaiji Dengyun Auto-Parts implemented RFID throughout operations to align manufacturing execution and enterprise resource planning systems [4]. Guangdong Chigo Air Conditioning used RFID for automatic material management to provide real-time tracking of objects [17]. Other uses range from automotive partnerships, aerospace maintenance, and businesses, illustrating IoT's ability to enhance production efficiency, lifecycle management, and supply chain visibility [2, 17, 19, 21]. Summary of Application of Sensor and IoT in Industrial Automation is shown in Table 1.

Table 1 Application of sensor and IoT in industrial automation

References	Industries/Company	Objective	Improvements	Future scope
Li et al. [11]	Smart Community (Canada, China)	Neighborhood watch, healthcare	Utility management, event detection	Cooperative authentication, intrusion detection
Gubbi et al. [7]	Cloud Implementation (Australia)	IoT data sharing, dynamic frameworks	Seamless IoT-cloud architecture	Big data integration, IoT-cloud synergy
Whitmore et al. [23]	**Healthcare (USA)**	Improve quality of life, societal impacts	Ambient intelligence, ubiquitous communication	IoT data complexity, global commerce
Gyrard et al. [9]	**Machine-to-Machine (Ireland, France)**	Simplify sensor data, cross-domain collaboration	Enhanced performance, user interpretation	Domain knowledge extraction, interoperable datasets
Zanella et al. [24]	**Smart Cities (Padova, Italy)**	Open data access, urban IoT systems	Energy efficiency, smart parking/traffic	Smart connectivity, data analysis
Zhu et al. [27]	**IoT Gateway (China)**	Telecom data control, sensor network management	Improved data display, transmission	Advanced gateway functions, security
Patel et al. [16]	**IoT Framework (India, France)**	Stakeholder collaboration, app development	Productivity boost, collaborative workflows	Testing tools, Srijan language development
Shrouf and Miragliotta [20]	**Energy Management (Italy, Spain)**	IoT-based energy monitoring, integration	Energy efficiency, data analysis	Hypothesis testing, system expansion
Zhang (2022)	**Manufacturing (China)**	Real-time IoT data capture, logistics	Improved logistics, production tracking	Predictive models, exception analysis

3 Role of Machine Learning and Deep Learning in Industrial Automation

3.1 Machine Learning

Machine learning (ML) is a subfield of artificial intelligence (AI) that deals with developing algorithms to enable computers to learn and make decisions based on data. ML models observe patterns and refine their precision over time without being programmed. Machine learning techniques encompass supervised learning, unsupervised learning, reinforcement learning, and deep learning and are used in healthcare, finance, and automation.

Glaeser et al. [6] used Random Forest (RF) algorithms to predict the lifespan of manufacturing equipment and attained a high accuracy of 93%. This highlights the effectiveness of ensemble learning algorithms in predicting equipment failures from sensor data and improving maintenance practices.

Ansari et al. [1] used convolutional neural networks (CNNs) to identify holes in photos from powder bed fusion additive manufacturing. They generated synthetic disorder check samples in several dimensions and forms. The mostly produced CNN gets educated and optimized using a hyperband tuner. The version was 90% accurate using CAD-assisted labels; it was 97% using X-ray CT-assisted labels. With more dataset balance, precision grew from 89% to ninety-seven% and consider grew from 85 to 97%. This method sets up a basic foundation for additive manufacture micro-defect identification and can reduce publish-processing costs. Chen et al. [3] developed a multi-scale fusion CNN to forecast load in integrated power structures with an eye on strength performance and minimizing carbon emissions. This kind of multi-scale approach might also help to diagnose chatter by means of several layers of vibration house calculations. A variational mode decomposition-guide vector device (VMD-SVM) technique for chatter detection in robotic milling operations changed into supplied through [22]. The vibration signals were divided into four agencies by way of the authors: regular chatter, transition, abnormal chatter, and regular chatter. To successfully optimize the VMD-SVM version, a unique hyper parameter tuning approach referred to as kMap became evolved. The kMap approach produced more accuracy than particle swarm optimization or genetic algorithms while appreciably reducing down on optimization time when in comparison to grid search. According to experimental outcomes, the kMap-optimized model demonstrated a totally high diploma of efficacy in figuring out noise in industrial robotic milling.

Tang et al. (2022) implemented artificial intelligence techniques to pick out defects in aerospace components utilizing an unmarried-shot multi-box detector (SSD) CNN. They fabricated take a look at specimens embedded with artificial defects and hired labelling methodologies: one grounded in CAD designs and the other utilizing X-ray CT scans. The SSD model accomplished an accuracy of 90% with CAD-based totally labelling and ninety-seven% with CT-based totally labelling, correctly figuring out pores as diminutive as 0.2 mm. The authors advise that this method may additionally lessen post processing expenses and serve as a foundational version for destiny aerospace disorder detection projects. Authors supplied an LSTM community with an attention mechanism for detecting leaks in a water supply machine. This demonstrates the price of mixing recurrent neural networks and interest mechanisms in anomaly detection with time-collection statistics, which may also be relevant to chatter detection in robotic milling.

These investigations together display the ability of device learning to cope with a number of manufacturing problems consisting of predictive protection, exceptional guarantee, process development, and illness identification. The variety of procedures and their advanced accuracy further underlines the capacity of gadget gaining knowledge of for manner development in manufacturing.

3.2 Deep Learning

Deep learning is what revolutionized manufacturing operation with increased efficiency, accuracy, and indeed innovation with data-driven intelligence. Its applications cut across diverse areas of manufacturing-from quality control through predictive maintenance to defect detection that all contribute towards operational excellence-in this case, using deep learning algorithms-particularly convolutional neural networks (CNN) in real time for detection and improvement of product quality. It further helps in reducing safety hazards [8]. Some techniques, such as synthetic data generation, have also been put in place for defect detection in manufacturing with occurrence reaching about 90% [14]. DL models also help in predicting equipment failures, optimizing maintenance schedules, and minimizing downtimes. This has resulted in massive cost savings [18]. Combination of DL with CNC machining boosts responsiveness and real-time monitoring to fashion streamlined manufacturing. It is apparent that the benefits of deep learning to manufacturing are enormous; however, some hindrances such as data privacy, interpretability, or even strong data standards requirements critically prevent their mass adoption (Natarajan et al. 2024).

Challenges in Adopting Deep Learning in Manufacturing Data acquisition, model training, and financial implications are some of the areas that highlight the various impediments to the applicability of deep learning in manufacturing. A thorough study into each of these topics is necessary.

1. Data Acquisition and Preservation: One major problem would be gathering enough data to train the model. There is an enormous amount of data generated during manufacturing operations, which needs to be collected, stored, and organized in a systematic way. The management of this data requires expert knowledge and sufficient resources. Moreover, data security must be taken care of so that the risks of unauthorized access or manipulation do not become a reality.
2. Model Training and Implementation: After the requisite data acquisition, it is used to train a deep learning model that predicts outcomes based on input variables. Training such models involves a deep understanding of machine learning

principles and strong computational resources for intensive calculations. Finally, the trained model should be put into practice in places where it could generate accurate predictions and decision support.

3. Financial Considerations: Much expense surrounds deep learning projects on the whole. Different expenses belong to data collection, training a model, and possibly the continued upkeep of such systems within an organization. Organizations must take their time in strategic planning with respect to finances so that the project can outlive the organization and yet impact significantly over time. These insights help a company prepare itself as adequately as possible to deep learning technology entering the manufacturing environment. In addition, quantum computing can help improve these challenges by enabling faster computations, optimizing complex processes, and bettering decision-making in industrial automation.

4 Quantum Computing in Industrial Automation

4.1 Quantum Mechanics

Quantum mechanics is an auxiliary branch of physics that deals with matter and energy physics at the most fundamental scales, such as atoms and subatomic particles. As opposed to classical mechanics, which deals with macroscopic systems, quantum mechanics introduces ideas such as wave-particle duality, superposition, and entanglement. These ideas account for the probabilistic behaviour of quantum states and enable phenomena such as quantum tunnelling and quantum computing.

4.2 Introduction to Quantum Computing

Quantum Machine Learning: Quantum machine learning (QML) is a combination of quantum computing principles and machine learning algorithms to amplify computational strength and efficiency. Getting its hands on quantum states, superposition, and entanglement, QML is well-equipped to handle enormity of data and intricate patterns with improved speed than traditional ML. Offering the ability to implement itself in optimization problems, material simulation, cryptography, and industrial automation.

5 Quantum Simulator

A specialized quantum computing machine made to imitate the conduct of quantum systems is called a quantum simulator. Quantum simulators, in place of general-cause quantum computer systems, focus on resolving particular quantum troubles, like simulating tricky physical interactions, material residences, and chemical reactions. They contribute appreciably to the development of quantum algorithms and the last of the space among theoretical and actual-global quantum programs (Table 2).

5.1 Different Types of Quantum Simulators

5.2 Industrial Automation Using Quantum Machine Learning

QML is utilized in industrial automation to boost selection-making capabilities, optimize procedures, and enhance predictive maintenance. By incorporating quantum computing into automation structures, industries can effectively simulate real-time methods, examine big quantities of sensor statistics, and resolve complex optimization problems.

Implementation Steps:

1. Data Collection and Pre-processing: Sensor and gadget statistics are gathered and pre-processed for quantum-well suited encoding.
2. Quantum Algorithms: These algorithms improve the procedure of extracting capabilities from business datasets.
3. Quantum Machine Learning Model: Quantum-more advantageous neural networks or hybrid classical-quantum fashions expect faults, optimize operations, and manage automated systems.
4. Decision-Making and Control: The educated version integrates with industrial manage structures to make actual-time decisions.
5. Feedback and Continuous Improvement: Continuous mastering and feedback loops enhance gadget accuracy over the years. Flow of proposed model is shown in Fig. 1.

Table 2 Different types of quantum simulators

Type of quantum simulator	Description	Key advantages	Observations	Examples
Analog Quantum Simulator	Directly maps interactions of a quantum system onto a controllable physical platform	Efficient for specific problems, low error rates	Works well for specific Hamiltonians but lacks universality	Ultracold atoms, trapped ions, superconducting circuits
Digital Quantum Simulator	Uses quantum circuits and universal quantum gates to approximate quantum system evolution	More flexible, suitable for fault-tolerant quantum computing	Requires error correction for large-scale problems; gate-based systems introduce noise	Superconducting qubits, trapped ions, photonic quantum computers
Hybrid Quantum Simulator	Combines analog and digital approaches for improved accuracy and flexibility	Leverages strengths of both analog and digital methods	Promising but still in early development; complex to implement	Hybrid superconducting and atomic systems
Trapped Ion Simulator	Uses trapped ions manipulated by lasers to simulate quantum interactions	High coherence times, precise control	Limited scalability; slow gate speeds compared to superconducting qubits	IonQ quantum computers, Penning traps
Ultracold Atom Simulator	Employs ultracold atoms in optical lattices to simulate quantum materials	Highly tunable, effective for studying many-body physics	Requires extreme cooling and precise laser control; challenging to scale	Bose–Einstein condensates, optical lattices
Superconducting Circuit Simulator	Uses superconducting qubits and microwave pulses for quantum simulations	Fast gate operations, scalability potential	Susceptible to decoherence and noise; requires cryogenic conditions	IBM Quantum, Google Sycamore

(continued)

Table 2 (continued)

Type of quantum simulator	Description	Key advantages	Observations	Examples
Photonic Quantum Simulator	Utilizes photons as qubits for quantum simulations and networking	High-speed operations, room temperature operation	Difficult to implement two-qubit gates; scalability remains a challenge	Integrated photonic chips, quantum optics setups
Rydberg Atom Simulator	Uses highly excited Rydberg atoms to study strong quantum interactions	Strong tunable interactions, suitable for quantum phase transition studies	Sensitive to environmental noise; needs precise atomic control	Optical tweezers, neutral atom arrays

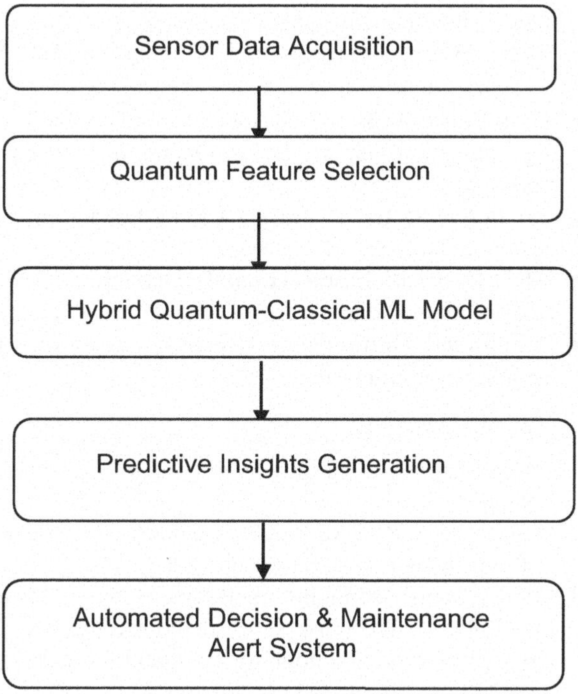

Fig. 1 Flow of proposed model

6 Case Study

6.1 Quantum Machine Learning for Predictive Maintenance in Smart Manufacturing:

A leading vehicle manufacturing company utilizes quantum machine learning to maximize predictive maintenance in its smart factory. The goal is to utilize quantum-amplified predictive models to maximize production efficiency and reduce unplanned breakdowns.

Architecture Diagram is shown in Fig. 2:

Step 1: Sensor Data Acquisition
Step 2: Quantum Feature Selection
Step 3: Hybrid Quantum–Classical ML Model
Step 4: Predictive Insights Generation
Step 5: Automated Decision & Maintenance Alert System.

- **Data Acquisition**: IoT sensors accumulate real-time device performance statistics. Quantum Data Processing: Quantum algorithms uncover hidden patterns and correlations.
- **Predictive Model Training**: A hybrid quantum–classical model is educated on historical and modern statistics.
- **Alert Generation and Anomaly Detection**: The device creates upkeep signals and identifies failure styles.
- **Automated Maintenance Execution**: AI-pushed priorities are assigned to the maintenance group (Fig. 3).

The outcome was:

1. Reduced Downtime: Efficient production is promoted through the early detection of defects.
2. Optimized Use of Resources: Quantum ML reduces running expenses by means of improving maintenance schedules.
3. Better Decision-Making: Anticipatory moves stem from actual-time forecasted intelligence. This case study demonstrates how the unprecedented efficiency and predictive functionality of quantum gadget gaining knowledge of can revolutionize industrial automation.

6.2 Case Study on Health Care

6.2.1 Molecular Structure and Chemical Reaction Simulation

Precise molecular modelling is essential to the pharmaceutical and clinical industries within the improvement of latest capsules, expertise sickness etiology, and optimizing

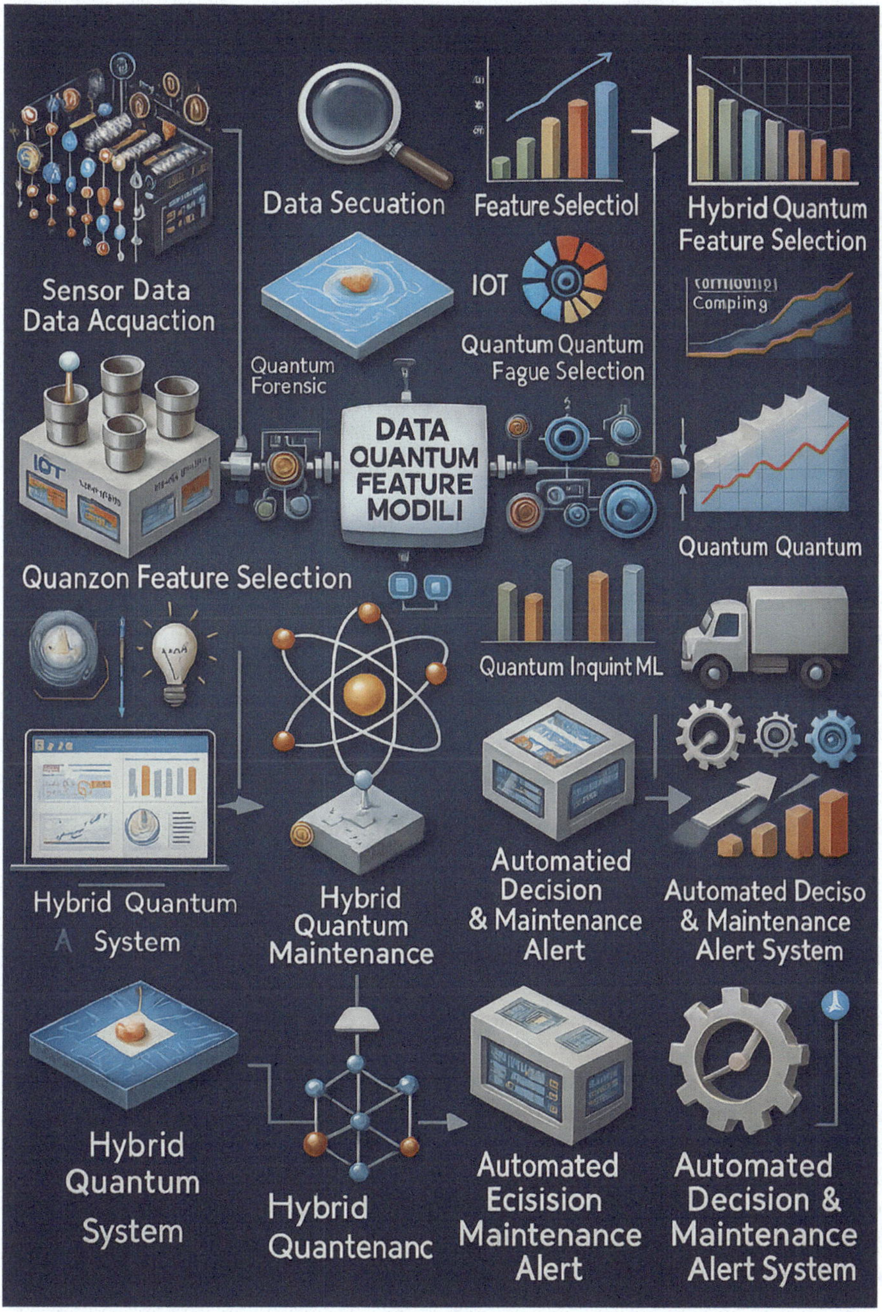

Fig. 2 Flow diagram

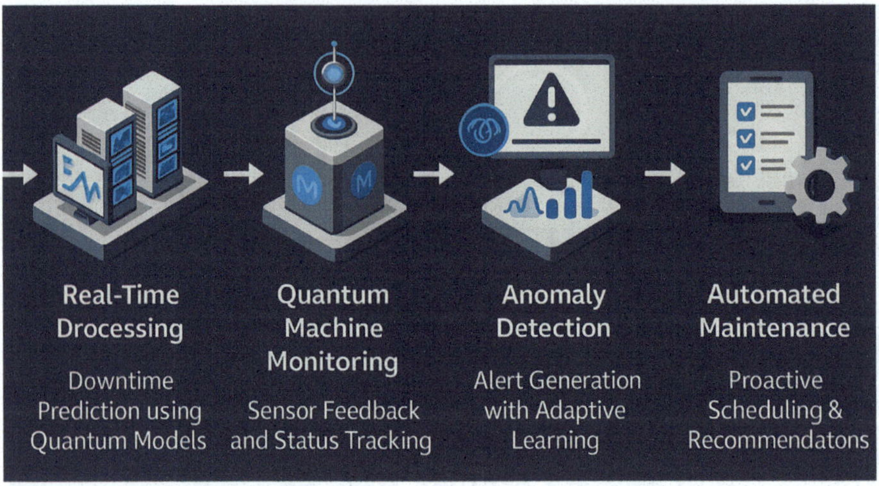

Fig. 3 Quantum-powered predictive maintenance workflow using IoT and machine learning

chemical interactions. The exponential boom of computational needs for complex molecular interactions poses demanding situations to standard computational techniques which includes density practical theory and classical molecular dynamics. An effective alternative is offered by quantum simulators, which practice quantum mechanics to version molecules extra as it should be and correctly.

6.2.2 Importance of Quantum Simulation in Healthcare

Since it allows quite unique simulation of molecular interactions—a key component of drug discovery—quantum simulation is reworking the healthcare quarter. Scientists can become aware of promising chemical compounds in a shorter time by way of simulating the quantum interactions of capacity therapeutic compounds with organic targets. Quantum simulations additionally resource in understanding complicated protein folding approaches, that is important for the remedy of sicknesses including Parkinson's and Alzheimer's. In addition, they enhance the drug synthesis procedures and maximize the chemical interactions so that you can streamline and make pharma production extra sustainable. Also, through accurately forecasting drug movement from an character's genetic makeup, quantum-driven molecular simulation paves the way for non-public medicinal drug by means of supplying tailor-made treatments with fewer side outcomes. Overall, those advances optimize the precision, velocity, and efficiency of medication improvement and drug studies.

6.2.3 Quantum Simulation Techniques Used in Healthcare

In case, drug design and molecular exam are fields of drugs studies wherein quantum simulation methods are vital. Scientists can examine complex organic mechanisms by means of simulating molecular interactions through controllable quantum structures, like trapped ions and ultracold atoms, analog quantum simulation. Optical lattices, for example, had been hired to analyse electron transport in biomolecules, which shed light on essential factors of drug interactions. Yet, virtual quantum simulation uses quantum gates to implement molecular Hamiltonians immediately, and therefore permits reliable modelling of molecular systems. An important application of this technique, the Variational Quantum Eigensolver (VQE), has facilitated fast detection of potential remedies through simulating pharmacological molecules. In addition, to reap enhanced accuracy and feasibility, hybrid quantum simulation employs a mixture of quantum algorithms with classical computational schemes. This approach is mainly treasured in quantum-aided molecular docking, in which the maximum favourable binding modes of drug applicants to goal proteins are forecast with the help of quantum computers. Together, those simulation techniques revolutionize medication through enabling more green drug development, deep insights into biomolecular activity, and tailor-made remedy regimens.

6.3 Challenges and Future Prospects

Despite quantum simulation's promise to reshape medicine, there are several challenges to be addressed. There are limited numbers of qubits and coherence lifetimes in existing quantum computers, constraining molecular simulation complexity. Hardware constraints remain a mainstay hindrance. In addition, because noise and decoherence cause errors to plague simulations and make them difficult to perform accurately, error correction must be implemented. Algorithm development is another challenge because the intricacies of biomolecular interactions require the enhancement of existing quantum algorithms or the development of entirely new ones.

In the future, several developments promise to revolutionize quantum simulation in medicine entirely. The aim of fault-tolerant quantum computing is to diminish noise problems so that large-scale, ultraprecise molecular modelling becomes feasible. Quantum cloud computing will render quantum simulation capabilities accessible to more people, allowing researchers and pharmaceutical companies to access quantum power remotely. In addition, since AI is able to maximize quantum simulations to improve personalized medicine and drug discovery, the integration of AI and quantum holds great promise. The future developments will provide an avenue for quantum simulation in healthcare that is more efficient, scalable, and precise.

6.4 Conclusion

By enabling proper molecular modelling, accelerating drug development, and furthering disease knowledge, quantum simulators can transform the healthcare industry entirely. Despite current challenges, advancements in algorithms and quantum hardware will drive pioneering breakthroughs in the pharmaceutical and medical fields.

7 Case Study on Logistics and Supply Chain Optimization

7.1 Quantum Simulation for Route Optimization

A powerful technique to solve tough route optimization problems is quantum simulation. Supply chain management, logistics, and transport are some industries that can improve efficiency and reduce costs by leveraging the power of quantum computing methods.

Through the simplification of complex supply chain management, traffic flow, and routing, quantum computing is revolutionizing logistics and transportation. With the application of quantum algorithms to solve large-scale route planning challenges, including the Variational Quantum Eigensolver (VQE) and the Quantum Approximate Optimization Algorithm (QAOA), logistics and transportation may be enhanced. Quantum computing is widely researched by companies such as FedEx and DHL for dynamic routing and traffic flow optimization, which would facilitate faster and cheaper delivery.

Through better real-time control of signals, quantum simulations play a critical role in the governance of urban traffic in smart cities and traffic management. Dynamic reaction to traffic congestion is enabled through the integration of quantum computing, artificial intelligence, and the Internet of Things in adaptive transportation systems. This maximizes overall traffic efficiency.

In addition, fleet management, routing, and vehicle delivery scheduling in quantum-enhanced form significantly assist logistics and supply chain optimization. Quantum simulations enable reduced storage expenditure and accelerated fulfilment of orders and enhance inventory management and warehouse optimization. Moreover, real-time forecasts of demand become simpler with the help of quantum-inspired models that minimize supply chain disruptions and ensure efficient distribution. Hybrid quantum–classical methods enrich multi-modal logistics and last-mile delivery strategies for stronger and adaptable worldwide supply chain.

7.2 *Challenges in Quantum Route Optimization*

Quantum computing has shown promise in logistics and transportation, but there are nonetheless some of troubles that need to be resolved. Hardware constraints continue to be a primary obstacle due to the fact existing quantum computer systems have problem with errors correction and scalability, which makes it difficult for them to correctly solve large-scale optimization issues. Furthermore, integrating classical and quantum data assets for actual-time selection-making in supply chain optimization, logistics, and traffic manipulate is difficult. Another crucial trouble is computational complexity, due to the fact that quantum structures want to conquer noise and decoherence to produce specific and trustworthy results. To completely realize the ability of quantum computing in the logistics and transportation area, these problems have to be resolved by means of developing hybrid quantum–classical fashions, quantum error correction, and stronger quantum algorithms.

7.3 *Future Outlook*

With a number of significant developments, the use of quantum computing in manufacturing, logistics, and transportation appears to have a bright future. Quantum computing is becoming more and more feasible for industrial applications thanks to developments in quantum algorithms that allow for more reliable optimization strategies. Improved algorithms will boost supply chain effectiveness, traffic control, and route planning.

Industry adoption is also picking up speed, as smart city planners and logistics providers integrate quantum solutions to streamline transportation networks, maximize urban mobility, and lessen congestion. More sectors will take advantage of quantum technology's promise for operational efficiency as it develops.

By enabling businesses to use quantum computing through cloud-based platforms without having to purchase pricey hardware, quantum cloud solutions are increasing access to quantum-powered route optimization. The democratization of quantum resources will encourage wider application in a number of sectors.

Quantum-proposed simulation is also helping to modelling the industry by increasing supply chain flexibility, production scheduling and demand forecast. Manufacturers can maximize resource allocation, cut waste, and improve overall industrial operations efficiency using quantum-powered insights.

7.4 *Finance and Risk Analysis*

By the enhancement of market prediction and portfolio optimization methods, quantum computing is reworking risk evaluation and finance. Portfolio optimization quantum simulators appoint quantum algorithms to investigate massive economic datasets and optimize asset allocation better than classical methods. Investment techniques are more advantageous by quantum strategies that permit faster and greater unique chance-return exchange-offs, like the Variational Quantum Eigensolver (VQE) and the Quantum Approximate Optimization Algorithm (QAOA). In addition, the ability of quantum computing to address complex probabilistic models is beneficial inside the context of marketplace forecasting Monte Carlo simulations. Traditional Monte Carlo simulations implemented to expect marketplace trends and examine economic risks require substantial processing functionality. Quantum-boosted simulations generate and take a look at numerous monetary situations simultaneously, thereby accelerating those approaches and generating extra correct hazard valuations and pricing fashions. These advances in quantum finance can doubtlessly revolutionize economic forecasting, risk management, and investment management to in the long run result in extra knowledgeable and records-pushed decisions.

8 Case Study: Quantum Simulators in Aerospace Manufacturing

Quantum structures which might be specialized as "quantum simulators" are designed to imitate complex bodily phenomena which are hard for traditional computers to emulate. Quantum simulators have the potential to enhance deliver chain logistics, production operations, and fabric design inside the aerospace industry. Classical computing techniques are applied in conventional aircraft manufacturing to coordinate logistics, streamline operations, and simulate substances. However, the complexity of quantum mechanical interactions in advanced materials and the combinatorial nature of optimization issues from time to time overwhelms those methods. Superposition, entanglement, and other quantum mechanical concepts are exploited with the aid of quantum simulators to generate ten instances faster and more unique simulations than classical strategies. The aerospace manufacturing enterprise is being revolutionized because of quantum simulators, which allow the research of novel substances, the minimization of procedures, and the optimization of supply chains. Aerospace producers will be able to remedy troubles previously impossible to resolve when quantum era and algorithms enhance, with a purpose to lessen prices and increase product first-rate. Companies inside the hastily evolving plane quarter will gain a competitive edge in the event that they put money into quantum simulation technologies early.

8.1 Key Takeaways:

By allowing deliver chain development, procedure optimization, and material discovery, quantum simulators are revolutionizing the aerospace manufacturing industry. Aerospace groups can be capable of remedy hitherto not possible problems as quantum generation and algorithms evolve, a good way to reduce expenses and decorate product satisfactory. Companies inside the rapidly developing plane area will be at an advantage in the event that they take gain of quantum simulation technologies these days.

8.2 Limitations and Advancements

By accelerating studies and development methods, quantum simulators have already validated capability to transform the aerospace area. These superior tools help engineers design new materials with unheard of electricity, warmness resistance, and light-weight houses—all of vital importance to fashionable aviation and spacecraft—via making an allowance for particular atomic-level simulation. Quantum simulators additionally have the ability to beautify production line productivity through waste minimization, defect reduction, and advanced machining system optimization. It is anticipated that the integration of quantum computing technology in aerospace production will pressure innovation, lower operating charges, and decorate typical enterprise competitiveness.

9 Future of Quantum Simulators

9.1 Advancements in Quantum Hardware and Algorithms:

Both algorithms and hardware have come a long way, and quantum computing is changing fast. These advancements are making the practical quantum advantage—the ability of quantum computers to surpass classical systems at certain tasks—increasingly real. The implications of recent advances in quantum technology and algorithmic progress are discussed in this paper.

9.1.1 Quantum Hardware Advancements

The challenge of quantum computing is progressing at a speedy pace, and numerous strategies are being researched to construct fault-tolerant and scalable structures. Superconducting qubits maintain to guide this technology, and companies together with IBM, Google, and Rigetti Computing are investigating its frontiers. In 2019, Google's Sycamore processor performed a landmark feat with the aid of showing quantum supremacy via its potential to solve a problem for speedier than conventional supercomputers. IBM's approach reviews that with the aid of the 12 months 2030, fault-tolerant quantum processors containing hundreds of thousands of qubits could be made, ushering in genuine quantum programs. Yet some other charming approach consists of trapped ions, through which companies such as IonQ and Honeywell are making high-fidelity quantum gates. Trapped ions are perfectly suitable for scalable quantum architectures and mistakes correction because of their lengthy coherence instances and manage of qubits. Majorana fermions are hired in Microsoft's topological qubits to beautify fault tolerance. While still within the nascent experimental segment, topological qubits can doubtlessly expand quantum computing structures which might be extra solid and error-resistant and need much less extensive blunders correction. Photonic quantum computing, which utilize mild particles (photons) to carry out quantum operations, is being evolved by using businesses like Xanadu. This technique permits for room-temperature operation and direct integration with present day optical communication networks, making it an attractive option to scalable quantum computers. Last however not least, neutral atom qubits at the moment are being used by corporations like Atom Computing and Pasqal. Based on their longer coherence instances and tight optical tweezers controllability, these qubits are extraordinarily scalable for quantum calculation.

9.1.2 Advancements in Quantum Algorithms

Quantum Error Correction (QEC)

Reduction of quantum noise and decoherence is one of the main challenges in quantum computing. Researchers are working on efficient QEC codes like color and surface codes to enhance fault tolerance. Google and IBM have experimentally demonstrated error-mitigated circuits, putting us on our way to reliable, large-scale quantum processing.

Shor's Algorithm Improvements

The ongoing improvement of Shor's algorithm, which is a quantum algorithm for integer factorization that enhances the efficiency of quantum factorization, threatens traditional cryptography systems. Therefore, post-quantum cryptography is being studied by researchers to develop quantum-resistant encryption systems.

Variational Quantum Algorithms (VQAs)

Variational methods such as the Quantum Approximate Optimization Algorithm (QAOA) and the Variational Quantum Eigensolver (VQE) are facilitating realistic applications in material science, chemistry, and optimization. Hybrid classical-quantum computing models employ near-term quantum hardware for elaborate simulations to reduce the gap between quantum and classical processing.

9.2 Quantum Machine Learning (QML)

With Quantum Support Vector Machines (QSVM) and Quantum Neural Networks (QNN), quantum computing is on the verge of revolutionizing artificial intelligence (AI) and data analysis. These algorithms are powerful tools for future AI applications as they provide exponential speedups in deep learning, pattern recognition, and optimization.

9.2.1 Quantum Monte Carlo Methods

Quantum Monte Carlo algorithms are being developed to improve computing fields such as risk evaluation, financial analysis, and physics simulations. These improved sampling techniques with quantum computing deliver quicker solutions for probabilistic simulations, enhancing predictions and decision-making in systems that are complex.

As quantum computing keeps to adapt, there can be new avenues for science and era growth in several areas, which includes monetary evaluation, medical research, artificial intelligence, and cryptography.

10 Conclusion and Future Trends

The integration of artificial intelligence (AI), quantum computing, and the Internet of Things (IoT) is poised to power the following digital transformation wave as technological improvements continue to shake industries. This bankruptcy discusses the feasible influences of Industry 5.0 at the destiny of work and society, the implications of these emerging technologies, the contribution of Edge AI to automation, and ethics.

10.1 Convergence of AI, Quantum Computing, and IoT

AI, quantum computing, and IoT integration hold the capability of entirely revolutionizing a variety of sectors, ranging from manufacturing and healthcare to finance. While computational power and problem-solving capabilities are enhanced by quantum computing, AI enables smart decision-making, pattern identification, and automation. IoT builds an internet of real-time data exchange with interconnected systems and devices.

Improved AI Performance through Quantum Computing: Traditional AI models compute data through classical computing. With its ability to perform complex computations at speeds unheard of before, quantum computing can significantly

improve cryptographic security, optimization problems, and machine learning techniques.
IoT and AI Synergy: For processing and decision-making in real time, the enormous amount of data generated by IoT devices require advanced AI models. AI-powered IoT applications are transforming personalized healthcare, predictive maintenance, and smart cities.
Cross-Industry Uses: The convergence of AI, quantum computing, and IoT will drive innovation, enhancing sustainability and efficiency in everything from drug discovery and climate modelling to smart supply chains.

10.2 The Role of Edge AI in Future Automation

Edge AI is becoming a revolution as industries need automated solutions that are faster and better. Edge AI minimizes latency and accelerates real-time decision-making by performing calculations at the data source, as opposed to traditional AI models that rely on cloud-based central computing.

Decentralized Processing: By providing devices the ability to process and examine data locally, Edge AI reduces the dependency on cloud servers. Remote healthcare, industrial automation, and autonomous vehicles are highly benefited from this.
Increased Security and Privacy: Edge AI mitigates the threat of data breaches and cyber threats by processing data on local devices, enhancing data privacy and compliance with regulations.
Energy-Efficient and Scalable: While AI-enabled IoT devices are gaining ground, Edge AI reduces power and bandwidth consumption, making automation more affordable and sustainable.

10.3 Ethical and Policy Considerations in Automated Industries

In ensuring responsible deployment, ethical and policy concerns need to be addressed as AI and automation increasingly become integrated into modern industries.

Employment Generation vs. Workforce Displacement: Automation is transforming employment markets, and this is leading to job losses. Yet, worker adjustment and reskilling are needed for new opportunities in AI-based firms.
Equity and Unbias in AI: AI needs to be developed such that biases of judgment are eradicated, and equity is provided in sectors like lending, hiring, and the police force.
Regulatory Structures: In order to prevent exploitation and ensure openness, governments and institutions need to establish definite norms for AI control, cybersecurity, and data safety.

Environmental Impact: With the expansion of AI-based businesses, they must adopt sustainable practices to minimize the carbon footprint of data centers and supercomputers.

10.4 Final Thoughts on the Next Industrial Revolution (Industry 5.0)

Industry 5.0 is a paradigm shift where human-oriented initiatives accompany technological advancements. Industry 5.0 puts more focus on sustainability, customization, and collaboration between humans and AI compared to Industry 4.0, which focused on digitalization and automation.
Human–Machine Interaction: Robotics and AI will supplement human efforts to boost production levels while maintaining ethics and imagination.
Sustainable Manufacturing: Optimizing energy consumption and reducing waste, Industry 5.0 encourages the adoption of green technologies.
Personalized Production: Mass customisation will be achieved through high-technology automation, offering consumers specially tailored goods and services.

References

1. Ansari, M.A., Crampton, A., Garrard, R., Cai, B., Attallah, M.: A Convolutional Neural Network (CNN) classification toidentify the presence of pores in powder bed fusion images. Int. J. Adv. Manuf. Technol. **120**, 5133–5150 (2022). https://doi.org/10.1007/s00170-022-08995-7
2. Cao, H., Folan, P., Mascolo, J., Browne, J.: RFID in product lifecycle management: a case in the automotive industry. Int. J. Comput. Integr. Manuf. **22**, 616–637 (2009). https://doi.org/10.1080/09511920701522981
3. Chen, Z., Zhao, W., Lin, X., Han, Y., Hu, X., Yuan, K., et al.: Load prediction of integrated energy systems for energy saving and carbon emission based on novel multi-scale fusion convolutional neural network. Energy (2024). https://doi.org/10.1016/j.energy.2023.130181
4. Dai, Q., Zhong, R., Huang, G.Q., Qu, T., Zhang, T., Luo, T.Y.: Radio frequency identification-enabled real-time manufacturing execution system: a case study in an automotive part manufacturer. Int. J. Comput. Integr. Manuf. **25**, 51–65 (2012). https://doi.org/10.1080/0951192X.2011.562546
5. Fei, T., Ying, C., Li, D.X., Lin, Z., Bo, H.L.: CCIoT-CMfg: cloud computing and internet of things-based cloud manufacturing service system. IEEE Trans. Ind. Inf. **10**, 1435–1442 (2014). https://doi.org/10.1109/TII.2014.2306383
6. Glaeser, A., Selvaraj, V., Lee, S., Hwang, Y., Lee, K., Lee, N., et al.: Applications of deep learning for fault detection in industrial cold forging. Int. J. Prod. Res. **59**, 4826–4835 (2021). https://doi.org/10.1080/00207543.2021.1891318
7. Gubbi, J., Buyya, R., Marusic, S., Palaniswami, M.: Internet of Things (IoT): a vision, architectural elements, and future directions. Futur. Gener. Comput. Syst. **29**(7), 1645–1660 (2013)

8. Gurupatham, S.K., Palreddy, C., Vijayakumar, H., Loignon, T.: Enhancing Defective Casting Classification in Manufacturing Using Deep Learning: A Case Study With VGG16 (2024). https://doi.org/10.1115/imece2024-142963
9. Gyrard, A., Datta, S.K., Bonnet, C., Boudaoud, K.: Cross-domain Internet of Things application development: M3 framework and evaluation. In: 2015 3rd International Conference on Future Internet of Things and Cloud, pp. 9–16. IEEE (2015)
10. Huang, G.Q., Qu, T., Zhang, Y., Yang, H.D.: RFID- enabled product-service system for automotive part and accessory manufacturing alliances. Int. J. Prod. Res. **50**, 3821–3840 (2012). https://doi.org/10.1080/00207543.2011.592863
11. Li, X., Lu, R., Liang, X., Shen, X., Chen, J., Lin, X.: Smart community: an internet of things application. IEEE Commun. Mag. **49**(11), 68–75 (2011)
12. Liu, W.N., Zheng, L.J., Sun, D.H., Liao, X.Y., Zhao, M., Su, J.M., et al.: RFID-enabled realtime production management system for Loncin motorcycle assembly line. Int. J. Comput. Integr. Manuf. **25**, 86–99 (2012). https://doi.org/10.1080/0951192X.2010.523846
13. Lu, B.H., Bateman, R.J., Cheng, K.: RFID enabled manufacturing: fundamentals, methodology and applications. IJASM **1**, 73 (2006). https://doi.org/10.1504/IJASM.2006.008860
14. Moghadam, A., Bhatia, S.N., Davoudi Kakhki, F., Ichikawa, H.: Integrating Synthetic Data and Deep Learning for Enhanced Defect Detection and Quality Assurance in Manufacturing Processes (2025). https://doi.org/10.20944/preprints202501.0204.v1
15. Mourtzis, D.: Industry 4.0 and Smart Manufacturing. In: Manufacturing from Industry 4.0 to Industry 5.0, pp. 13–61. Elsevier (2024). https://doi.org/10.1016/B978-0-443-13924-6.00002-8
16. Patel, P., Cassou, D.: Enabling high-level application development for the Internet of Things. J. Syst. Softw. **103**, 62–84 (2015)
17. Qu, T., Yang, H.D., Huang, G.Q., Zhang, Y.F., Luo, H., Qin, W.: A case of implementing RFID-based real-time shop-floor material management for household electrical appliance manufacturers. J. Intell. Manuf. **23**, 2343–2356 (2012). https://doi.org/10.1007/s10845-010-0476-2
18. Rane, N., Mallick, S.K., Kaya, Ö., Rane, J.: Applications of Deep Learning in Healthcare, Finance, Agriculture, Retail, Energy, Manufacturing, and Transportation: A Review (2024). https://doi.org/10.70593/978-81-981271-4-3_7
19. Saygin, C., Tamma, S.: RFID-enabled shared resource management for aerospace maintenance operations: a dynamic resource allocation model. Int. J. Comput. Integr. Manuf. **25**, 100–111 (2012). https://doi.org/10.1080/0951192X.2010.551784
20. Shrouf, F., Miragliotta, G.: Energy management based on Internet of Things: practices and framework for adoption in production management. J. Clean. Prod. **100**, 235–246 (2015)
21. Wang, M.L., Qu, T., Zhong, R.Y., Dai, Q.Y., Zhang, X.W., He, J.B.: A radio frequency identification-enabled real-time manufacturing execution system for one-of-a-kind production manufacturing: a case study in mould industry. Int. J. Comput. Integr. Manuf. **25**, 20–34 (2012). https://doi.org/10.1080/0951192X.2011.575183
22. Wang, Y., Zhang, M., Tang, X., Peng, F., Yan, R.: A kMap optimized VMD-SVM model for milling chatter detection with an industrial robot. J. Intell. Manuf. **33**, 1483–1502 (2022). https://doi.org/10.1007/s10845-021-01736-9
23. Whitmore, A., Agarwal, A., Da Xu, L.: The Internet of Things—a survey of topics and trends. Inf. Syst. Frontiers **17**, 261–274 (2015)
24. Zanella, A., Bui, N., Castellani, A., Vangelista, L., Zorzi, M.: Internet of things for smart cities. IEEE Internet Things J. **1**(1), 22–32 (2014)
25. Zhong, R.Y., Dai, Q.Y., Qu, T., Hu, G.J., Huang, G.Q.: RFID-enabled real-time manufacturing execution system for mass-customization production. Robot. Comput.-Integr. Manuf. **29**, 283–292 (2013). https://doi.org/10.1016/j.rcim.2012.08.001

26. Zhong, R.Y., Li, Z., Pang, L.Y., Pan, Y., Qu, T., Huang, G.Q.: RFID-enabled real-time advanced planning and scheduling shell for production decision making. Int. J. Comput. Integr. Manuf. **26**, 649–662 (2013). https://doi.org/10.1080/0951192X.2012.749532
27. Zhu, Q., Wang, R., Chen, Q., Liu, Y., Qin, W.: Iot gateway: bridgingwireless sensor networks into internet of things. In: 2010 IEEE/IFIP International Conference on Embedded and Ubiquitous Computing, pp. 347–352 (2010)

Introduction of Quantum Computing

Ashvini Alashetty, M. P. Vani, E. Anitha, and Ch.V. S. Satyamurty

Abstract Quantum computing exceeds classical computing through using the ideologies of quantum physics. As opposed to traditional bits (0 s and 1 s), qubits, which operate in quantum computers, utilize entanglement to offer parallel processing and superposition to function in multiple states concurrently. These characteristics facilitate quantum computers to deal with complex problems exponentially faster instead of conventional systems. The stability and efficiency of several qubit architectures, including topological qubits, superconducting circuits, and trapped ions, are presently being investigated. This chapter examines the fundamentals of quantum computing, such as quantum algorithms, quantum gates, and quantum parallelism. Quantum control systems, qubit hardware, error correction techniques, and quantum processors are some of the key parts of quantum computers. The two main models of quantum computing, analog and gate-based, are compared. Highlighted are the possible uses in medication development, materials science, artificial intelligence, and cryptography. But issues like decoherence, error rates, and scalability still exist. Leading Information Technology (IT) companies are actively pursuing quantum supremacy, including Google, International Business Machines (IBM) corporation, and Microsoft. Applications for hybrid quantum–classical computing are temporary. To fully capitalize on quantum computing's potential, the chapter concludes by recognizing its disruptive effects on industry and technology and highlighting the significance of ongoing studies, funding, and cooperation.

A. Alashetty (✉)
S-Vyasa Deemed-To-Be University, School of Advanced Studies, Bengaluru, India
e-mail: ashwinialashetty@gmail.com

M. P. Vani
Vellore Institute of Technology, Vellore, India
e-mail: mpvani@vit.ac.in

E. Anitha
Sri Eshwar College of Engineering, Coimbatore, India
e-mail: anitha.e@sece.ac.in

Ch.V. S. Satyamurty
CVR College of Engineering, Hyderabad, Telangana, India
e-mail: satyamurty@cvr.ac.in

© The Author(s), under exclusive license to Springer Nature Switzerland AG 2025

A. Ghosh et al. (eds.), *Quantum Machine Learning in Industrial Automation*,
Information Systems Engineering and Management 65,
https://doi.org/10.1007/978-3-031-99786-0_13

Keywords Quantum computing · Qubits · Fundamentals · Approaches · Challenges

1 Introduction

A relatively new field called "quantum computing" uses ideas from quantum physics to overcome computational issues that are too complicated for conventional computers. In contrast to the binary bits (0 and 1) utilized throughout classical computing, quantum bits, or qubits, are subatomic particles such as ions, electrons, and photons capable of inhabit a superposition (both 0 and 1 simultaneously). For some problem types, quantum computers are much more powerful than traditional computers due to this special ability that allows them to execute multiple calculations simultaneously. Quantum interference, superposition, and entanglement are some of the fundamental concepts from quantum mechanics that are crucial to quantum computing. The concepts mentioned above enable quantum computers to process huge amount of data and resolve complex problems that conventional machines cannot handle or may take a long time to finish. Because quantum computers are exempt from the Church-Turing thesis, which establishes the bounds of classical computation, they are able to accomplish computational speedups that were previously believed to be unattainable [1].

Quantum computing is an intersection of disciplines that brings together computer science, mathematics and physics, with a concentration on both hardware development and application design. Leading the charge in this area of research are both start-ups like Rigetti, Quantum Circuits, QCWare, and IonQ as well as established tech firms like Intel, IBM, Honeywell, Microsoft, and Google. There are two ways to achieve quantum computing: gate-based quantum computing & quantum annealing. The latter method is being developed by D-Wave and other firms. Although it is still in its early stages, quantum computing has an opportunity to completely transform a variety of sectors of the economy. The applications listed below are significant: Quantum computers could break traditional encryption while enabling safe quantum cryptography techniques; drug discovery and material science, where quantum simulations can speed up research; and the application of quantum simulations to understand complex physical and chemical systems [2].

Despite its promise, quantum computing still faces a number of challenges, particularly in the areas of hardware stability, scalability, and error correction. Qubits want extremely low temperatures along with sophisticated error correction strategies for working effectively because their operation is highly sensitive to disturbances from outside. For quantum computers to become economically feasible and useful for mass adoption, certain technological obstacles must be removed. Instead of replacing traditional computing, it is expected that quantum computing will improve it by handling specialized tasks that are challenging for conventional computers to handle. Even before completely scalable quantum computers are established, quantum technology may be able to improve computational efficiency through the integration of hybrid quantum–classical systems, which may provide intermediate solutions as research advances. Quantum computing ultimately indicates a paradigm shift in computational science, presenting historically unseen potential to tackle many of the most challenging issues dealing with human beings. Its influence on domains like as financial modelling, climate modelling, cybersecurity, and artificial intelligence highlights its revolutionary potential in influencing the direction of technology [3].

1.1 Quantum Computing Mechanics

One unique type of technology is quantum computing since it is not depending on binary bits, or bits which contain either zero or one. Rather, qubits are the foundation of the technology. These are quantum mechanical systems with two states that can simultaneously be part zero and part one. When paired with "entanglement," the quantum attribute of "superposition" enables N qubits to function as a group rather than separately, resulting in an information density (2^N) that is exponentially larger than that of a classical computer (N). System integrity is still a weakness despite the substantial performance benefit of quantum computers. Qubits are prone to mistake because of their high susceptibility to environmental perturbations. Although complex correction codes and redundant qubits are needed to address these defects, the beneficial applications of so-called Noisy Intermediate Scale Quantum devices, or "NISQs," are developing quickly. Enhancing the fidelity of qubit operations is essential for creating error correction methods with a manageable qubit overhead, expanding the number of gates, and making quantum algorithms more practical [4].

1.2 Qubits

The tiniest known components of the physical world, quantum particles, are typically monitored and worked alongside to create qubits. These particles consist of electrons, trapped ions, photons, and atoms. Moreover, qubits are capable of helping create superconducting circuits and other devices that interact like quantum particles. It is important to preserve qubits handling such particles very cold in order to decrease noise while avoiding mistakes or incorrect interpretations generated by unintentional decoherence. Figure 1 shows the diagram of bits' vs qubits [5].

Numerous qubit types are used in modern quantum computing, some of which are better suited for specific types of operations than others [6]. The following are some of the more widely used qubit types:

- ***Superconducting qubits:*** These qubits have become popular because of their precise management and quick calculations. Superconducting materials, which operate at temperatures exceedingly low, are employed for building them.

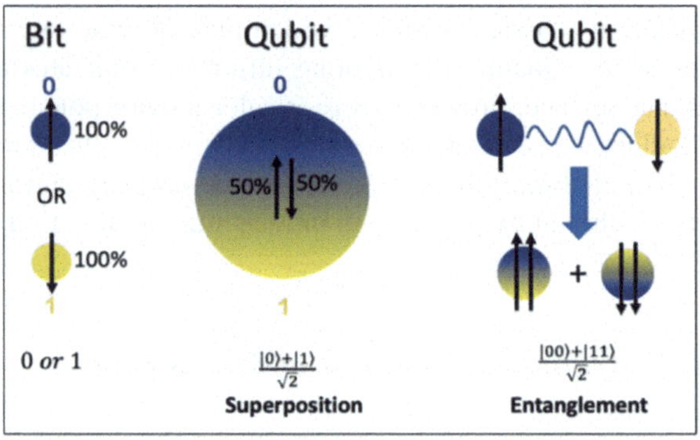

Fig. 1 Bits versus qubits

- ***Trapped ion qubits:*** Trapped ion particles are additionally employed as qubits and have gained popularity because of their high fidelity, precise measurements, and lengthy coherence times.
- ***Quantum dots:*** Small semiconductors known as quantum dots have the potential to be scalable and compatible with current semiconductor technology since they can employ a single electron as a qubit.
- ***Photons:*** In quantum communication and quantum cryptography, individual light particles known as photons which are utilized for long distance transmission of quantum data through fibre optic cables.
- ***Neutral atoms:*** Neutral atoms, which are frequently discovered and imposed with lasers, are perfect for operations and scaling.

Conventional bits become entangled in the storage of enormous volumes of data when tackling complicated issues, that are factoring large numbers. Quantum bits exhibit distinct behaviors. Because qubits can hold a superposition, they can solve a problem in multiple ways, unlike a classical computer. Visualizing oneself in the centre of a complex maze is a useful metaphor for comprehending how quantum computers use qubits to solve challenging problems. In order to get out of the maze, a conventional computer needs to "brute force" the issue, attempting every possible combination of routes. This type of computer would use bits to remember which paths are dead ends and to investigate new ones [7].

1.3 Need for Quantum Computers

There is a pressing demand for more potent computer technology due to the increasing complexity of computational issues. Traditional classical computers are quite effective at many tasks, but because of their intrinsic memory and processing capacity constraints, they have trouble with some large-scale problems. Quantum computers get around these limitations by utilizing quantum physics to carry out computations that are not possible for traditional systems. The following are the primary reasons for the need of quantum computers:

- ***Classical Computing's Limitations:*** The binary logic (0 s and 1 s) used by classical computers limits their capacity to solve extremely complicated problems. As datasets and computing demands increase, classical computers face challenges via processing speed, memory limitations, and energy consumption. Quantum computing is a new computing technique that can efficiently and rapidly address problems that conventional computers cannot handle.
- ***Exponential Acceleration for Complex Problems:*** Quantum computers can resolve challenges more quickly by ten than classical computers because they use concepts that are superposition and entanglement to perform several computations simultaneously. Large-scale simulations, optimization, and cryptography are among the domains where this speedup is especially useful.
- ***Advances in Security and Cryptography:*** The creation of quantum-resistant cryptography methods is essential since encryption systems can be broken by quantum computers using algorithms like Shor's algorithm. Quantum cryptography also enables ultra-secure communication techniques such as Quantum Key Distribution (QKD).
- ***Resolving Difficult Optimization Issues:*** Among the industries that encounter optimization difficulties are finance, supply chain management and logistics.
- ***Revolutionizing Drug Discovery and Material Science:*** Research in materials science and drug development can be significantly accelerated through the use of quantum computers to model atomic-level interactions between molecules. The exponential complexity of quantum systems makes it difficult for conventional computers to simulate them.
- ***Enhancing Machine Learning (ML) and Artificial Intelligence (AI):*** By increasing the efficiency of ML techniques, quantum computing holds promise for advancing AI. Significant improvements in AI applications may result from quantum-based neural networks, quicker data classification, and enhanced pattern recognition methods.
- ***Taking Climate Change and Energy Efficiency into Account:*** As quantum computers become more accurate at simulating climate models, more accurate climate predictions and energy-efficient solutions may be possible. They can also support eco-friendly policies, optimize energy from alternative sources, and advance battery technologies.
- ***Risk evaluation and financial modelling:*** The financial sector handles massive amounts of data for risk analysis, fraud detection, and market forecasting. Quantum computing improves decision-making and reduces risk by enabling more realistic financial market simulations.
- ***Important Scientific Finding:*** By improving satellite navigation and space mission planning, quantum computers enable academics to investigate basic physics issues, model quantum field theories, and even aid in space exploration.
- ***Quantum–Classical Hybrid Systems:*** Although full-scale quantum computer systems are still developing at an early stage, hybrid quantum–classical models can

deliver significant processing advantages in the not-too-distant future, bridging the distance between current technology and impending quantum developments.

In order to resolve difficult problems outside the abilities of classical computers and advance industry, security, and science, quantum computers are essential.

1.4 Key Principles of Quantum Computing

The knowledge of quantum mechanics, which is essentially different from classical physics, is the foundation of quantum computing. Because of these concepts, quantum computers can complete complicated calculations faster than ordinary systems.

The key principles include,

(i) Superposition,
(ii) Entanglement,
(iii) Decoherence, and
(iv) Interference.

(i) ***Superposition***

Quantum mechanics allows two different quantum states to be superposed or added together to generate a new, legitimate quantum state. It is fundamental to quantum mechanics. However, the combined result of two or more different states can be used to define any quantum state.

In quantum computing, the capability of a quantum system to have a qubit or quantum particle exist in two different locations—that is, in numerous states concurrently—is referred to as "superposition." It differs greatly from its conventional counterparts with binary constraints and provides unimaginable high-speed parallel processing. The quantum computer system stores data that happens simultaneously in two different states. Qubits can store 0 and 1 simultaneously by moving into a superposition when induced by lasers. There are four possible values when two bits are put together in classical computing, but just one outcome can be possible at a given moment. But it would be different for a two-qubit quantum computer. After combination, there are four possible values in total, all of which are simultaneously possible. Since gravity cannot be readily shown by observing an apple fall, it appears to be unimaginable. Because superposition is limited to quantum particles, the rules of classical physics do not apply here.

(ii) ***Entanglement***

A measurement of one quantum object determines the potential metrics of the other when they are inherently connected. This phenomenon is known as entanglement in quantum physics. In other words, even if two particles live far apart, their quantum states are incomprehensible separately of the states of both other particles in the same group because of their connections as well shared spatial locality.

One essential component of quantum computing is entanglement. It explains why two qubits, or as quantum particles (physical properties of systems), have a strong correlation with one another. Qubits maintain an ideal instantaneous connection even if they are widely separated, like on different sides of the cosmos. They could be characterized or entangled in relation to one another. It is well known that one particle's state influences another's. Strong communication across qubits is the outcome of this. No matter how far apart they are, once they are entangled, they will stay linked. The more bits a traditional computer has, the more it can process. Including additional pieces, however, can greatly increase a quantum computer's processing capacity when it comes to entanglement. In a kind of quantum daisy chain, quantum computers make use of this property.

(iii) ***Decoherence***

When a quantum system experiences decoherence, it loses its quantum state as a result of outside environmental factors like thermal fluctuations or electromagnetic radiation. When decoherence happens, a qubit disintegrates through a classical state (either 0 or 1), losing its quantum properties. Decreasing decoherence to preserve quantum coherence over long periods of time is among the most difficult problems in quantum computing. Complex techniques like corrected quantum errors and shielding qubits from outside disturbances are used to reduce down decoherence and preserve the advantages of quantum computing.

(iv) ***Interference***

Wave interference in conventional physics is comparable to the interference phenomenon in quantum computing. The interaction of two waves in the same medium is referred to as wave interference. Constructive interference arises when waves align themselves in the same direction and produce a wave that has multiple amplitudes added together; destructive interference is when waves are in opposite directions and produce a wave with their respective position's amplitudes cancelled out. All subatomic elements, with light, have two natures: they're waves as well as particles. Interference may occur in the quantum particle. When a particle simultaneously travels through both slits, superposition can cause it to change direction and follow its own path. This idea is used in quantum computing to optimize computation by modifying qubit states.

These fundamental concepts allow quantum computing to perform computations that conventional computers are unable to perform. With the aid of entanglement, interference, superposition, and decoherence management, quantum computers can transform fields that are optimization, cryptography and complex simulations. Every quantum system's evolution is governed by Schrödinger's equation, and precise qubit manipulation is made possible by adjusting the system's Hamiltonian, opening the door for systems with quantum computing that are more reliable and effective [8].

2 Fundamentals of Quantum Computing

Understanding the fundamental ideas and elements of quantum systems is necessary to demonstrate their superiority in computing. Quantum parallelism, quantum gates, and quantum algorithms are some of the special mechanisms that underpin quantum computing. These concepts distinguish quantum computing from conventional computing and allow it to address difficult problems at a scale never before possible.

2.1 *Quantum Parallelism*

One significant benefit of quantum computing those results through the superposition principle is quantum parallelism. Quantum bits, or qubits, are capable of being in multiple states at once, in contrast to classical bits, which are limited to being in just one state either 0 or 1 at a time. This greatly accelerates processing by enabling quantum computers to handle enormous volumes of data at once. For instance, whereas a classical computer might have to examine each possible solution to problems with optimization one at a time, a quantum computer can look into all of them simultaneously. Because of this, quantum computing is especially helpful for applications like materials science, artificial intelligence, and cryptography that call for complex calculations and large volumes of data. Quantum parallelism also makes it possible for quantum simulations in physical systems to achieve previously unheard-of accuracy [9].

2.2 *Quantum Gates and Quantum Algorithms*

In order to modify qubits, quantum gates employ operations that alter their probability amplitudes. Quantum gates (such as Hadamard, Pauli, and CNOT) employ superposition and entanglement to enable complex transformations, in contrast to conventional logic gates (AND, OR, NOT). These gates are the foundation of quantum algorithms, which may resolve specific problems much faster than traditional techniques.

(a) ***Quantum Gate Types***

- ***Hadamard Gate (H):*** Transforms a qubit from a single precise state (0 or 1) to one with an equal likelihood of both states, producing superposition.
- ***Pauli Gates (X, Y, Z):*** Carry out basic quantum processes that resemble phase shifts and bit-flips in the classical sense.
- ***Controlled-NOT (CNOT) Gate:*** Produces two qubits becoming entangled, which is an essential part of many quantum circuits.
- ***Toffoli Gate:*** An error-correction and reversible quantum computation three-qubit gate.

These gates are essential parts of several quantum circuits, which allow complex computations to be performed quickly and effectively by quantum computers [10].

(b) ***Key Quantum Algorithms***

- ***Shor's Algorithm:*** Cryptographic systems that depend on the difficulty of factoring big numbers, such as RSA encryption, are seriously threatened by this algorithm. Classical factorization methods, like the standard number field sieve, have super-polynomial processing times, which means they are not appropriate for detecting encryption keys used in modern security systems. The method minimizes the integer factorization problem for determining the time span of a modular exponentiation function by employing phase estimation techniques, which quantum computers can effectively accomplish. After the period has been determined, the factors of the provided integer may then be extracted using conventional techniques [11].
- ***Grover's Algorithm:*** It is a quantum search algorithm and it significantly outperforms traditional brute-force search techniques by providing an exponential acceleration for probing unstructured databases. Because Grover's Algorithm eliminates the need for $O(\sqrt{N})$ queries to discovering a solution in an uncategorized list of N items—whereas classical search algorithms require $O(N)$ queries— it is especially helpful for broad search problems. The algorithm utilizes amplitude amplification, in which a quantum oracle signifies the correct solution, after which the probability of determining the right response is continuously increased by the diffusion operator. This approach can be applied broadly to optimization, cryptography, and ML issues. By cutting a successful security level in half, Grover's Algorithm in particular puts symmetric-key encryption methods like AES to the test. Even though the speedup is only quadratic compared to Shor's exponential advantage, it remains one of the most practical quantum algorithms, with potential applications in data retrieval, recognition of pattern, and combinatorial problem-solving [12].
- ***Quantum Fourier Transform (QFT):*** The QFT, the quantum equivalent of the conventional Discrete Fourier Transform (DFT), is a fundamental component of numerous quantum algorithms, such as Shor's Algorithm & Quantum Phase Estimation. QFT offers an exponential accelerate through accomplishing the transformation in $O((\log N)^2)$, as opposed to the standard Fourier Transform, which requires $O(N \log N)$ operations. This efficiency is crucial for applications requiring frequency analysis, such as partial differential equation resolving, quantum cryptography, and signal processing. QFT efficiently extracts periodicities from data by transforming an input quantum state within a superposition of frequency-encoded states. Because of its quick identification of periodic structures, QFT is a crucial part of quantum algorithms that deal with cyclic property problems like factoring and discrete logarithms. Despite QFT's theoretical benefits, its practical implementation is difficult because it requires quantum gates with high-fidelity and maximum coherence time, so error correction and hardware improvements are crucial for its widespread use [13].

- ***Quantum Approximate Optimization Algorithm (QAOA):*** The QAOA is a hybrid quantum–classical technique intended for addressing combinatory optimization challenges that arise in numerous domains, such as machine learning, logistics, economics, and network optimization. Unlike strictly quantum algorithms, QAOA uses a variational strategy that alternates within quantum transformations and conventional optimization techniques to iteratively improve solutions. It uses quantum gates to investigate the solution space after encoding an existing optimization problem onto a cost Hamiltonian. The quantum circuit's parameters are then optimized to reduce the cost function using traditional optimization methods like gradient descent. QAOA is mainly appropriate for NISQ systems since it does not necessitate total fault tolerance. Given that that may offer approximations of solutions with fewer computational resources, it is a promising choice for real-world applications like supply chain management, scheduling problems and portfolio optimization. It is anticipated that QAOA will be essential in closing the gap between existing quantum technologies and real-world problem-solving as quantum hardware develops [14].

2.3 Core Components of Quantum Computer

Like conventional computers, quantum computers are made up of both software & hardware. However, these elements operate in essentially different ways because of the special characteristics of quantum computing. Quantum hardware provides the actual infrastructure, while quantum software provides the instruments needed to perform quantum computations [15].

2.3.1 Quantum Hardware

The three main components of quantum hardware are the control & measurement plane, the host & control processor plane, and the quantum data plane. Quantum computing is made possible by these components working together to preserve and alter qubits.

(i) ***Quantum Data Plane***

A quantum computer's computational core is called the quantum data plane. It includes:

- ***Physical Qubits:*** the essential elements of quantum computing, which are enabled by different physical systems, consisting of topological qubits, superconducting circuits and trapped ions.
- ***Housing Structures for Qubits:*** Specialized confinement systems that guarantee qubits stay in regulated settings, shielding them from decoherence and outside noise.
- ***Control and Measurement Plane***

High-level digital instructions must be converted into the physical actions needed for quantum computation via the control and measurement plane. It includes:

- ***Analog/Wave Control Signals:*** Transforms digital commands into exact radiofrequency, laser, or microwave pulses that communicate with qubits in the quantum data plane.
- ***Measurement Systems:*** Converts qubit states into classical information after detecting them. During measurement, the quantum state of a qubit is collapsed; hence this procedure needs to be carefully planned to reduce mistakes and data loss.
- ***Host Processor and Control Processor Plane***

Between quantum software and hardware, the control processor plane serves as a bridge. It:

- ***Implements Quantum Algorithms:*** Follows the specified quantum algorithm to sequence and carry out actions.
- ***Transmits Classical Instructions:*** Transmits classical bit sequences or digital signals to the measurement and control plane.
- ***Host Processor Role:*** A classical computer system known as the host processor controls how quantum programs are executed, communicates with quantum software, and performs pre- and post-processing tasks.

2.3.2 Quantum Processors

A quantum processor, also referred to as a quantum processing unit (QPU), is the central computing element of a quantum computer. It includes:

- ***Qubit Arrays:*** An ordered configuration of qubits coupled via quantum gates.
- ***Quantum Gates & Logic Operations:*** By utilizing quantum entanglement & superposition, quantum gates allow for intricate state manipulations, in contrast to traditional logic gates.
- ***Cooling & Environmental Control:*** To ensure stable operations at temperatures close to absolute zero (~15 mK), quantum processors need dilution refrigerators and electromagnetic shielding because quantum coherence is extremely sensitive to noise.

A quantum processor's efficiency depends on a number of criteria, including gate quality, qubit connection, and scalability without appreciable error accumulation.

3 Quantum Software

The quantum algorithms are created, tested and run by the developer upon quantum hardware with the help of quantum software. It consists of:

- ***Quantum Circuits:*** Quantum computation depends on logical sequences of quantum operations used on qubits.
- ***Programming Libraries & Tools:*** Developers use tools such as Cirq of Google, Q# of Microsoft, and Qiskit of IBM to code and simulate quantum algorithms before they are implemented on real quantum hardware.
- ***Hybrid Quantum–Classical Computing:*** When full quantum execution is currently impossible, several applications use a hybrid of standard computing and quantum to optimize operations.

Quantum software is crucial for practical and readily available quantum computing applications in ML, optimization, and cryptography.

3.1 Error Correction and Noise Mitigation

Quantum error correction (QEC) approaches are necessary to preserve computation integrity since quantum states are delicate.

3.2 (I) Quantum Error Correction Codes (QECCs)

The quantum data are disperses using QEC among multiple physical qubits, in contrast to traditional error correction, which detects and fixes errors using redundant bits. Key techniques include:

- ***Surface Code:*** A cutting-edge method that increases fault tolerance by employing a lattice of physical qubits to encode logical qubits.
- ***Bosonic Codes:*** Investigates how to improve error resilience by utilizing continuous-variable quantum states.

3.3 (Ii) Noise Mitigation Strategies

Since research on full-scale QEC is still ongoing, quantum computers need techniques like these to reduce noise:

- ***Dynamical Decoupling:*** Applying pulses on a regular basis to offset interference from the environment.
- ***Techniques for Error Suppression:*** Circuit designs that are optimized to reduce decoherence and gate noise.

Among the more challenging issues with quantum computing is still creating fault-tolerant systems, which calls for ongoing improvements in software error-handling and hardware stability.

3.4 Quantum Computing Approaches

Analog quantum computing & gate-based quantum computing are the two important categories into which quantum computing can be generally divided. Both approaches utilize the unique concepts of quantum physics, yet they differ in how calculations are handled and executed [16].

3.4.1 Analog Quantum Computing

Analog quantum computing is dependent on ongoing quantum state changes, as opposed to digital quantum computing, and this employs discrete gate-based operations. The quantum system is guided toward its ideal ultimate state by gradually altering its energy landscape, which is determined by the Hamiltonian. As opposed to digital quantum computer technology, where qubits represent binary states (0 and 1), analog quantum computing allows qubits to remain in continuous states. This makes error correction considerably more difficult but provides for a more natural description of some issues.

For some applications, this method uses the physical interactions between quantum particles to carry out calculations in a more straightforward and effective manner. Key techniques in analog quantum computing include the following:

3.5 (I) Quantum Annealing

A particular type of analog quantum computing called quantum annealing was created mainly to address optimization issues. It effectively explores energy landscapes and determines the lowest-energy (optimal) solution to a given problem by utilizing quantum physics.

3.6 **Working Principle***:*

- A quantum system is started with a simple ground state.
- The Hamiltonian gradually changes from an initial state to the problem-specific Hamiltonian.
- The state that best resolves the problem is the final one at the end of the process.

3.7 Applications:

- ***Combinatorial optimization:*** It includes issues such as schedule optimization and the Traveling Salesman Problem (TSP).
- ***Machine learning:*** Neural network and clustering task optimization.
- ***Finance:*** Risk assessment and portfolio optimization.
- ***Supply chain and logistics:*** issues with routing and resource allocation.

3.8 *(Ii) Quantum Simulation*

One of the foremost important uses of analog quantum computing is quantum simulation. It enables researchers to examine and simulate intricate quantum processes that might be difficult or impossible to depict on standard computers.

3.9 Working Principle:

- A controllable quantum system, like superconducting circuits, ultra cold atoms, and trapped ions, imitates the characteristics of a target quantum system.
- To comprehend the target system's behaviour, the quantum system's evolution is monitored.
- The analog system offers extremely accurate insights because it adheres to the same physical principles as the simulated system.

3.10 Applications:

- ***Material Science:*** Knowledge of quantum phase transitions, high-energy materials, and superconductors.
- ***Chemistry:*** Molecular interaction simulation for catalysis and drug development.
- ***Nuclear Physics:*** Investigating quark-gluon plasma and unusual matter states.

3.11 *(Iii) Adiabatic Quantum Computing (AQC)*

AQC is deeply related to quantum annealing, but it has a wider range of applications. It provides a framework for solving problems beyond optimization by exploiting the gradual change in a quantum system's Hamiltonian.

3.12 Working Principle:

- The initialization of the system is done by the basic Hamiltonian's ground state.
- The Hamiltonian gradually changes into one that represents the answer to a computational issue.
- The system stays in the beginning state for the duration if the evolution is sufficiently slow, indicating that the last stage contains the right answer.

3.13 Applications:

- ***Graph theory:*** Resolving issues with connection and network architecture.
- ***Machine Learning*:** Training models and feature selection boosted by quantum mechanics.
- ***Artificial Intelligence:*** AI algorithms that use probabilistic inference and optimization.

3.13.1 Gate-Based Quantum Computing

Using quantum circuits made up of quantum gates that regulate qubits through unitary transformations, gate-based quantum computing employs a methodical approach to carrying out quantum computations. Unlike ordinary computers that comprehend bits as 0 or 1, quantum computers utilize superposition to allow qubits to exist in multiple states simultaneously. This allows for parallel computation, which significantly boosts processing power for some tasks. The final results, that disintegrates qubits into classical values, is retrieved through measurement after calculations are performed using a sequence of quantum gates.

The model depends on three fundamental quantum principles: superposition, which enables parallel computations; entanglement, which enables qubits to exhibit correlated contends regardless of distance and is integral to communication and error correction; as well as quantum interference, that enhances perfect results while cancelling out incorrect ones. These features work together to give gate-based quantum computing the ability to deal with issues in ways that are not possible with classical systems, especially in complex simulations, cryptography, and optimization. Quantum circuits are built using both single-qubit & multi-qubit gates. Single-qubit gates create superposition and change quantum phases, whereas multi-qubit gates enable entanglement and controlled operations. When combined, these gates allow quantum algorithms to be executed efficiently.

3.14 Quantum Algorithm Execution Process*:*

The following procedures are usually involved in executing a quantum algorithm:

Qubit Initialization: This involves setting up qubits in a specific initial state, typically $|0\rangle$.
Application of Quantum Gates: Putting a sequence of gate operations into practice to change qubit states.
Quantum Computation and Operations: This step uses entanglement and interference to guide the system toward the right answer.
Measurement: To obtain the final results, the quantum states are collapsed toward typical binary values (0 s and 1 s).

3.15 Challenges and Opportunities for the Future*:*

Gate-based quantum computing has a number of obstacles in spite of its revolutionary potential:

- ***Scalability Issues*** – One major challenge is to increase the overall amount of qubits while preserving their coherence and stability.
- ***Quantum Decoherence*** – Qubits eventually lose information because of their high sensitivity to background noise.
- ***Quantum Error Correction*** – Because of decoherence and noise, fault-tolerant quantum computing requires robust error correction mechanisms.

In order to overcome these constraints, scientists are currently investigating:

- QEC codes, like the Surface Code and Shor Code, enable **fault-tolerant quantum computing**.
- **Developments in quantum hardware**, including photonic quantum processors, superconducting qubits and trapped ions.

3.16 Advantages of Quantum Computing

Utilizing ideas from quantum physics to challenging computational problems that are outside the capabilities of traditional computers is the new field of "quantum computing". It has tremendous potential in many areas, such as cryptography, AI, optimization, and sustainability [17]. Figure 2 shows the main benefits of quantum computing, which are described in more detail below.

(i) ***Exponential Computational Power***
- Security and encryption are impacted by quantum algorithms' exponentially faster factoring of large numbers than standard algorithms, such as Shor's algorithm.

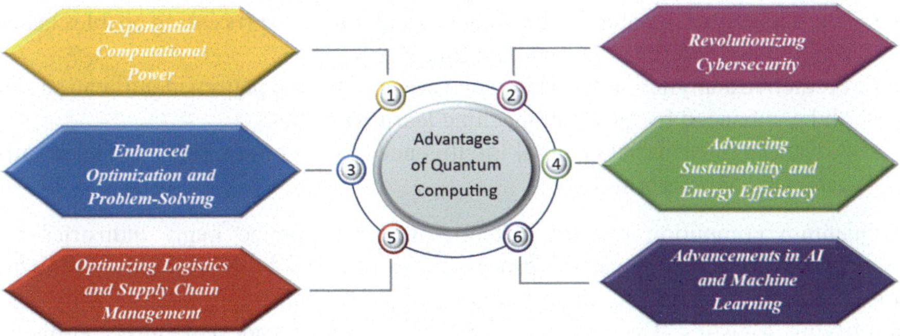

Fig. 2 Advantages of quantum computing

- Quantum system simulation is crucial in domains where classical simulations are inadequate, that are drug development, chemistry and materials science.

(ii) ***Enhanced Optimization and Problem-Solving***
- Solving intricate optimization problems is necessary for many real-world issues, including scheduling, logistics, and financial portfolio optimization.
- Large solution spaces may be efficiently explored by quantum computers thanks to quantum techniques like quantum annealing.
- ML and AI techniques can be enhanced by quantum computing, producing models for classification and regression tasks that are quicker and more accurate.

(iii) ***Revolutionizing Cybersecurity***
- Traditional cryptography could be threatened by quantum computing since it could potentially break elliptic curve and RSA encryption using Shor's method.
- Communication is made incredibly secure by quantum cryptography, such as QKD, that utilizes the laws of quantum physics to identify eavesdropping attempts.

(iv) ***Advancing Sustainability and Energy Efficiency***
- Compared to classical supercomputers, quantum computers need a lot less energy to do the same tasks.
- Quantum simulations possess the capacity to get better fuel efficiency, battery technology, energy storage, and material research.
- Quantum computing can improve resource allocation, farming sustainability, water management, and climate modeling.

(v) ***Optimizing Logistics & Supply Chain Management***
- By determining the best routes, timetables, and inventory control techniques, quantum algorithms can simplify logistical processes.
- More effective supply chain operations are advantageous for sectors including manufacturing, retail, and transportation.
- By modelling and forecasting supply chain disruptions, quantum computing improves resilience and catastrophe response.

(vi) ***Advancements in AI and ML***

- The capacity of quantum computing to accelerate data processing enables faster training and inference in deep learning (DL).
- Predictive analytics, natural language processing, and pattern identification can all be enhanced by quantum machine learning (QML).
- Advances in Artificial General Intelligence (AGI) could result from quantum-enhanced ML algorithms.

Quantum computing has the potential to revolutionize many industries by providing previously unheard-of processing power, resolving complex optimization problems, safeguarding digital communications, enhancing sustainability, and simplifying logistics. Even though hardware stability and scalability problems persist, ongoing research and development is expanding the potential applications of quantum computing.

4 Drawbacks of Quantum Computing

Although quantum computing is a ground-breaking technology with enormous promise, it also has a number of significant drawbacks that prevent its widespread use and useful application [18]. Figure 3 shows the main drawbacks that are described below.

(i) ***Hardware Fragility and Extreme Operating Conditions***
   - Cryogenic cooling systems are therefore required, which raises the expense and complexity.
   - The loss of a quantum state, or decoherence, occurs as a result of qubits' extreme sensitivity to external perturbations such as temperature changes, vibrations, and electromagnetic interference.
   - Controlling heat dissipation is a major problem since quantum systems produce heat that can affect dependability and performance.

(ii) ***Scalability and Stability Challenges***
   - Even though businesses like Intel, IBM, and Google have produced quantum devices with dozens of qubits (like Google's processor with 72 qubits), millions of qubits are likely required for notable quantum advantage in fields like ML, encryption, and chemical simulations.
   - Due to its intrinsic instability and proneness to errors, qubits have intricate error correcting techniques. For real error correction, current coherence times (such IBM's claimed 47 microseconds for a 27-qubit system) are insufficient.
   - Because they need exact calibration and specific conditions, quantum processors are challenging to test and maintain.

(iii) ***Limited Accessibility and High Cost***
   - Currently, only government agencies, research institutes, and significant technological companies have access to quantum computing resources. Innovation is hampered by the high expenses and limited access faced by smaller businesses and individuals.

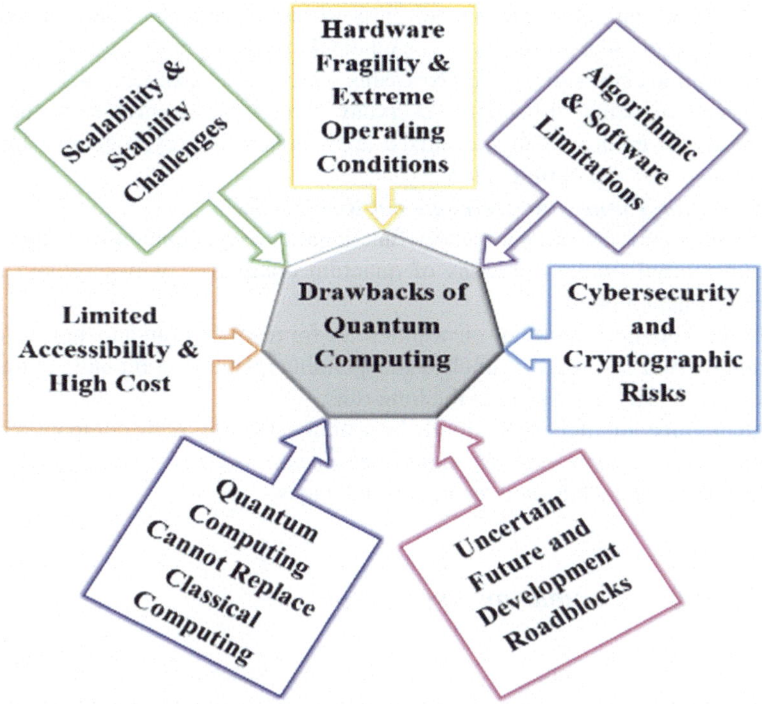

Fig. 3 Drawbacks of quantum computing

- Developing consumer-grade quantum computers is difficult due to the complex engineering and resource-intensive methods required to build and scale quantum hardware.

(iv) ***Algorithmic and Software Limitations***

- Creating algorithms for quantum computers necessitates a thorough comprehension of the extremely intricate and specialized field of quantum physics.
- Because there are no standard quantum programming languages or software tools, software development on different quantum computing platforms is uneven.
- Because quantum systems are non-intuitive, it is difficult to debug and enhance quantum algorithms.

(v) ***Cybersecurity and Cryptographic Risks***

- Quantum computers makes defense networks, government databases, banking institutions, and IoT security vulnerable.
- To mitigate these concerns, PQC technologies must be developed and put into use.

(vi) ***Quantum Computing Cannot Replace Classical Computing***

- Quantum computers are not all-purpose tools; while they are excellent at solving particular problems, they cannot take the place of traditional computers for routine chores like word processing, spreadsheets, and email.
- Instead of being utilized for everyday computing requirements, they will mostly be employed in specialized applications like cryptography, intricate simulations, and optimization.

(vii) ***Uncertain Future and Development Roadblocks***

- There is currently no completely functional universal quantum computer and only partial implementations of quantum computers, which remain in the experimental phase.
- Since there is currently no clear timescale for resolving fundamental hardware and software issues, it is unclear if expanding quantum computers to millions of qubits will be feasible in the long run.
- Quantum computers will initially be available to large-scale companies before making their way into the consumer market because to their high costs, infrastructure needs, and technical challenges.

4.1 Quantum Computing Applications

Using concepts via quantum mechanics like superposition as well entanglement, quantum computing has the possibility of completely transform a number of varied industries. AI, intricate chemical simulations, industrial optimization, and secure communication are just a few of the areas it can significantly enhance [19]. The quantum computing's main applications are showed in Fig. 4.

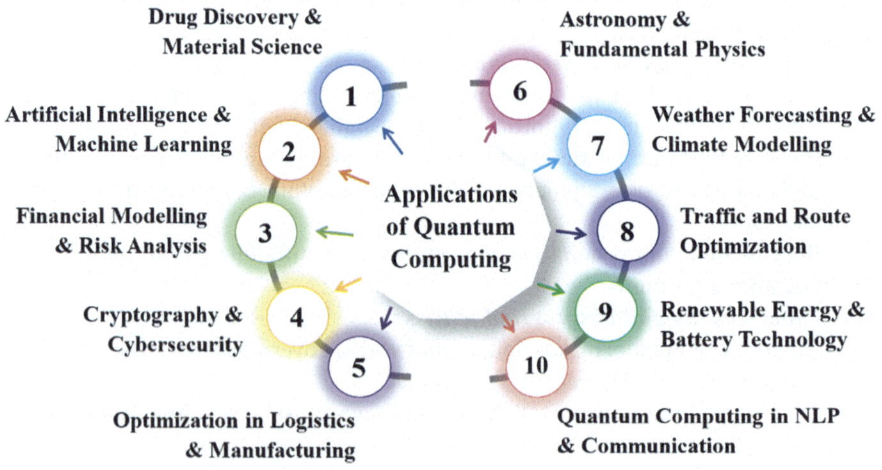

Fig. 4 Quantum computing applications

4.2 (I) Drug Discovery & Material Science

Through precise molecular interaction simulation, quantum computing can effectively simulate quantum systems, aiding in materials science and drug development. Businesses that investigate QC's potential in protein modelling and molecular behaviour analysis include Google, IBM, and AstraZeneca.

- ***Example:*** IBM's research helped with chemical and petroleum applications by demonstrating quantum modelling of beryllium hydride (BeH2) and lithium hydride (LiH).
- ***Impact on the Industry:*** ExxonMobil uses quantum computing to improve environmental modelling, optimize power grids, and create carbon capture materials.

4.3 (Ii) Artificial Intelligence and Machine Learning

By using statistical distributions more effectively than classical systems, quantum computers can speed up ML tasks.

- ***Example:*** IBM created a hybrid ML classification algorithm that combines quantum and classical methods.
- ***Research Breakthrough:*** When compared to traditional methods, Google's Sycamore quantum processor showed exponential gains in quantum learning agents.
- ***Application Areas:*** AI optimization, NLP developments, and automated decision-making in the healthcare and financial sectors.

4.4 (Iii) Financial Modelling and Risk Analysis

It is anticipated that quantum computing will improve risk management, portfolio optimization, and financial modelling.

- ***Example:*** A collaborative study by Multiverse Computing, Pasqal, and Crédit Agricole enhanced credit risk assessments and financial product valuations through the use of quantum-inspired methodologies.
- ***Industry Adoption:*** To predict markets and detect fraud, JPMorgan Chase and other international banks are investigating quantum computing.

4.5 (Iv) Cryptography and Cybersecurity

Quantum-enabled encryption ensures secure data transfer through methods like super dense coding and QKD.

- ***Prospects for the Future:*** Reducing the threats that quantum computing poses to established cryptography methods is the aim of PQC development.

4.6 (V) Optimization in Logistics and Manufacturing

Production procedures, resource optimization, & supply chain management are all enhanced by quantum computing.

- ***Example:*** Quantum-South used quantum annealers via Amazon Braket to optimize cargo loading for passenger aircraft.
- ***Automotive Industry:*** Businesses such as Volkswagen are investigating quantum computing for battery material research and traffic optimization.

4.7 (Vi) Astronomy and Fundamental Physics

By simulating molecular interactions, quantum computing can support astrophysical research on the Big Bang, black holes, and cosmology.

- ***Possible Findings:*** Refining space exploration technologies, forecasting the behavior of celestial objects, and comprehending dark matter and energy.

4.8 (Vii) Quantum Computing in NLP and Communication

AI-driven communication systems benefit from Quantum Natural Language Processing (QNLP), which improves text analysis and speech recognition.

- ***Example:*** QNLP toolkits for automated dialogue, bioinformatics applications, and language translation were created by Cambridge Quantum (now Quantinuum).
- ***Industry Collaboration:*** Merck Group used QNLP to enhance healthcare explainable AI and text classification.

4.9 (Viii) Weather Forecasting and Climate Modelling

Large-scale climate datasets can be analysed by quantum computers, improving weather forecasting and climate change mitigation techniques.

- ***Example:*** IBM and NASA are investigating quantum computing to improve atmospheric modeling and forecast disaster response.

4.10 (Ix) Renewable Energy and Battery Technology

Optimizing energy storage and integrating renewable energy sources is made easier by quantum computing.

- ***Example***: Manufacturers can improve the longevity and efficiency of lithium-ion batteries for electric vehicles by using quantum models.
- ***Impact***: Less dependence on fossil fuels and improved solar capture technologies.

4.11 (X) Traffic and Route Optimization

Real-time traffic data is processed by quantum algorithms to enhance logistics and route planning.

- ***Example:*** Volkswagen's research on quantum computing improves traffic flow control throughout the city.

There are still problems in quantum computing with QEC and scalability, ongoing research and collaborations between university researchers and industry executives are accelerating advancements and ensuring that quantum computing will be applied in practical settings.

4.12 Major Challenges in Quantum Computing

Since quantum computing may address difficulties that conventional computers cannot, it has the possibility to entirely alter computing. However, there are significant barriers preventing its actual use. These difficulties fall into four categories: workforce-related, algorithmic, software, and hardware-related [20]. Figure 5 shows the major challenges in quantum computing.

(i) ***Hardware Challenges***

(a) ***Quantum Decoherence and Noise Sensitivity***

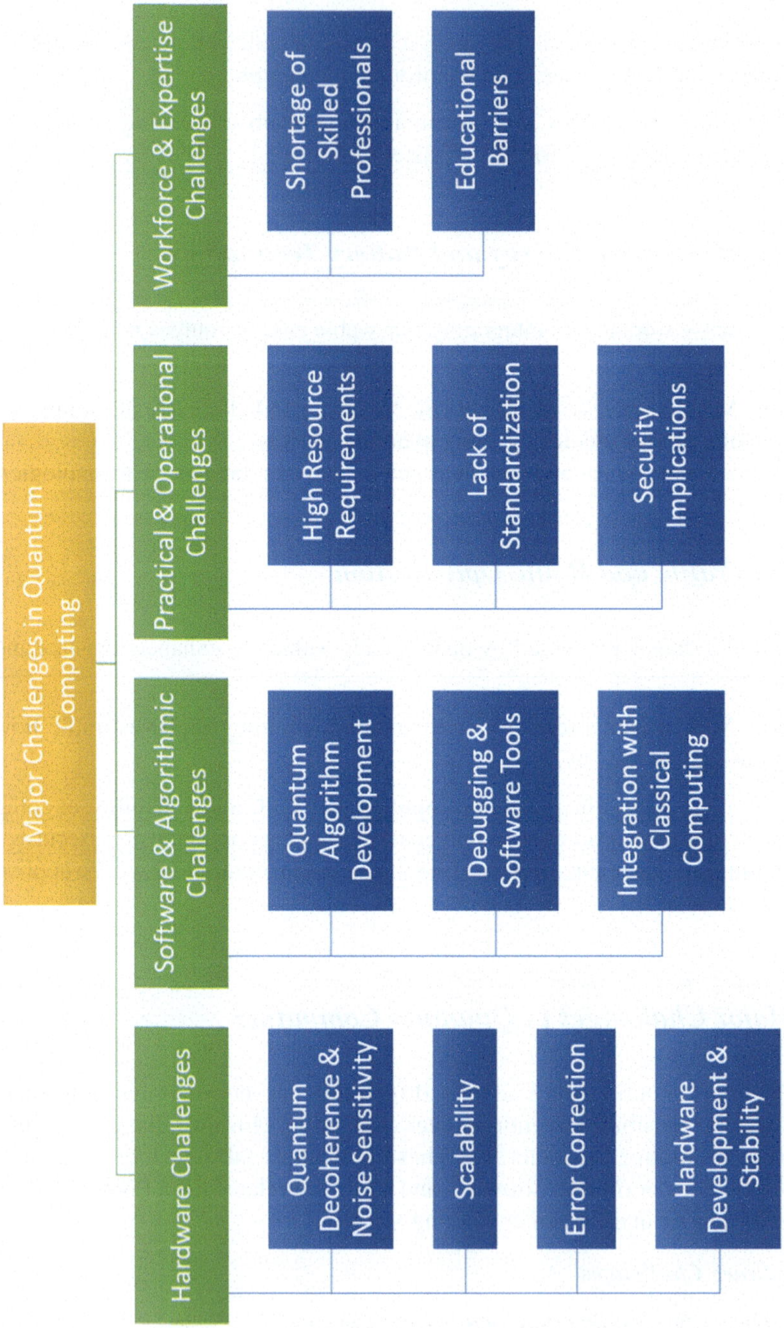

Fig. 5 Major challenges in quantum computing

- Qubits can decohere, or lose their quantum state, because of their fragility and heightened sensitivity to environmental noise.
- Qubits are more prone to errors in computations than classical bits because they are not inherently able to ignore noise.

(b) ***Scalability***

- There are only so many qubits available for use in modern quantum computers.
- Scaling is expensive and difficult because it requires specialized equipment and extremely low temperatures.

(c) ***Error Correction***

- Due to qubits' high error susceptibility, QEC is crucial.
- QEC requires extra qubits and resources, which makes implementation difficult and resource-intensive. In contrast, redundant bits in classical systems can detect and correct errors.

(d) ***Hardware Development and Stability***

- It is extremely difficult to create reliable, high-quality qubits because of the limitations of different quantum computing models.
- To ensure qubit stability, cooling systems, vacuum chambers, and exact control mechanisms are required, which raises the price and complexity.

(ii) ***Software and Algorithmic Challenges***

(a) ***Quantum Algorithm Development***

- Many real-world problems lack feasible quantum solutions, even with the existence of quantum algorithms.
- In contrast to classical methods, creating quantum algorithms necessitates a new paradigm of computation.

(b) ***Debugging and Software Tools***

- Quantum states cannot be directly observed or replicated without collapsing their superposition, in contrast to classical computers. Debugging becomes very difficult as a result.
- To test and improve quantum algorithms, new programming languages, simulators, and software tools must be created.

(c) ***Integration with Classical Computing***

- Real-world applications require a smooth transition between quantum and classical systems.
- To close the gap between these two computing models, frameworks for hybrid computing and effective data transfer must be created.

(iii) ***Practical and Operational Challenges***

(a) ***High Resource Requirements***

- Quantum computers are expensive and energy-intensive because they require extreme conditions, like laser-controlled ion traps or superconducting environments (near absolute zero).
- Most organizations cannot afford the infrastructure required to house and run quantum systems.

(b) ***Lack of Standardization***
- Interoperability and broad adoption of quantum computing are challenging due to the lack of universal standards or protocols.
- Different quantum computing models, like gate-based and annealing, make it more challenging to develop a coherent plan.

(c) ***Security Implications***
- Current cryptographic systems, especially those based on RSA and ECC encryption, are at risk from quantum computing since these systems could be cracked by quantum algorithms.
- The development of PQC will necessitate a global effort to secure digital communications during the transition.

(iv) ***Workforce and Expertise Challenges***

(a) ***Shortage of Skilled Professionals***
- Proficiency in computer science, electrical engineering, mathematics, and quantum mechanics is necessary for quantum computing.
- Research and development are being slowed down by the severe lack of professionals with training in these interdisciplinary fields.

(b) ***Educational Barriers***
- Due to its complexity, quantum computing is not frequently covered in traditional school curricula.
- Open-source frameworks and hands-on training programs must be developed to close the knowledge gap and prepare the quantum engineers' future generation.

Researchers and organizations around the world are actively trying to overcome these obstacles. For quantum computing to become feasible and widely available, developments in qubit stability, error correction, algorithm design, and workforce training will be essential.

5 Conclusion

A thorough explanation of quantum computing is given in this chapter, with a focus on its fundamental ideas, required parts, and uses. It describes how entanglement, superposition, and qubits allow for quicker problem-solving than traditional computers. Real-world uses in AI, cryptography, and drug discovery are discussed, along with a comparison between gate-based quantum computing with quantum annealing. Along with continuing research by large IT businesses, issues like scalability and mistake rates are addressed. The chapter emphasizes the need for consistent investment and collaboration to adequately explain the massive growth of quantum computing.

References

1. Sood, Vaishali, and Rishi Pal Chauhan. "Archives of Quantum Computing: Research Progress and Challenges." Archives of Computational Methods in Engineering. State of the Art Reviews, vol. 31, no. 1, 2024, pp. 73–91, https://doi.org/10.1007/s11831-023-09973-2.
2. Sidak Bawa. "Exploring Quantum Computing: Principles and Applications." Journal of Quantum Science and Technology, vol. 1, no. 3, 2024, pp. 57–69, https://doi.org/10.36676/jqst.v1.i3.27.
3. Oyeniran, Oyekunle Claudius, et al. "Advancements in Quantum Computing and Their Implications for Software Development." Computer Science & IT Research Journal, vol. 4, no. 3, 2023, pp. 577–593, https://doi.org/10.51594/csitrj.v4i3.1558.
4. Yang, Zebo, et al. "A Survey of Important Issues in Quantum Computing and Communications." IEEE Communications Surveys & Tutorials, vol. 25, no. 2, Secondquarter 2023, pp. 1059–1094, https://doi.org/10.1109/comst.2023.3254481.
5. Gill, Sukhpal Singh, et al. "Quantum Computing: A Taxonomy, Systematic Review and Future Directions." Software: Practice & Experience, vol. 52, no. 1, 2022, pp. 66–114, https://doi.org/10.1002/spe.3039.
6. Dejpasand, Mohamad Taghi, and Morteza Sasani Ghamsari. "Research Trends in Quantum Computers by Focusing on Qubits as Their Building Blocks." Quantum Reports, vol. 5, no. 3, 2023, pp. 597–608, https://doi.org/10.3390/quantum5030039.
7. Chae, Eunmi, et al. "An Elementary Review on Basic Principles and Developments of Qubits for Quantum Computing." Nano Convergence, vol. 11, no. 1, 2024, p. 11, https://doi.org/10.1186/s40580-024-00418-5.
8. Bhat, Hilal Ahmad, et al. "Quantum Computing: Fundamentals, Implementations and Applications." IEEE Open Journal of Nanotechnology, vol. 3, 2022, pp. 61–77, https://doi.org/10.1109/ojnano.2022.3178545.
9. Wang, Yan, et al. "Opportunities and Challenges of Quantum Computing for Engineering Optimization." Journal of Computing and Information Science in Engineering, vol. 23, no. 6, 2023, https://doi.org/10.1115/1.4062969.
10. Tripathi, Vinay, et al. "Benchmarking Quantum Gates and Circuits." arXiv [Quant-Ph], 2024, http://arxiv.org/abs/2407.09942.
11. Dsouza, Vishal, et al. "A Systematic Review on Quantum Computing: Concepts, Algorithms, Hardware and Applications." Algorithms, Hardware and Applications, 2024.
12. Krizhanovskyi, D.: An Overview of Quantum Algorithms: Detailed Analysis of the Deutsch-Jozsa and Grover's Algorithms. (2024). https://doi.org/10.2139/ssrn.4930685
13. Ebrahimpour-Komleh, H., Hadipour, A.: Quantum Computing in Image Processing: A Comprehensive Survey of Applications, Challenges, and Future Direction. Artificial Intelligence Applications and Innovations **1**(2), 9–25 (2024)
14. Blekos, Kostas, et al. "A Review on Quantum Approximate Optimization Algorithm and Its Variants." Physics Reports, vol. 1068, 2024, pp. 1–66, https://doi.org/10.1016/j.physrep.2024.03.002.
15. Kumar, Vinay, et al. "Quantum Internet: Technologies, Protocols, and Research Challenges." arXiv [Quant-Ph], 2025, http://arxiv.org/abs/2502.01653.
16. Rietsche, Roman, et al. "Quantum Computing." Electronic Markets, vol. 32, no. 4, 2022, pp. 2525–2536, https://doi.org/10.1007/s12525-022-00570-y.
17. Scholten, Travis L., et al. "Assessing the Benefits and Risks of Quantum Computers." arXiv [Quant-Ph], 2024, http://arxiv.org/abs/2401.16317.
18. Di Meglio, Alberto, et al. "Quantum Computing for High-Energy Physics: State of the Art and Challenges." PRX Quantum: A Physical Review Journal, vol. 5, no. 3, 2024, https://doi.org/10.1103/prxquantum.5.037001.

19. Bova, Francesco, et al. "Commercial Applications of Quantum Computing." EPJ Quantum Technology, vol. 8, no. 1, 2021, p. 2, https://doi.org/10.1140/epjqt/s40507-021-00091-1.
20. Natarajan, Gnanasankaran, et al. "Quantum Computers—Real-world Applications and Challenges." Quantum Computing and Artificial Intelligence, Wiley, 21 Mar. 2025, pp. 47–75, https://doi.org/10.1002/9781394242399.ch1.

Advancing AI with Quantum Computing: Synergies Between Quantum Neural Networks and Quantum Reinforcement Learning

Aanal Raval and Ruchi Oza

Abstract Quantum-enhanced computation enables decision-making and new paradigms for learning consequently transforming artificial intelligence (AI). This chapter explores the individual capabilities of Quantum Neural Networks (QNNs) and Quantum Reinforcement Learning (QRL) before delving into their synergies. QNNs enables advancements in deep learning applications leveraging quantum circuits to improve pattern recognition, optimization, and generative modeling. On other hand, QRL integrates quantum mechanics into reinforcement learning. This integration improves exploration, accelerate policy optimization, and improve decision-making in complex environments. First we establish foundational principles of QNNs and QRL and then provide a structured analysis of how these two fields can be combined. We examine theoretical synergies, such as using QNNs as function approximators in QRL and leveraging quantum-enhanced exploration strategies. Additionally, practical implementations, real-world applications, and current research frameworks are discussed. Finally, we highlight key challenges, including hardware limitations, scalability issues, and algorithmic constraints, while outlining future research directions. Furthermore, we explore practical implementations, including existing frameworks such as TensorFlow Quantum, Qiskit, and PennyLane, and discuss real-world applications in robotics, finance, autonomous systems, and drug discovery. A comparative analysis of classical and quantum reinforcement learning is presented, evaluating computation time, accuracy, and scalability. Additionally, we address key challenges such as quantum noise, decoherence, hardware limitations, and scalability issues, which hinder the widespread adoption of quantum AI. Finally, we outline future research directions, emphasizing advancements in error correction, hybrid quantum–classical models, and the long-term vision of generalized quantum AI. This chapter provides a comprehensive overview

A. Raval (✉)
Department of Computer Engineering, Gujarat Technological University, Ahmedabad, India
e-mail: aanalraval@gtu.edu.in

R. Oza
Independent Researcher, Ahmedabad, India
e-mail: ruchikoza@gmail.com

© The Author(s), under exclusive license to Springer Nature Switzerland AG 2025

A. Ghosh et al. (eds.), *Quantum Machine Learning in Industrial Automation*,
Information Systems Engineering and Management 65,
https://doi.org/10.1007/978-3-031-99786-0_14

of QNNs and QRL, highlighting their transformative potential in revolutionizing artificial intelligence.

Keywords Quantum computing · Quantum neural networks (QNNs) · Quantum reinforcement learning (QRL) · Quantum machine learning (QML) · Variational quantum circuits (VQCs) · Quantum AI · Quantum policy optimization · Quantum feature representation · Hybrid quantum–classical models · Quantum decision-making

1 Introduction

1.1 Overview of Quantum Computing in AI

Quantum computing has emerged as a transformative technology with significant potential across various domains along with AI. Quantum computing (QC) offers advantages in specific problems in handling complex optimization and learning tasks compared to classical computers. Leveraging on quantum mechanical principles such as superposition, entanglement, and quantum parallelism, QC can perform computations that may be infeasible for classical systems [1]. This synergy between quantum computing and AI has given rise to Quantum Artificial Intelligence (Quantum AI), encompassing areas such as quantum machine learning (QML), quantum neural networks (QNNs), and quantum reinforcement learning (QRL). It has been recently observed that quantum computing-related technologies, such as security with quantum key distribution (QKD), quantum communications, quantum optimization, quantum artificial intelligence (Quantum AI), quantum machine learning (QML), and quantum neural networks (QNNs), are attracting significant attention. Among the various AI and deep learning algorithms, reinforcement learning (RL) and deep reinforcement learning (DRL) have been actively considered for various autonomous control applications. The benefits of using QNN-based learning algorithms in RL models includes fast training, high scalability, and efficient learning parameter utilization. Integration of QNN based learning algorithms in RL models can efficiently utilize fast training, high scalability, and learning parameter.

Integration of quantum computing to machine learning (ML) models aim at high-dimensional data processing and optimization tasks [2]. A key class of QML algorithms relies on variational quantum algorithms (VQAs). VQAs are hybrid quantum–classical models are used to optimize learning processes facilitating classification, sequential learning, and reinforcement learning that harness quantum computational advantages while mitigating hardware limitations [3]. Quantum Approximate Optimization Algorithms (QAOA) is one of the notable QML approaches that can solve combinatorial optimization problems [4]. Leveraging quantum feature maps, quantum-enhanced kernel methods are being explored to improve supervised and unsupervised learning models [5]. QNN integrates parametrized quantum circuits

(PQCs)for processing and learning patterns from data [6]. QNNs utilizes quantum parallelism to improve training efficiency and scalability [7]. This makes QNNs highly suitable for applications such as pattern recognition, anomaly detection, and optimization tasks where classical methods face exponential scaling challenges. To automate the design of QNN architectures, Quantum Neural Network Architecture Search (QNNAS) is recent advancement in QNN research. QNNAS efficiently explores quantum circuit structures to improve model adaptability for various applications [8]. Quantum Reinforcement Learning (QRL) incorporates quantum computing to reinforcement learning to enhance learning efficiency and generalization capabilities [9, 10]. Leveraging quantum-enhanced policy optimization techniques, QRL models such as Quantum Centralized-Critic and Multiple-Actor (Q-CCMA) networks provide scalable solutions for multi-agent RL scenarios [11]. These models offer improved training speed and robustness against noisy data. Entangled Quantum Multi-Agent Reinforcement Learning (EQMARL) has emerged as a promising framework for distributed cooperation in quantum networks [12, 13]. The extensions of QNN-based RL models to multi-agent cooperation and coordination lead to direction of distributed autonomous multi mobility control. Multi-agent QNN-based RL models includes quantum centralized-critic and multiple-actor (Q-CCMA) network. The realization for need of modern noisy intermediate-scale quantum (NISQ) era arises from limitations of modern quantum computing platforms while implementing multi-agent RL.

1.2 Importance of Quantum Neural Networks (QNNs) and Quantum Reinforcement Learning (QRL)

Models like QNN and QRL depend on quantum principles such as superposition, entanglement, and quantum parallelism, making them highly efficient for tasks involving large-scale optimization and high-dimensional data processing [6, 7]. QNNs use parametrized quantum circuits (PQCs) which makes traditional neural network computations faster, generalized with reduced complexity. In contrast to NN which rely on gradient-based optimization, QNNs use quantum properties to enhance learning capabilities in high-dimensional search spaces [11]. Integration of quantum computing enables quantum parallelism to process multiple computations simultaneously accelerating learning processes [2]. To enhance model efficiency for tasks like image processing and NLP, quantum states can encode vast amounts of information using fewer qubits [13, 14]. QNNs exhibit notably good performance in noise-tolerant and probabilistic learning scenarios which make them ideal for anomaly detection and cryptographic applications [2]. Quantum Neural Network Architecture Search (QNNAS): a development of QNN, enables automated optimization of QNN and dynamic adaption to various data domains [8].

QRL extends capabilities of RL for faster convergence and improved exploration–exploitation trade-offs through policy optimization, value estimation, and multi-agent cooperation [9]. For faster policy optimization, quantum superposition enables agents to evaluate multiple actions simultaneously [11]. To enhance decision making, quantum entanglement allows reinforcement learning agents to consider interconnected states more efficiently [12, 15]. A framework in QRL called Quantum Centralized-Critic and Multiple-Actor (Q-CCMA) model enables distributed learning in quantum-enhanced reinforcement learning tasks. These are helpful in scenarios like autonomous mobility, quantum networking, and complex game-theoretic decision-making [11].

1.3 Motivation for Exploring Synergies Between QNNs and QRL

Key motivations for exploring synergies between QNN and QRl are outlined below:

- Real time computations: Traditional DL and RL algorithms face scalability and computational challenges in high-dimensional optimization problems. QNNS utilizes quantum parallelism and entanglement for complex AI tasks like sequential learning [2, 3, 14].
- Efficiency: Unlike RL that rely on iterative updates for optimization, QRL accelerates learning through quantum superposition and faster optimization techniques [9, 11, 15]. Incorporation of QNNs into QRL frameworks enables quantum feature extraction and faster state-space exploration [7, 8].
- Training Bottlenecks in QNNs: A major challenge in QNNs is the presence of barren plateaus—regions in the optimization landscape where gradients vanish, making training difficult [16]. This issue can be mitigated through QRL's policy based approaches that optimizes reward-driven training strategies. These strategies allow QNNs to escape poor local minima and improve generalization capabilities [5, 12].
- Realization of Multi-Agent and Distributed AI: Integration of QNNs with QRL can enhance enables quantum-enhanced coordination, communication, and decision-making to enhance multi agent RL [9, 12]. Noteworthy scalability and robustness has been observed with entangled quantum multi-agent RL (EQMARL) that exhibits the potential for distributed cooperation over quantum channels [9].
- Quantum-Assisted Neural Architecture Search (QNNAS): Neural Architecture Search (NAS) has been critically emerging area for automated discovery of optimal network architectures [8]. This optimizes the search space and computational cost. Integrating this with QRL can dynamically adapt QNN architectures for specific learning tasks [8, 13].
- Secure AI: Integration of QNN with QRL can ensure robust cryptographic protocols, secure multi-agent interactions, and improved generalization of AI models in adversarial environments [6, 9].

2 Fundamentals of Quantum Neural Networks (QNNs)

2.1 Introduction to Quantum Neural Networks

QNNs extend classical neural networks through usage of quantum states and quantum gates to process and encode information. Classical neurons rely on weighted sum operations followed by activation functions. QNNs in order to use unitary transformations, employ quantum circuits through qubits [14]. These circuits can now process information in a high-dimensional Hilbert space with increased expressiveness and computational speedups [3]. A high-dimensional Hilbert space mathematically describes quantum states in quantum mechanics and quantum computing. A Hilbert space is a generalization of Euclidean space with infinitely many dimensions where each quantum state is represented as a vector. Quantum states or n qubit can exist in 2^n dimensional space which can superposition as linear combinations of basis states, can have an inner product to allow measurement and probability calculations, and a tensor product which can create high dimensional spaces for quantum entanglement. A typical QNN consists of:

1. Quantum Encoding Layer: To translate normal data to quantum states using feature maps or quantum embeddings.
2. Quantum Processing Layer: to apply unitary transformations like quantum gatesto process encoded data.
3. Quantum Measurement Layer: To measure quantum state in order to extract relevant information.

2.2 Quantum Circuit-Based Neural Architectures

The smallest building block of a quantum neural network is the quantum perceptron [17]. A quantum perceptron can be defined as an arbitrary unitary operator with i input qubits and o output qubits. An arbitrary unitary is applied to the $i + o$ input and output qubits depending on $(2^{i+o})^2$ - 1 parameters. The input qubits are initialised to a random mixed state ρ^{in}. The output qubits assume the state given by $|O...O\rangle_{out}$. QNN architecture is composed of perceptron unitaries applied from top to bottom layer wise [17]. The mixed output state ρ^{out} corresponding to a probabilistic mixture of pure states is defined as per Eq. (1). U is the quantum NN given by Eq. (2). U^1 are layer unitaries. These layer unitaries are made up of product of quantum perceptrons in layers l-1 and l. Figure 1 shows the architecture of QNN.

$$\rho^{out} \equiv \text{train}_{in}, \text{hidden}\big(\text{U}\big(\rho^{in}|\text{O}\ldots\text{O}\rangle\text{out}, \text{hidden}\,\text{O}\ldots\text{O}|\big)U+\big)[17] \tag{1}$$

$$U \equiv U^{out}U^{L}U^{L-1}U^{1} \tag{2}$$

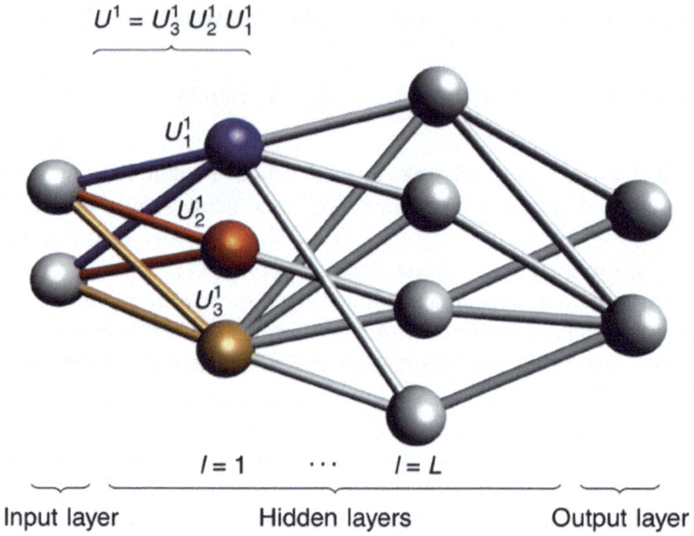

Fig. 1 Architecture of QNN perceptron unitaries layer wise from top to bottom. Sequence of unitary application is from violet, followed by orange and then yellow [17].

2.3 *Variational Quantum Circuits (VQCs) and Hybrid Quantum–Classical Models*

2.3.1 Variational Quantum Circuits (VQCs)

VQCs are parameterized quantum circuits designed to solve optimization problems. Variational quantum circuits employ rotation operator gates with free parameters. All quantum gate operations are reversible linear operations, which means a neural network's activation function can be replaced by entanglement layers. VQC starts with encoding input values into an appropriate qubit state followed by transformation of qubit by parametrized rotation gates and entangling gates and finally measuring transformed qubit state through expected value of the system's Hamiltonian operator. This expected values is decoded to an appropriate output and parameters of QNN are updated by relevant classical optimizer [18]. Figure 2 shows VQC structure. The workflow of a VQC can be summarized as:

1. Initialization Phase: Initial set of parameters are defined and a quantum state is prepared.
2. Parameterized Evolution: Perform transformations through quantum gates depending on trainable parameters
3. Extract: The final quantum state is measured to extract relevant information.
4. Optimization: A classical optimizer such as gradient descent or evolutionary algorithms updates the circuit parameters based on a cost function.

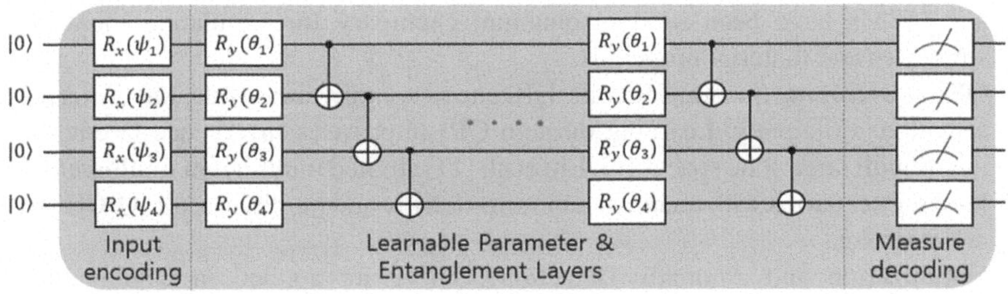

Fig. 2 VQC working structure showing data processing [18]

2.3.2 Hybrid Quantum–Classical Models

Pure quantum algorithms are impractical for large-scale applications as limitations exist for quantum hardware. Hybrid quantum–classical models address this challenge by dividing computation between quantum and classical systems. These models utilize both quantum circuits for parallelism and classical processors for efficient parameter optimization as well as data processing. A hybrid model consists of:

1. Quantum component: For processing quantum data and quantum feature transformations, VQC is used.
2. Classical Component: A classical neural network or optimizer updates the quantum circuit parameters based on a loss function.

2.4 Applications of QNNs in Machine Learning

Integrating high-dimensional Hilbert space for efficiency, scalability, and expressivity [6], QNNs offer advantages over classical neural networks. QNNs are increasingly being explored for various machine learning applications, including classification, generative modeling, reinforcement learning, and optimization.

- Classification: QNNs use quantum feature maps that enhance the separability of data in a quantum Hilbert space providing advantageous support to classification tasks. VQCs can learn complex decision boundaries in high-dimensional data spaces. Chen [9, 11] explored Quantum convolutional neural networks (QCNNs) for medical image classification and remote sensing. Park and Kim [15] tested Hybrid quantum–classical architectures for NLP tasks like sentiment analysis and speech recognition.
- Generative modelling: Here quantum generative adversarial networks (QGANs) are used [19]. These models use quantum circuits to generate data distributions that mimic real-world datasets. QGANs can generate high-quality synthetic images with fewer parameters than classical GANs [12]. Chen [9] explored Quantum Boltzmann Machines (QBMs) to create complex distributions efficiently. QGANs

and QBMs have been used in quantum chemistry for simulating molecular structures and material properties.

- QRL: Leveraging quantum circuits, QRl encodes and processes state-action representations efficiently. Learning through QRl improves convergence in environments with large state spaces [6]. Liu et al. [11] studied multi-agent reinforcement learning scenarios with quantum centralized-critic and multiple-actor (Q-CCMA) architectures.
- Optimization and Anomaly Detection: QNNs are applied in combinatorial optimization tasks for supply chain optimization and portfolio management.

2.5 Challenges and Limitations of QNNs

QNNs has challenges and limitations that hinder their practical deployment. These challenges span hardware constraints, algorithmic limitations, and theoretical uncertainties.

- Data Encoding Bottleneck: QNNs face challenges in encoding classical data into quantum states. To encode data quantum features map needs to be created that introduce overhead and require complex circuit implementations [2].
- Hardware limitations: Despite significant advancements, quantum computers remain highly susceptible to noise and decoherence. The complexity of QNN models is affected by number of qubits, fidelity of quantum gates and coherence times [20]. Recent area of research and challenge includes fault-tolerant quantum computing and QNN scaling [2].
- Optimization in VQC: VQCs require optimization over a high-dimensional parameter space. Training these models suffers from issues such as barren plateaus, where the gradients of the cost function vanish exponentially with the system size, making training inefficient. This problem increases with depth of NN and limits ability to scale for larger tasks [19].
- Hybrid Quantum–Classical Integration Issues: In hybrid quantum–classical models, quantum circuits handle specific computations while classical optimizers adjust parameters. However, this integration introduces inefficiencies due to quantum–classical communication bottlenecks. Additionally, mapping classical problems into quantum-representable forms requires domain expertise and careful algorithmic design [19].
- Generalization and Interpretability: Classical neural networks use training heuristics, regularization techniques, and interpretability tools. Whereas QNNs lack comprehensive methods for understanding their decision-making processes and behavior across different datasets reducing generalizability [20].
- Scalability Concerns: To scale a network, exponential growth is required for qubits in larger networks. This scaling can impose hardware constraints [2].

3 Foundations of Quantum Reinforcement Learning (QRL)

3.1 Basics of Classical Reinforcement Learning (RL)

Reinforcement Learning (RL) is a machine learning paradigm. The core part here are the agents which makes sequential decisions through interaction with an environment. The goal is to learn decision making by trial and error maximizing cumulative rewards. These have wider application domains like robotics, game playing, autonomous systems, and various optimization problems [10]. RL framework consists of the following key components [21]. RL learns an optimal policy to maximize the expected cumulative reward as per Eq. (3). γ $(0 \leq \gamma \leq 1)$ indicates the discount factor showing importance of future rewards.

$$\mathrm{G_t} = \sum_{k=0}^{\infty} \gamma^k R_{t+k} \tag{3}$$

- Agent: the entity that learns and makes decisions to maximize rewards.
- Environment: The external system with which the agent interacts and percepts.
- State(s): Representation of environmental situation in given time period.
- Action (a): Decision made by agent to maximize rewards that affect environment.
- Reward (r): feedback evaluating the agent's action
- Policy (π): Strategic mapping of states to actions
- Value Functions (V): Estimated long term rewards from a given sate.
- Q function (Q): An expected reward corresponding to given state and action.

Classical RL methods can be broadly categorized into three types:

1. Value-Based Methods: In this type, the action selection is done through value functions. Q learning is one of most popular algorithm updates Q-values iteratively using the Bellman equation as per Eq. (4). α is the learning rate and s′ is the next state after taking action a.

$$\mathrm{Q(s, a) \leftarrow Q(s, a) + \alpha[r + \gamma a' max_{\mathit{a}} Q\big(s', a'\big) - Q(s, a)]} \tag{4}$$

2. Policy-Based Methods: Instead of using value functions. Policy-based methods directly optimize the policy. These methods use gradient-based optimization to improve performance
3. Actor-Critic Methods: Actor-Critic methods combine value-based and policy-based approaches. The actor updates the policy, while the critic evaluates it using a value function. Notable examples include Advantage Actor-Critic (A2C) and Proximal Policy Optimization (PPO) [22].

3.2 *Quantum Principles in RL (Quantum Superposition, Quantum Parallelism)*

The development of Quantum Reinforcement Learning (QRL) through integration of reinforcement learning with quantum principles leverages the unique properties of quantum mechanics to enhance learning efficiency and scalability.

1. Quantum Superposition in RL

In RL, the state of agent is represented as a definite value in a high-dimensional space (Fig. 3). The principle of quantum superposition allows a quantum system to exist in multiple states simultaneously. If a measurement collapses, the state assumes a single outcome. Basic units of quantum information are qubits. Classical bits can assume value either 0 or 1, whereas qubits can exist in multiple states at once due to a property called superposition. This makes quantum computer more powerful than traditional ones by handling vast amount of data at same time. Mathematically, a qubit (quantum bit) can be in a superposition of two basis states $|0\rangle$ and $|1\rangle$: $|\psi\rangle = \alpha|0\rangle + \beta|1\rangle$, where $|\alpha|^2 + |\beta^2 = 1$. A quantum agent uses Hadamard gates to create superposition. This enables the agent to explore across multiple states at once: $H|0\rangle = 1/\sqrt{2}\ (|0\rangle + |1\rangle)$. This enables faster policy updates and a more efficient search for optimal actions. This approach has been explored in Quantum Q-learning and Quantum Policy Gradient methods, where superposition enables faster convergence [9, 15].

2. Quantum Parallelism in RL

Classical RL uses sequential computation in which an agent updates its value functions one step at a time. In contrast, quantum parallelism allows a quantum system to process multiple computations at once by leveraging quantum gates that operate on superposed states. A quantum system applies a unitary transformation U to a superposed state: $U|\psi\rangle = U(\alpha|0\rangle + \beta|1\rangle) = \alpha U|0\rangle + \beta U|1\rangle$. This enables simultaneous evaluation of multiple state-action transitions in QRL. Quantum parallelism improves RL efficiency by accelerating key operations like updation of Q values, parallel policy evaluation and bat6ch learning acceleration. Quantum circuits can process multiple Q-values at the same time, thus evaluating multiple policies in parallel. This enables faster policy optimization. Traditional RL uses large mini-batches for training. Quantum algorithms such as Quantum Generative Adversarial

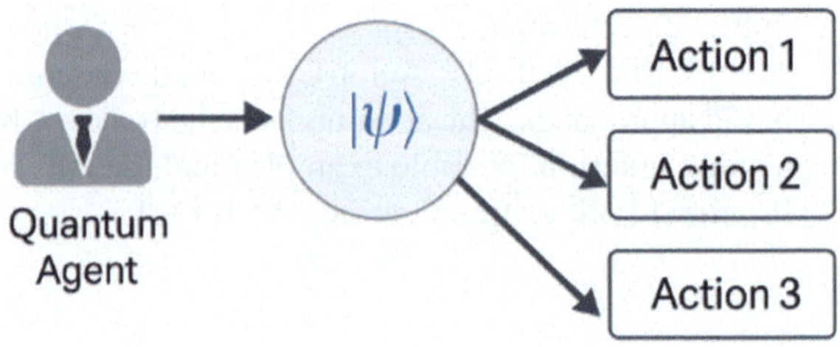

Fig. 3 Superposition concept for agents' multiple actions

Imitation Learning (QGAIL) exploit parallelism to process entire batches of experience simultaneously. A key example is Quantum Grover Search, which provides quadratic speedup in finding optimal actions: $O(\sqrt{N})$ vs. O(N) (classical search time complexity).

3. Quantum Entanglement and Quantum gates:

   Quantum Entanglement is a phenomenon in which qubits become interconnected. Due to interconnection of qubits, the state of one qubit spontaneously affects the state of another one, irrelevant of how far they are. This interconnectedness provides a more efficient as well as faster processing of information. The basic module of quantum circuits are quantum gates. These manipulate qubits for further calculations. Some of them are: Pauli-X flips the state of a qubit, Hadamard creates superposition and CNOT entangles two qubits [23].

3.3 Quantum Policies and Quantum Value Functions

For policy optimization and value function estimation, quantum computing has been integrated to reinforcement learning (RL). Quantum principles such as superposition, entanglement, and quantum parallelism are used to enhance decision-making efficiency in Quantum Policies and Quantum Value Functions. These quantum enhancements with quantum states and unitary operations allow agents to evaluate multiple state-action pairs simultaneously for faster learning and improved generalization.

Quantum Policies: In reinforcement learning, the policy is a function that maps states to actions. This decides the behavior of agent in an environment. In classical RL, policies are represented as either deterministic ($\pi(s) = a$) or probabilistic mappings ($\pi(a|s) = P(A = a|S = s)$). As Quantum Reinforcement Learning (QRL) policies use quantum states and unitary operations, the agents can explore multiple actions simultaneously. Using superposition, agent maintains a quantum state that represents a probability distribution over multiple actions by Eq. (5). U_π is a unitary operation to transform the policy state: ($|\psi_\pi'\rangle = U_\pi|\psi_\pi\rangle$.) This enables the agent to evaluate and sample multiple actions in parallel, resulting in an improved exploration and reduced sample complexity. To enable correlated action selections across different states, quantum entanglement is used. Especially when multi agent scenarios enter with Quantum Multi-Agent Reinforcement Learning (QMARL) [12], entanglement can achieve faster convergence and better coordination in complex environments.

$$|\psi_\pi\rangle = a \sum c_a|a\rangle. \tag{5}$$

For implementation of quantum policies, Variational Quantum Circuits (VQCs) are applied. With optimized quantum gates, a classical state is encoded as a quantum state with input encoding layer. Unitary transformations $U(\theta)$ modify quantum states based on learnable parameters through parameterized quantum gates. And finally, the measurement layer determines the action to be taken by the collapsed quantum state.

These quantum circuits are trained using gradient-based optimization techniques like Quantum Policy Gradient (QPG). The policy updates are handled by quantum gradients [11].

Quantum Value Functions: Value function gives the expected value reward from a given state or state-action pair. In traditional RL, this function is defined by Eq. (6). Using action- value function it can be redefined as per Eq. (7). For evaluation of multiple state-action values simultaneously, quantum value functions use quantum parallelism. Thus instead of estimating Q(s,a) for single action at a time, a quantum processor can estimate values for all actions in parallel. This speeds up the policy evaluation. As stated by [9], Bellman equation can be updated through quantum computing as: $Q(s,a) \approx UQ|s,a\rangle$, where U_Q is unitary matrix representing the transition dynamics. This in turn enables quantum circuits to efficiently approximate Q-values across multiple state-action pairs in a single step. To find best action, Quantum algorithms like Grover's search are explored. Grover's search provides a quadratic speedup in finding the best action by amplifying the probability of the optimal action. Unlike classical evaluation of Q values, Grover's algorithm helps identify the action with the highest expected reward in $O(\sqrt{N})$ time complexity [6]. This approach can accelerate value function computations in large action spaces as per Eq. (8).

$$V(s) = E\left[\sum_{t=0}^{\infty} \gamma^t r_t | S_o = s\right] \tag{6}$$

$$Q(s, a) = E\left[\sum_{t=0}^{\infty} \gamma^t r_t | S_o = s, A_0 = a\right] \tag{7}$$

$$a* = \arg\max_a Q(s, a). \tag{8}$$

Quantum Agents and Environment Interactions: A quantum agent's state $|\psi_A\rangle$ can be represented by Eq. (9), where $|si\rangle$ are possible states the agent can occupy, ci are complex probability amplitudes satisfying $\sum|c_i|^2 = 1$. The agent evolves via unitary transformations based on quantum gates along with dynamic policy updates: $|\psi A'\rangle = U|\psi A\rangle$. U is a quantum gate matrix representing learning updates. Similar to feedback to agent's actions, in quantum enhanced environments, both state transitions and reward functions can be quantum-encoded. This allows faster computation and more efficient optimization. A quantum environment can be modelled as per Eq. (10). This modelling uses density matrices instead of classical Markov decision processes (MDPs). ρ_{env} represents the mixed quantum state of the environment and p_i is the probability of the environment being in state $|s_i\rangle$. Here this interaction is monitored by a quantum evolution in Eq. (11). U_{env} is a quantum unitary matrix for evolution of environmental states based on the agent's actions. Instead of scalar values of reward, QRL store them as quantum registers. This allows multiple potential reward values to be evaluated in parallel as per Eq. (12). This enables faster reward evaluation and policy updates via quantum measurements.

$$|\psi_A\rangle = \sum i c_i |s_i\rangle \tag{9}$$

$$\rho_{env} = \sum i p_i |s_i\rangle\langle s_i| \tag{10}$$

$$\left|\psi'_{env}\right\rangle = U_{env}|\psi_{env}\rangle. \tag{11}$$

$$|\psi_R\rangle = \sum i r_i |r_i\rangle \tag{12}$$

The agent-environment interaction in a quantum setting follows a modified Quantum Markov Decision Process (QMDP) framework. This interaction cycle involves:

1. Quantum State Preparation: The agent and environment begin with superposition states. The quantum circuit encodes initial conditions into quantum registers.
2. Quantum Action Selection: The agent applies quantum gates to evolve its state, selecting actions probabilistically. Then Actions are encoded as quantum operations.
3. Quantum Environment Response: The environment processes the agent's action using quantum unitary transformations. State transitions are modeled as quantum state collapses.
4. Quantum Measurement & Reward Processing: Measurement is applied to determine the resulting environment state and reward. The reward function is quantum-evaluated in parallel for multiple action choices.
5. Quantum Policy Updates: The agent updates its policy using a quantum back-propagation method, such as Quantum Policy Gradient (QPG) or Quantum Q-Learning (QQ-Learning).

3.4 Applications of QRL in Decision-Making, Robotics, and Finance

With enhanced problem-solving capabilities and better handling of complex environments, applications of QRL can be seen in multiple domains, including autonomous systems, financial modeling, and strategic decision-making (Fig. 4).

1. Quantum RL for Decision-Making: The property of quantum parallelism can be particularly useful in Supply chain optimization (such as optimizing inventory management), Healthcare decision-making (such as personalized treatment planning) and Strategic planning in dynamic environments like military logistics and resource allocation.
2. Quantum RL in Robotics: With enhanced decision making, applications in RL include faster route planning / obstacle avoidance in self-driving cars, assembly line automation in industrial robots and efficient search-and-rescue missions in

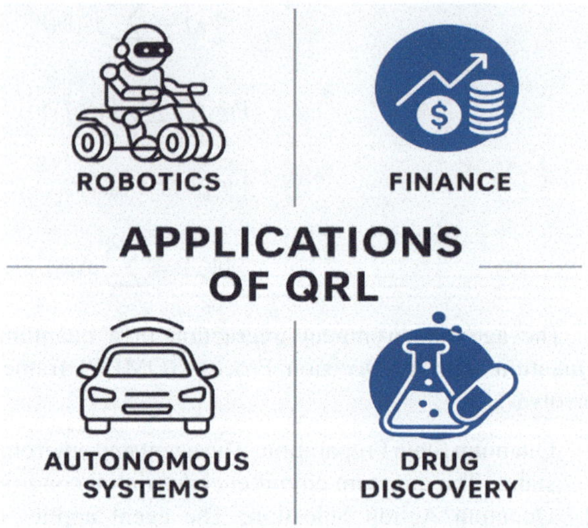

Fig. 4 Application domain of QRL

drones. Using a quantum Hamiltonian formulation, multiple flight paths can be exploring simultaneously in real time. Quantum-enhanced trajectory planning allows for adaptive responses in unpredictable environments, improving robustness.

3. Quantum RL in Finance: Quantum Grover's search achieves a quadratic speedup in finding and evaluating the optimal asset allocation in parallel. This enables faster financial decision-making by optimizing investment strategies under uncertainty. Applications include Portfolio optimization (selecting the best asset mix), Fraud detection (identifying anomalies in real-time) and Algorithmic trading (optimizing trading strategies using quantum-enhanced risk assessment).

3.5 Challenges in QRL: Noisy Quantum Hardware, Scalability Issues

1. Quantum Decoherence: Decoherence is a phenomenon where quantum states degrade due to interactions with the environment. This is caused generally due to gate errors that introduce noise in quantum circuits. This leads to incorrect QRL policy updates. As a part of its possible solutions, techniques like surface codes help reduce noise. This includes Quantum Error Correction (QEC) for stable learning. Another solution includes utilization of Hybrid Quantum–Classical Models. Integrating quantum processing with classical reinforcement learning mitigates errors.
2. Scalability of qubits: Current quantum computers have low qubit counts which in turn restricts the complexity of QRL models. Large-scale QRL applications require thousands of qubits, which are not yet available. The usage

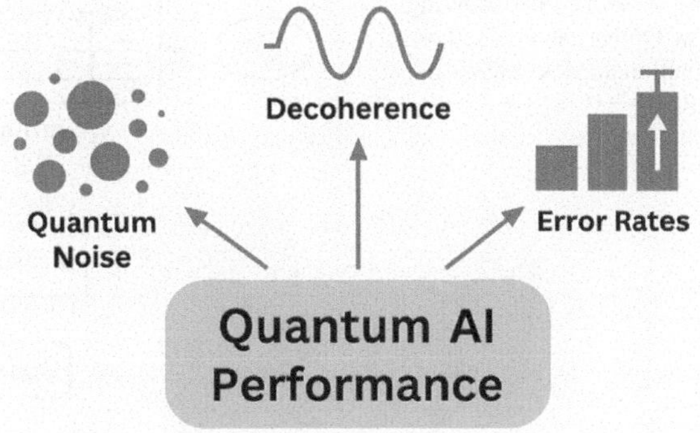

Fig. 5 Parameters affecting quantum AI performance

of Variational Quantum Circuits (VQCs) here reduces qubit requirements by optimizing quantum resources. Apart, if qubits are efficiently encoded, state space complexity can be reduced, these are explored as Quantum Compression Techniques.

QRL presents groundbreaking applications in decision-making, robotics, and finance. Figure 5 shows parameters affecting quantum AI performance. Though QRL offers speedups over classical RL, hardware limitations and scalability issues remain key obstacles. As highlighted, these issues can be solved by focusing on improving quantum error resilience. This requires the development of hybrid QRL architectures and quantum kernel methods. For this, the exploration for synergies between QNN and QRL remains an inevitable and challenging topic with its complexities.

4 Bridging QNNs and QRL: Theoretical Synergies

4.1 How QNNs Can Be Used as Function Approximators in QRL

Quantum Neural Networks (QNNs) can efficiently approximate value functions, policies, and reward functions in quantum environments. This can assist with crucial role in Quantum Reinforcement Learning (QRL) where complex and high-dimensional nature of quantum state spaces exists. QNNs provide an efficient way to model and generalize the relationship between quantum states and optimal actions.

As its known traditional RL function approximators such as Deep Neural Networks (DNNs) includes Q-values (Q(s,a)) (with Q learning), Value functions (V(s)) (in policy-based RL) and Policy functions (π(s,a)) (in policy gradient methods). QRL face challenge of extending these approximations to quantum states ($|\psi\rangle$) and quantum operations. This could consist of parametrized quantum circuits (PQCs)

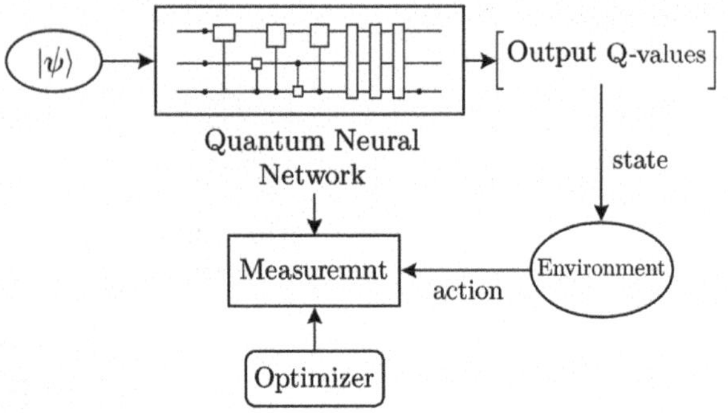

Fig. 6 Work flow showing how QNNs can be used as function approximators in QRL

which serve as quantum-native function approximators. These leverages on principles of quantum superposition and entanglement for more efficient learning. QNNs approximate three key functions in QRL: Q function, policy and value function. Yang et al. [24] provides critical insights into their theoretical foundation and implementation. QNNs in QRL are implemented using Variational Quantum Circuits (VQCs), which consist of parameterized quantum gates optimized through classical gradient-based methods. These circuits serve as function approximators by learning quantum representations of policies, value functions, and Q-tables [24]. Figure 6 shows the work flow.

Universal Approximation Theorem for QNNs: The work [24] studied QNN and derived that like classical neural networks, QNNs can approximate any continuous function with arbitrary precision. For this task, a sufficient number of qubits and circuit depth is used. The mapping from input states to output values can be expressed as Eq. (13). U_θ is a parameterized quantum unitary transformation. ρ_x is the quantum state encoding the input. O is the measurement observable.

$$f_\theta(x) = \mathrm{Tr}\left(U_\theta \rho_x U_\theta^\dagger O\right) \tag{13}$$

Quantum Feature Embedding for Function Approximation: Instead of classical activation functions, QNNs rely on quantum feature maps and kernel methods to encode data into high-dimensional Hilbert spaces. Feature encoding uses quantum unitary transformations such as: $U_\phi|0\rangle^{\otimes n} = |\psi_{enc}(x)\rangle$. This transformation ensures non-linearity and expressivity, crucial for approximating complex functions.

1. Q-Function Approximation: In Q learning, the Q-function estimates the expected future reward in order to action a in state s. A QNN can be used to approximate the Q-function by encoding states and actions into a quantum circuit as per Eq. (15). F_θ is a QNN with parameters θ. Quantum gates learn the optimal mapping between quantum states and Q-values. Introducing QNN here introduces the advantage of quantum parallelism where QNNs evaluate multiple state-action pairs simultaneously. Along with parallelism, quantum states naturally encode high-dimensional data. This leads to better generalization. Equation (14) can be modified and rewritten to Eq. (15) showing the Q value function. $|\psi_{enc}(s,a)\rangle$ is the

quantum feature encoding of state-action pairs. Quantum Bellman Update can be implemented using variational quantum gradient descent. This quantum formulation reduces the number of required parameters while enhancing efficiency, exploiting quantum parallelism for faster function approximation.

$$Q(s, a) \approx f_{\theta}(|\psi_s\rangle, |a\rangle) \tag{14}$$

$$Q\theta(s, a) = \langle\psi_{enc}(s, a)|U_{\theta}^{\dagger} O U_{\theta}|\psi_{enc}(s, a)\rangle \tag{15}$$

2. Policy Approximation: The classical softmax function uses policy based QRL where the policy function π(s,a) determines the probability of choosing action an in state s. A QNN-based policy model can be formulated as per Eq. (16). U_θ is the unitary operator of quantum circuit which encodes the policy. This quantum policy approximation enables efficient action selection using quantum measurement. This exponentially speed ups the process as quantum states store information more efficiently. Apart, due to superposition, sampling across multiple actions can be done simultaneously. Again this principle enhances the exploration. Yang et al. [24] approximated the policy function as a parameterized quantum circuit by Eq. (17). U_θ is a variational circuit with tunable quantum gates. O is an observable for measurement. ρ_s is the quantum state representation of the classical state. This approach leverages quantum entanglement to encode correlations between actions and states, leading to better exploration strategies in RL [24].

$$\pi(a|s) = |\langle a|U_{\theta}|\psi_s\rangle|^2 \tag{16}$$

$$\pi_{\theta}(s) = \text{Softmax}\left(\text{Tr}\left(O U_{\theta} \rho_s U_{\theta}^{\dagger}\right)\right) \tag{17}$$

3. Value Function Approximation: The expected cumulative reward from state s gives the value function V(s). QNNs can approximate this function using parametrized quantum circuits with Eq. (18). H_θ is a Hamiltonian operator learned during training. Here the rewards are quantum encoded which allows efficient learning in quantum environments. Modelling value functions through quantum gates makes model function more smoothly than classical networks.

$$V(s) \approx \langle\psi_s|H_{\theta}|\psi_s\rangle \tag{18}$$

4. Quantum Gradients in Training and Optimization: The optimization of QNN-based function approximators is achieved via hybrid quantum–classical training techniques. Quantum Gradient Descent (QGD) uses the parameter shift rule to compute gradients efficiently as per Eq. (19). This technique avoids computationally expensive full wave function simulation. The classical optimizer like Adam

or SGD updates parameters θ based on quantum-executed loss function evaluations. Training involves a quantum–classical feedback loop, where gradients are computed on a quantum processor, while updates are performed classically.

$$(\partial/\partial\theta i)\langle O\rangle = 1/2\left[\langle O\rangle_{\theta i+2\pi} - \langle O\rangle_{\theta i+2\pi}\right] \quad (19)$$

Implementation:

Following describes the details that QNN-based Q-learning algorithm models:

Implementation starts with conversion of classical states into quantum states using quantum feature maps. This is Quantum state encoding stage. This stage is followed by Quantum Circuit Execution. This utilizes a variational quantum circuit (VQC) to approximate Q-values. Next for Quantum Measurement, action values are extracted from the quantum state. Finally, the QNN parameters are updated using a quantum gradient descent method. This approach is implemented using Qiskit's variational quantum classifiers, optimizing a hybrid quantum–classical Q-learning model. To meet the hardware limitations, Noisy Intermediate-Scale Quantum (NISQ) are applied. Due to limited qubits, low-depth ansatz circuits are used for practical implementation. Error mitigation techniques such as zero-noise extrapolation are employed to reduce decoherence effects. Quantum gates are optimized to minimize depth while retaining expressivity. Quantum Hardware Platforms include IBM Q, Google Sycamore and IonQ. While practical limitations exist due to NISQ hardware constraints, advances in quantum algorithms, error mitigation, and variational training methods are paving the way for scalable QRL implementations.

4.2 *Quantum Variational Circuits for Policy Optimization*

Quantum Variational Circuits (QVCs) play a crucial role in Quantum Reinforcement Learning (QRL) by optimizing policies using parameterized quantum gates. These circuits serve as trainable quantum function approximators that leverage quantum superposition, entanglement, and parallelism to efficiently model decision-making processes in reinforcement learning.

1. VQCs in Policy Optimization:

Generally, policy is represented as a classical probability distribution over actions. Unlike this, quantum policies are encoded as a quantum state in a Hilbert space by Eq. (20). U_θ is a variational circuit with tunable quantum gates. $|0\rangle^{\otimes n}$ represents an initial state of n qubits. The quantum measurement collapses the wavefunction resulting to an action probability distribution. The output action is determined by measuring an observable O by Eq. (21). O represents the measurement operator for action selection. The Born rule determines the probability of selecting action a.

$$\langle\psi_\theta(s)| = U_\theta|0\rangle^{\otimes n} \quad (20)$$

$$\pi_\theta(s, a) = \langle \psi_\theta(s)|O|\psi_\theta(s)\rangle \tag{21}$$

2. Quantum Policy Gradient Optimization:

Normally reinforcement learning uses gradient-based methods such as REINFORCE or Proximal Policy Optimization (PPO) to optimize policy. In QRL, policy gradients are computed using the parameter shift rule to update quantum gate parameters. The quantum gradient of the policy loss function is approximated as per Eq. (22). $J(\theta)$ is the expected reward. e_i is a unit vector corresponding to parameter θ_i. This gradient estimate is used in Stochastic Gradient Descent (SGD) or Adam optimizers. The issue of small step sizes gets solved here which exists in finite difference approximations. It is more computationally stable on Noisy Intermediate-Scale Quantum (NISQ) hardware.

$$\frac{\partial J(\theta)}{\partial \theta_i} = \frac{1}{2}\left[J\left(\theta + \frac{\pi}{2}e_i\right) - J\left(\theta - \frac{\pi}{2}e_i\right)\right] \tag{22}$$

3. Hybrid Quantum–Classical Optimization Loop:

Though its applications and advantages, quantum hardware has limited depth and qubit count. This forces policy optimization through a hybrid approach. It starts with quantum execution where a quantum device computes the policy function $\pi_\theta(s)$. Followed by Quantum Execution is Classical Optimization. A classical optimizer such as Adam, RMSProp are used here to update the parameters θ using computed gradients. Finally, the feedback loop feeds back the updated parameters to the quantum circuit to refine the policy. This hybrid quantum–classical approach ensures that QRL models remain feasible on existing quantum devices.

4. Quantum Variational Advantage Actor-Critic (QVA2C)

Integrating VQCs with Actor-Critic framework gives a Quantum Variational Advantage Actor-Critic (QVA2C) model. In this model, the actor uses a VQC to represent the policy function. The Critic approximates the value function $V(s)$ using another quantum circuit. The policy update or actor is modelled by Eq. (23). $A(s,a)$ is the advantage function. The Value Function Update or critic module is given by Eq. (24). w is the critic parameters. r is the immediate reward. γ is the discount factor. This architecture allows for efficient policy refinement using quantum circuits.

$$\theta \leftarrow \theta + \alpha \nabla\theta \log \pi_\theta(s, a)A(s, a) \tag{23}$$

$$w \leftarrow w - \beta\nabla_w\left(r + \gamma V_w(s') - V_w(s)\right) \tag{24}$$

4.3 Enhancing QRL with Quantum Feature Representations

One key aspect of improving QRL models is the effective representation of features in quantum-enhanced environments. Quantum Feature Representations (QFRs) transform classical data into high-dimensional quantum Hilbert spaces, improving learning efficiency and generalization capabilities.

1. Quantum Feature Representations in QRL

Feature representations have been a critical part in defining how an agent perceives its state space. Traditional feature engineering methods rely on hand-crafted embedding or deep learning-based feature extractors. In QRl, quantum states can be used to encode richer representations via Quantum Feature Embeddings (QFEs). Mathematically, a classical state s is mapped into a quantum state $|\psi(s)\rangle$ using a feature encoding unitary U_ϕ. This is shown in Eq. (25), where $U_\phi(s)$ is a quantum feature map. The feature map embeds classical states into quantum Hilbert space. Unlike complex feature interactions were not easily captured by classical models, the higher-dimensional quantum space enables the same. QFRs allow for implicit representation in exponentially large Hilbert spaces with complex function approximation complex function approximation. This improves generalization by reducing overfitting through encoding entangled correlations across states.

$$|\psi(s)\rangle = U_\phi(s)|0\rangle^{\otimes n} \tag{25}$$

2. Quantum Kernel Methods in QRL:

Quantum kernels map classical data into a quantum space. Here the state similarity is computed via an inner product given by Eq. (26). $K(s_i,s_j)$ measures the similarity between states. The feature space is nonlinear and implicitly large. These arrangements make QRL adaptable to complex environments. Quantum kernel embeddings improve RL tasks such as policy learning, function approximation, and reward prediction [2].

$$K\left(s_i, s_j\right) = |\psi(s_i)\rangle\left|\psi\left(s_j\right)\right\rangle^2 \tag{26}$$

3. Quantum Fourier Features for Function Approximation:

Quantum Fourier Features (QFFs) encode state-action representations using periodic quantum transformations [25]. QRL benefits from this. The function approximation is given by Eq. (27), where e $i\omega_k s$ represents a Fourier frequency encoding of the state. Quantum phase embeddings enhance long-term dependencies in decision-making. This technique improves function approximation, particularly in value function estimation in QRL.

$$|\psi(s, a)\rangle = \sum_k ei\omega_k s|k\rangle \tag{27}$$

4. VQCs for Feature Learning:

Variational Quantum Circuits (VQCs) are commonly used to learn feature representations dynamically in QRL. A hybrid quantum–classical QRL model integrates a VQC-based encoder for state-action representations. For quantum state encoding, a classical state sss is mapped into a quantum feature space using a parametrized encoding circuit U_{ϕ}. The circuit is trained to extract optimal state representations. Next, the quantum state is measured, and the outcome determines the policy probability distribution. Finally, the Parameter Shift Rule is used to optimize the circuit parameters for optimal decision-making [6].

5. Hybrid Quantum–Classical Feature Learning in QRL

As discussed the current limitation of Noisy Intermediate-Scale Quantum (NISQ) in Sect. 4.1, hybrid approaches combine classical deep learning with quantum feature encodings. Classical Neural Networks extract raw features from state observations. Quantum Circuits encode and process these features using quantum transformations. Quantum-enhanced embeddings are used to improve value function approximation and policy selection. Such hybrid models have demonstrated faster convergence rates in QRL tasks, particularly in robotic control and finance [9].

4.4 Quantum Speedup in Exploration–Exploitation Tradeoff

The exploration–exploitation tradeoff is a fundamental challenge in reinforcement learning (RL), requiring agents to balance exploring new actions to gather information and exploiting known actions to maximize cumulative rewards. In classical RL, this tradeoff is managed using techniques like epsilon-greedy, Upper Confidence Bound (UCB), and Thompson Sampling. However, Quantum Reinforcement Learning (QRL) introduces new mechanisms, leveraging quantum superposition, entanglement, and quantum parallelism to accelerate the exploration–exploitation balance, leading to faster learning convergence and improved decision-making efficiency.

1. Quantum Speedup in Exploration

- Quantum Superposition for Parallel Exploration: Quantum superposition is one of key advantages that allows the agent to explore multiple states simultaneously rather than sequentially, as in classical RL. Mathematically, a classical agent selects a single action a_t at each time step t, whereas a quantum agent can represent all possible actions in a superposed state: $|\Psi\rangle = \sum_{\alpha} \alpha_{\alpha} |a\rangle$. $|a\rangle$ represents an action state. α_a denotes the probability amplitude of selecting action a. This enables a QRL agent to evaluate multiple actions in parallel, significantly reducing the time required to explore an environment.

Quantum walks are an essential mechanism used in QRL to enhance exploration. Unlike classical random walks, which take O(N) steps to explore a space of size N, quantum walks propagate quadratically faster requiring only $O(\sqrt{N})$ steps [26].

- Quantum Search Algorithms for Exploration: Quantum search algorithms, such as Grover's search, can improve the efficiency of exploration. Classical methods require O(N) evaluations to find an optimal action, whereas Grover's search reduces it to $O(\sqrt{N})$, accelerating policy learning. A QRL agent using Grover's search applies the unitary transformation: $G = (2|\psi\rangle\langle\psi| - I)\,O$, where $|\psi\rangle$ is the superposition of all actions. O marks the optimal action. G amplifies the probability of selecting the best action. This results in faster convergence to optimal policies, particularly in high-dimensional state-action spaces [27].
- Quantum Speedup in Exploitation
- Quantum Policy Evaluation with Entanglement: Once promising actions are identified, QRL exploits the best policy using quantum entanglement, which maintains correlations between state-action pairs. This enables faster value function propagation across the state space. More efficient policy evaluation using quantum-enhanced Bell measurements. Entanglement ensures that decisions in one part of the state space influence others, reducing the number of updates needed in Q-learning algorithms [9].
- Variational Quantum Circuits (VQCs) for Policy Optimization: Variational Quantum Circuits (VQCs) serve as function approximators for policy optimization, reducing the computational overhead in learning optimal policies. Classical policies $\pi(a|s)$ are mapped into quantum states using parameterized quantum circuits. QRL leverages the Parameter Shift Rule for optimizing quantum circuits, which allows efficient training of quantum policies.
- Quantum Parallelism for Balancing Exploration–Exploitation:
- Hybrid Quantum–Classical UCB for Optimized Action Selection: Classical Upper Confidence Bound (UCB) methods balance exploration and exploitation by selecting actions with the highest confidence interval as per Eq. (28). Q(a) is the estimated reward for action a. N(a) is the number of times action a has been selected. c is a confidence parameter. A quantum-enhanced version of UCB (Quantum-UCB) replaces Q(a) with quantum amplitude estimates, reducing variance and accelerating decision-making [11].

$$A_t = \arg\max_a\left(Q(a) + c\sqrt{\frac{lnt}{N(a)}}\right. \tag{28}$$

- Quantum Bayesian Inference for Adaptive Exploration: Quantum Bayesian inference refines exploration strategies by using quantum-enhanced probability distributions to model uncertainty in decision-making. This is particularly useful in non-stationary environments where optimal actions change over time [28].

5 Practical Implementations and Use Cases

5.1 Existing Frameworks and Toolkits

Several quantum computing frameworks and toolkits facilitate the development and deployment of Quantum Neural Networks (QNNs) and Quantum Reinforcement Learning (QRL) models. These platforms provide quantum circuit simulation, hybrid quantum–classical computing support, and machine learning integration. Developed by Google, TFQ integrates TensorFlow with quantum computing for implementing QNNs and QRL models using variational quantum circuits (VQCs). Qiskit is an open-source framework from IBM that supports quantum circuit design, noise modeling, and hybrid algorithms for running quantum AI experiments on IBM Quantum hardware and simulators. Pennylane is a powerful tool for quantum differentiable programming, enabling the training of QNNs and quantum policy networks with automatic differentiation and hybrid quantum–classical optimization. Developed by Google, Cirq provides fine-grained quantum circuit control, making it suitable for QRL experiments and variational quantum algorithms.

5.2 Implementing QNNs in QRL: Case Studies

Several case studies highlight the implementation and evaluation of QNN-based function approximators in Quantum Reinforcement Learning:

- Quantum Policy Networks for Grid-World Navigation: A hybrid QNN-based RL model was trained to navigate a grid-world environment using quantum-enhanced value function approximation. Results showed faster convergence and improved generalization over classical deep RL models.
- Quantum-Assisted Portfolio Optimization: A QRL-based financial model was developed to optimize stock portfolio allocations. The use of quantum variational circuits led to better risk-adjusted returns compared to traditional RL-based portfolio strategies.
- Quantum Robotics: Autonomous Path Planning: QNNs were employed for policy learning in robotic motion planning, demonstrating improved performance in noisy and uncertain environments. The ability of quantum superposition to evaluate multiple paths in parallel resulted in faster decision-making.

6 Challenges and Future Directions

Challenges include hardware limitations, scalability issues, and the need for robust error correction. Additionally, future research must focus on developing hybrid quantum–classical algorithms, improving quantum optimization techniques, and expanding real-world applications.

6.1 Limitations of Current Quantum Hardware

The success of QNNs and QRL relies heavily on quantum hardware capabilities, which are still in their early stages. The primary limitations include:

1. Noisy Intermediate-Scale Quantum (NISQ) Devices: Current quantum processors operate in the NISQ era, characterized by high decoherence rates and noise. QNNs and QRL models require deep quantum circuits, but limited qubit coherence time restricts their usability. Noise-induced gate errors reduce the accuracy of quantum policy optimization and function approximation in QRL.
2. Limited Qubit Connectivity and Fidelity: Quantum circuits require efficient qubit connectivity to execute entanglement-based operations. Many current superconducting and trapped-ion quantum processors suffer from restricted connectivity, impacting the scalability of QNN architectures [6]. Gate fidelity constraints affect the accuracy of quantum backpropagation and Q-learning algorithms.
3. Quantum Measurement Overhead: Unlike classical models, which allow rapid access to network weights, QNNs require quantum measurement, introducing a significant computational bottleneck [9]. Frequent measurements collapse quantum states, limiting the efficiency of QRL training loops.

6.2 Scalability and Error Correction in Quantum AI

Scaling QNNs and QRL to practical problems requires significant advancements in quantum error correction (QEC) and fault-tolerant quantum computing. Quantum Error Correction QEC methods, such as Surface Codes and Topological Codes, require thousands of physical qubits per logical qubit, making large-scale QNNs infeasible on current hardware. Near-term quantum systems must rely on error-mitigating variational algorithms, which still suffer from hardware limitations. Future QNNs should integrate quantum submodules into classical neural networks, offloading computation to quantum processors only where beneficial. Hybrid QRL approaches, combining classical deep Q-networks (DQNs) with quantum-enhanced search, can provide near-term scalability. Fault-tolerant quantum computing requires logical qubits with error rates below 10^{-4}, which is not yet achievable in practical settings. Scalable QNN architectures must be designed with low-depth circuits and error-aware optimizers to function on NISQ devices.

6.3 Future Research Directions for QNNs and QRL Integration

To advance the integration of QNNs and QRL, research must focus on:

1. Efficient Variational Quantum Circuits for QRL: Variational Quantum Circuits (VQCs) can approximate QNN policies with shallow, trainable quantum gates. Research should explore adaptive circuit depth optimization, balancing expressivity and noise resilience [5].
2. Quantum Kernel Methods for Function Approximation: Quantum kernels can be used for function approximation in QRL, allowing higher-dimensional representations with polynomial speedups [2]. Future research should investigate quantum-enhanced Bellman operators for faster Q-value propagation.
3. Entanglement-Driven Policy Optimization: QRL can leverage multi-qubit entanglement to encode action correlations, leading to faster policy learning. Open research questions involve optimizing entanglement structures in reinforcement learning tasks.
4. Quantum Meta-Learning for Autonomous Adaptation: Classical deep RL struggles with transfer learning and adaptation across different environments. Quantum meta-learning algorithms could enable QNNs to generalize policies across tasks, reducing retraining overhead.
5. Long-Term Vision: Towards Generalized Quantum AI-The ultimate goal of quantum-enhanced artificial intelligence is to build generalized Quantum AI systems capable of:Universal Quantum Function Approximators and Quantum Decision-Making Agents
6. Universal Quantum Function Approximators: Developing QNNs that can approximate any function in quantum-enhanced Hilbert spaces with optimal computational resources.
7. Quantum Decision-Making Agents: Building autonomous quantum agents that leverage quantum-enhanced exploration for real-time decision-making in fields like robotics, healthcare, and finance.
8. Fully Fault-Tolerant Quantum AI: Transitioning from NISQ-era heuristics to full-fledged, fault-tolerant quantum learning models that surpass classical AI in both efficiency and accuracy.

7 Conclusion

Quantum Neural Networks (QNNs) and Quantum Reinforcement Learning (QRL) represent a transformative paradigm in artificial intelligence by leveraging quantum mechanics to enhance computation, learning efficiency, and decision-making. Throughout this discussion, several key insights emerged:

- QNNs enable quantum-enhanced function approximation, allowing superior representations in high-dimensional Hilbert spaces.

- QRL leverages quantum superposition, entanglement, and parallelism to improve policy optimization, exploration–exploitation balance, and scalability in complex decision-making tasks.
- Variational Quantum Circuits (VQCs) and hybrid quantum–classical models provide practical solutions for near-term quantum AI applications.
- Despite these advantages, challenges persist, including hardware noise, limited qubit connectivity, error correction difficulties, and scalability constraints.
- Future advancements in fault-tolerant quantum computing, quantum-enhanced feature representations, and quantum–classical hybrid architectures will be critical for realizing the full potential of QNNs and QRL.

The Potential of QNNs and QRL in Advancing AI:

Quantum AI holds immense potential to redefine various fields, including machine learning, optimization, robotics, and financial modeling:

- Speedup in learning algorithms: Quantum models can process information in exponentially large state spaces, making them ideal for complex AI tasks.
- Enhanced decision-making: QRL has the potential to outperform classical RL in sequential decision problems, leading to breakthroughs in autonomous systems, drug discovery, and high-frequency trading.
- Generative modeling and creativity: QNNs can revolutionize data generation, quantum-enhanced GANs, and AI-driven creativity.
- Real-time adaptive learning: Future quantum systems could enable real-time learning and adaptation, leading to more intelligent, self-improving AI agents.

While quantum AI remains in its infancy, early research demonstrates its tangible advantages over classical approaches. Hybrid quantum–classical architectures will bridge the gap until fully fault-tolerant quantum computers become available.

Final Thoughts on Quantum AI Evolution: The road to practical Quantum AI is filled with challenges, but the potential rewards are game-changing. Quantum AI will not replace classical AI but will augment it, enabling solutions that were previously intractable. Cross-disciplinary collaboration between quantum computing, machine learning, and applied AI fields will be essential. As quantum hardware matures, QNNs and QRL will transition from theoretical constructs to real-world applications, reshaping AI-driven industries. Ultimately, Quantum AI represents the next frontier of intelligent computing, pushing the boundaries of efficiency, learning, and decision-making. The continued development of QNNs, QRL, and hybrid models will be pivotal in unlocking this potential, paving the way for a new era of quantum-enhanced artificial intelligence.

References

1. Biamonte, J., Wittek, P., Pancotti, N., Rebentrost, P., Wiebe, N., Lloyd, S.: Quantum machine learning. Nature **549**(7671), 195–202 (2017)
2. Schuld, M., Sinayskiy, I., Petruccione, F.: The quest for a quantum neural network. Quantum Inf. Process. **13**(11), 2567–2586 (2014)
3. Cerezo, M., Arrasmith, A., Babbush, R., Benjamin, S.C., Endo, S., Fujii, K., Coles, P.J.: Variational quantum algorithms. Nat. Rev. Phys. **3**(9), 625–644 (2021)
4. Farhi, E., Goldstone, J., Gutmann, S.: A quantum approximate optimization algorithm. arXiv preprint arXiv:1411.4028 (2014)
5. Zhou, L., Wang, S.T., Choi, S., Pichler, H., Lukin, M.D.: Quantum approximate optimization algorithm: Performance, mechanism, and implementation on near-term devices. Phys. Rev. X **10**(2), 021067 (2020)
6. Behera, B. K., Al-Kuwari, S., Farouk, A.: QNN-QRL: Quantum neural network integrated with quantum reinforcement learning for quantum key distribution. arXiv preprint arXiv:2501.18188 (2025)
7. Muscalagiu, A.I.: Quantum neural network design via quantum deep reinforcement learning
8. Elsken, T., Metzen, J.H., Hutter, F.: Neural architecture search: A survey. J. Mach. Learn. Res. **20**(55), 1–21 (2019)
9. Chen, S.Y.C.: An Introduction to Quantum Reinforcement Learning (QRL). In 2024 15th International Conference on Information and Communication Technology Convergence (ICTC), pp. 1139–1144. IEEE (2024, Oct)
10. Sutton, R.S., Barto, A.G.: Reinforcement learning: An introduction. MIT press (2018)
11. Liu, C.Y., Lin, C.H.A., Yang, C.H.H., Chen, K.C., Hsieh, M.H.: Qtrl: Toward practical quantum reinforcement learning via quantum-train. In 2024 IEEE International Conference on Quantum Computing and Engineering (QCE), Vol. 2, pp. 317–322. IEEE (2024, Sept)
12. DeRieux, A., Saad, W.: Eqmarl: Entangled quantum multi-agent reinforcement learning for distributed cooperation over quantum channels. arXiv preprint arXiv:2405.17486 (2024)
13. Zhao, Y., Zhang, X., Huang, H.: A review on quantum neural networks: Methods and applications. IEEE Access **10**, 12870–12888 (2022). https://doi.org/10.1109/ACCESS.2022.3149024
14. Zhao, H., Li, J., Wang, Y.: Quantum-enhanced deep learning: A survey on advances and future directions. IEEE Trans. Neural Netw. Learn. Syst. **34**(1), 234–251 (2022). https://doi.org/10.1109/TNNLS.2022.3147653
15. Park, S., Kim, J.: Trends in quantum reinforcement learning: State-of-the-arts and the road ahead. ETRI J. **46**(5), 748–758 (2024)
16. McClean, J.R., Boixo, S., Smelyanskiy, V.N., Babbush, R., Neven, H.: Barren plateaus in quantum neural network training landscapes. Nat. Commun.Commun. **9**, 4812 (2020). https://doi.org/10.1038/s41467-018-07090-4
17. Beer, K., Bondarenko, D., Farrelly, T., et al.: Training deep quantum neural networks. Nat. Commun.Commun. **11**, 808 (2020). https://doi.org/10.1038/s41467-020-14454-2
18. Amzhao, E.C., Zhao, A.: Quantum neural networks. Accessed on: 23rd march (2025). Available At:https://medium.com/mit-6-s089-intro-to-quantum-computing/quantum-neural-networks-7b5bc469d984
19. Benedetti, M., Garcia-Pintos, D., Perdomo, O., Leyton-Ortega, V., Nam, Y., Perdomo-Ortiz, A.: A generative modeling approach for benchmarking and training shallow quantum circuits. NPJ Quantum Inf. **5**(1), 45 (2019). https://doi.org/10.1038/s41534-019-0157-8
20. Huang, H.Y., Kueng, R., Torlai, G., Albert, V.V., Preskill, J.: Predicting many properties of a quantum system from very few measurements. Nat. Phys. **17**(7), 711–715 (2021). https://doi.org/10.1038/s41567-021-01287-2
21. François-Lavet, V., Henderson, P., Islam, R., Bellemare, M.G., Pineau, J.: An introduction to deep reinforcement learning. Found. Trends Mach. Learn. **11**(3–4), 219–354 (2018). https://doi.org/10.1561/2200000071

22. Schulman, J., Wolski, F., Dhariwal, P., Radford, A., Klimov, O.: Proximal policy optimization algorithms. arXiv preprint arXiv:1707.06347 (2017)
23. Hassan, A.: Quantum Neural Networks (QNN): The Next Frontier in AI", Medium. accessed on: 23rd March (2025). Available at: https://medium.com/@dmanikhs/quantum-neural-networks-qnn-the-next-frontier-in-ai-bb05056217d0
24. Yang, Z., Hou, X., Wang, X.: Quantum neural networks as universal function approximators: Theory and implementation. In: 2024 14th Asian Control Conference (ASCC), Dalian, China, pp. 1952–1956 (2024)
25. Chalumuri, R., Subramanian, S., Jha, A.K.: Quantum Fourier Feature Encodings in Reinforcement Learning. IEEE Trans. Quantum Eng. **4**, 2512347 (2023)
26. Venegas-Andraca, S.E.: Quantum walks: A comprehensive review. Quantum Inf. Process. **11**(5), 1015–1106 (2012)
27. Sweke, R., Wilde, F., Mayer, K., Eisert, J.: Speeding up reinforcement learning with quantum search. Phys. Rev. Res. **2**(3), 033446 (2020)
28. Das, S., Roy, R., Chakrabarti, A.: Quantum Bayesian inference for adaptive learning in QRL. J. Quantum Artif. Intell. **14**(3), 211–225 (2023)

Revolutionizing Pattern Recognition with Quantum Machine Learning

K. Uma and T. Gopu

Abstract Pattern recognition is a fundamental issue in machine learning, which forms the basis of applications such as image processing, financial forecasting, medical diagnostics, and speech recognition. Traditional machine learning (ML) approaches have made significant strides in pattern recognition, despite the fact that the computational efficiency of classical computing is occasionally constrained. As datasets grow in size and complexity, traditional machine learning algorithms are challenged by high-dimensional feature spaces and complex connections. Combining ML techniques with the concepts of quantum computing, Quantum Machine Learning (QML) is an exciting approach to overcome these limitations. By leveraging the fundamental characteristics of quantum physics, including quantum parallelism, entanglement, and superposition, QML processes data in novel ways. Compared to classical methods, these properties allow quantum-enhanced machine learning algorithms to more efficiently explore feature spaces and identify intricate patterns. Quantum computers may be able to handle high-dimensional data well because they can use quantum states to encode and modify data. This benefit is essential for improving the efficacy of classification, clustering, and anomaly detection methods in a wide range of fields, such as financial modelling, bioinformatics, autonomous systems, and cryptography. Key areas where quantum machine learning (QML) advances pattern recognition include quantum neural networks (QNN), quantum-enhanced support vector machines (QSVM), and quantum Boltzmann machines (QBM). QSVMs use quantum kernels to calculate higher-order correlations, which enables faster and more efficient classification. QBMs provide improved convergence properties by using quantum annealing to optimise deep learning architectures. Quantum neurones are combined in QNNs to improve feature extraction, which increases learning and generalisation. Such quantum-enhanced models could outperform conventional methods in terms of speed and accuracy, particularly when handling large and complex data. Furthermore, using quantum feature mappings to

K. Uma (✉) · T. Gopu
Sasi Institute of Technology and Engineering, Tadepalligudem, Andhra Pradesh, India
e-mail: uma.k@sasi.ac.in

T. Gopu
e-mail: gopu.t@sasi.ac.in

© The Author(s), under exclusive license to Springer Nature Switzerland AG 2025
A. Ghosh et al. (eds.), *Quantum Machine Learning in Industrial Automation*,
Information Systems Engineering and Management 65,
https://doi.org/10.1007/978-3-031-99786-0_15

convert classical data into a kind of quantum Hilbert space can make it simpler to identify complex structures. In fields like fraud detection, medical imaging, and natural language processing, where even the smallest changes in patterns are crucial, this transformation enables superior performance. Quantum generative adversarial networks (QGANs) further demonstrate how QML can enhance pattern recognition by generating synthetic data that enhances training processes for deep learning applications. Despite these promising developments, QML still has several issues, including noise sensitivity, hardware constraints, and the requirement for error correction. There aren't many immediate applications for quantum processors because they are still in their infancy. However, the integration of QML into practical systems is becoming simpler due to continuous developments in quantum hardware, error-reduction strategies, and blended quantum–classical algorithms. In conclusion, pattern recognition could be revolutionised by quantum machine learning, which offers faster and more efficient solutions for complex data analysis. In key applications, QML-based pattern recognition models might outperform traditional techniques as quantum technology advances, driving advancements in artificial intelligence and data-driven decision-making. The combination of machine learning and quantum computational methods has led to a paradigm shift in computational intelligence that has enabled unprecedented advances in scientific research and technological innovation.

Keywords Pattern recognition · Conventional machine learning · Quantum machine learning · Quantum generative adversarial networks

1 Introduction- Quantum Algorithm for Pattern Recognition

A key component of machine learning is pattern recognition, and as quantum computing develops, scientists are investigating quantum algorithms that could greatly improve pattern recognition jobs. Quantum algorithms use The principles of superposition, quantum parallelism, and entanglement to handle enormous volumes of data exponentially quicker than classical machine learning algorithms, which depend on conventional computing resources [1, 2].

A thorough analysis of the key distinctions, processing techniques, and real-world applications between both quantum and conventional machine learning (CML and QML) is shown in Fig. 1. CML is shown to rely on classical data in the top portion, where data is sequentially processed using binary bits (0 s and 1 s). After passing via mathematical methods and CPU/GPU processing, this data is utilised for supervised or unsupervised learning tasks that involve pattern recognition [3]. On the other hand, QML makes use of quantum data, it can exist in multiple states simultaneously

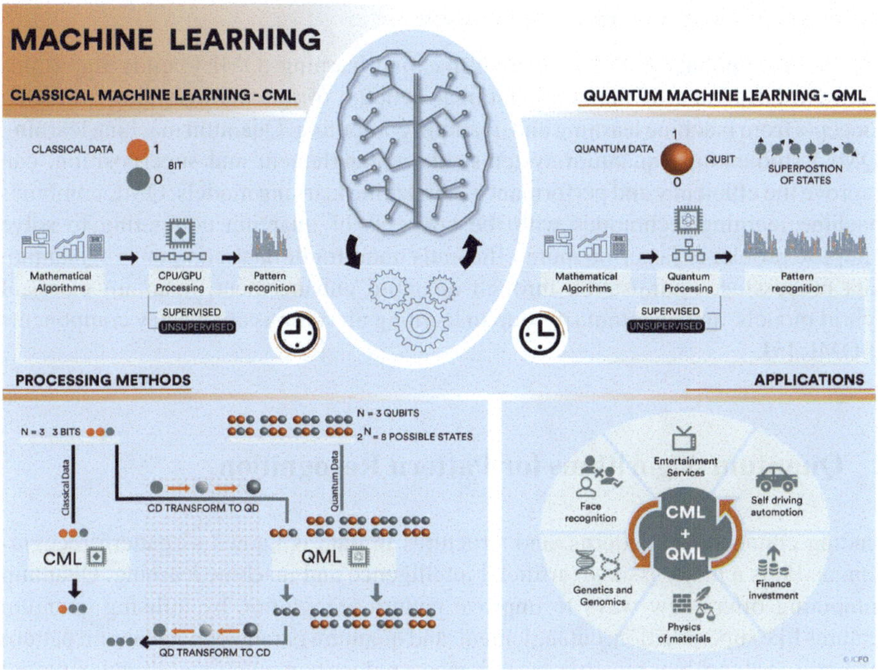

Fig. 1 Relationship between quantum and classical machine learning

due to superposition and is represented by qubits. Pattern recognition and data analysis are greatly accelerated by quantum computing concepts, which enable parallel processing in place of traditional CPU/GPU processing [1].

The middle section of the figure highlights the processing power of CML vs. QML. Computational efficiency is limited in classical computing because three bits can illustrate just one of the eight potential states at one time. However, QML enables quantum systems to process many possibilities simultaneously by allowing three All eight states can be superposed by qubits. One of QML's major edge for managing intricate calculations and high-dimensional datasets because of its intrinsic parallelism.

The bottom section illustrates the significant practical applications of combining CML and QML. This hybrid approach is expected to revolutionise several fields, including face recognition, where QML can enhance image and signal processing; genetics and genomics, where quantum computing expedites the analysis of biological data; and entertainment services, where it optimises personalised recommendations [4]. Additionally, self-driving automation benefits from better real-time decision-making, and financial investments benefit from advancements in risk management and fraud detection. Moreover, quantum simulations in material science could lead to improvements in the physics of materials.

Definition of Quantum Machine Learning

The interdisciplinary field of quantum machine learning (QML) builds algorithms that take advantage of the computational benefits of quantum physics by fusing concepts from machine learning and quantum computing. Quantum machine learning (QML) studies how quantum systems, like entanglement and superposition, can improve the efficiency and performance of machine learning models. QML combines machine learning techniques with the concepts of quantum computing to solve complex problems faster or more efficiently than traditional computers. Quantum data processing, optimisation through quantum enhancement, quantum–classical hybrid models, and quantum speedup in learning algorithms are the key components of QML [5].

2 Quantum Algorithms for Pattern Recognition

Finding connections, patterns, and structures in data is known as pattern recognition, and it is a major issue in artificial intelligence and machine learning. Quantum computing offers new ways to improve pattern recognition by utilising quantum features like superposition, entanglement, and quantum parallelism. Quantum pattern recognition algorithms aim to boost efficiency, reduce computational complexity, and create new applications that aren't achievable with conventional methods [6]. The objective of quantum pattern recognition is to use quantum algorithms to process and classify data patterns faster than with conventional methods. This approach works best with large and complex datasets, where traditional methods might run into computing limitations.

Quantum computing enables:

- Faster feature extraction through quantum superposition.
- More efficient similarity measurements using quantum distance metrics.
- Scalability improvements via quantum parallelism.

Several quantum algorithms have been developed to improve classical pattern recognition tasks include feature selection, grouping, and classification.

2.1 *Quantum k-Nearest Neighbours (QkNN)*

A data point is classified using the k-Nearest Neighbours (kNN) method by looking at the majority class of its k closest neighbours. Quantum kNN (QkNN) uses quantum computing to speed up the nearest neighbour search. In comparison to classical kNN and Grover's search method or Polar Distance, it achieves a quadratic speedup through the use of quantum distance estimation [7–9]. It can also be applied to superconducting processors in a quantum feature space and is used for image classification applications. Two quantum kNN algorithms have been proposed that incorporate a memory-efficient quantum subroutine for pattern distance computation. The subroutine maintains overall algorithmic complexity and performance while reducing qubit usage by more than 50%. It works with amplitude-encoded features and superconducting processors. Evaluations on 13 datasets show consistent qubit savings, with significant performance gains seen in Cryotherapy, Balance Scale, Caesarian, and Immunotherapy datasets. The datasets for Iris, Raisin, and Glass showed some accuracy trade-offs. All things considered, the method makes quantum kNN classification more effective and scalable [10].

Key Components of QkNN:

1. **Quantum Superposition**: A quantum computer processes numerous data points simultaneously.
2. **Quantum Distance Computation**: Distances between the query and dataset points are computed using quantum circuits.
3. **Grover's Algorithm for Search**: Identifies the k-nearest neighbors efficiently.
4. **Quantum Measurement**: The final state collapses to reveal the classification.

2.1.1 QkNN Algorithm

Step 1: Encoding of Data in Quantum States

Using quantum feature maps, traditional data points are converted into quantum states. A quantum state |xi⟩ is used to represent each data point xi.

Step 2: Distance Calculation Using Quantum Circuits

Quantum circuits compute the Euclidean or Hamming distance between the query point |q⟩ and all dataset points[11].

Step 3: Quantum-Based Search with Grover's Algorithm

Grover's algorithm efficiently searches the dataset for the k-nearest neighbours.

Step 4: Classification by Majority Vote

Once the k-nearest neighbours are found, the most common label determines the class of the query point (Fig. 2).

Example: Handwritten Digit Classification

Suppose if we want to classify a new digit (e.g., "7") using QkNN:

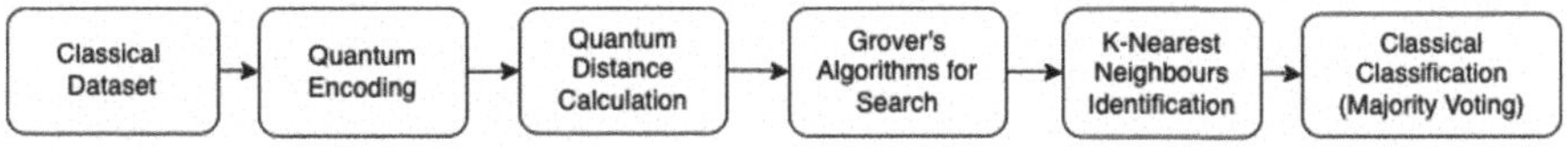

Fig. 2 Illustration of QKNN

1. **Dataset**: A quantum-encoded dataset of handwritten digits (0–9) [12].
2. **Query Point**: A new handwritten "7" digit.
3. **Quantum Distance Calculation**: Computes the similarity between the new digit and stored examples.
4. **Grover's Search**: Identifies k-nearest neighbors.
5. **Final Classification**: The query is categorized as "7" if the majority of the k closest neighbors are "7."

2.1.2 Advantages of QkNN

- **Speedup**: Quadratic improvement over classical kNN.
- **Efficient for Large Datasets**: Handles high-dimensional data better.
- **Optimized Search**: Number of comparisons are reduced by Grover's algorithm.

2.1.3 Challenges

- **Quantum Hardware Constraints**: Current quantum computers have limited qubits.
- **Quantum Error Rates**: Noisy computations can affect accuracy.

2.1.4 Implementation of QKNN

Here is a basic implementation that makes use of NumPy and PennyLane, a library for quantum computing. The code specifies a fundamental QKNN model in which data points are encoded and compared using quantum circuits. This algorithm computes quantum distances, encodes data into quantum states, and predicts labels using a k-NN classifier.

```
import pennylane as qml
import numpy as np
def encode_data(quantum_device, data):
    """Encodes classical data into a quantum state."""
    n_qubits = len(data)
    for i in range(n_qubits):
        qml.RY(data[i], wires=i)
def quantum_distance(quantum_device, x, y):
    """Computes quantum similarity/distance between two data points."""
    n_qubits = len(x)
    # Prepare the device
    @qml.qnode(quantum_device)
    def circuit():
        encode_data(quantum_device, x)
        qml.adjoint(encode_data)(quantum_device, y) # Compute inner product
        return qml.probs(wires=range(n_qubits))
    probabilities = circuit()
    return 1 - probabilities[0] # Distance measure
def qknn_classification(train_data, train_labels, test_point, k=3):
    """Quantum k-NN classification."""
    num_qubits = len(test_point)
    quantum_device = qml.device("default.qubit", wires=num_qubits)
    distances = [quantum_distance(quantum_device, test_point, x) for x in train_data]
    # Get the indices of the k nearest neighbors
    nearest_indices = np.argsort(distances)[:k]
    # Get the majority label
    nearest_labels = [train_labels[i] for i in nearest_indices]
    predicted_label = max(set(nearest_labels), key=nearest_labels.count)
    return predicted_label
# Example Usage:
train_data = [
    [0.1, 0.5],
    [0.2, 0.4],
    [0.9, 0.7],
    [0.8, 0.6]
]
train_labels = [0, 0, 1, 1]
test_point = [0.15, 0.45]
k = 3
predicted_label = qknn_classification(train_data, train_labels, test_point, k)
print(f"Predicted Label: {predicted_label}")
```

3 Quantum Support Vector Machines (QSVM)

A prominent learning with supervision strategy for applications involving regression and classification, an enhanced version of traditional support vector machines (SVMs) are called quantum support vector machines (QSVMs). In order to outperform classical SVMs computationally, especially in high-dimensional feature spaces, Quantum kernel estimation, entanglement, and superposition are some of the quantum computing ideas that QSVM uses.

Classical SVM versus Quantum SVM

An SVM in traditional machine learning finds the best hyperplane to divide a dataset into distinct classes [13]. Support vectors, or the data points nearest to the margin, determine the decision boundary. SVMs employ the kernel method to translate data into a higher-dimensional space where it can be separable in situations where the data is not linearly divisible. Although it uses quantum feature maps and quantum kernels to improve its performance, QSVM operates on a similar principle. QSVM enables significantly faster kernel computations by processing and converting classical data into quantum states inside a quantum circuit rather than utilizing classical transformations.

3.1 QSVM (Quantum Support Vector Machine) Algorithm

The Quantum Support Vector Machine (QSVM), an extension of the traditional Support Vector Machine (SVM), increases classification efficiency by utilizing quantum computing, particularly when dealing with non-linearly separable data. By processing data in an exponentially larger Hilbert space using quantum feature encoding and quantum kernel estimates, QSVM improves pattern identification with possible quantum speedups in contrast to classical SVMs, which rely on kernel functions calculated in a classical high-dimensional environment. Models of quantum machine learning based on quantum support vector regression (QSVR) and classification (QSVC), which are assessed using the IonQ Harmony processor and simulators with and without noise has also been implemented. On image and credit card fraud datasets, QSVC models with four qubits performed similarly to noiseless simulations. Using low-rank approximations and fine-tuning ε-SVR enhanced results for QSVR tasks on noisy devices. The alignment between ideal and noisy kernels using the Frobenius inner product was linked to prediction accuracy [14].

3.1.1 Quantum Feature Encoding

Before a machine learning model can process data, it must be mapped into a mathematical space where learning operations can be performed. Feature transformations, which transform data into a higher-dimensional space that allows linear separability, are used by classical SVMs to accomplish this mapping. Quantum feature mappings, which convert regular data points into quantum states that exist in a Hilbert space, are used by QSVM to achieve this. In a high-dimensional Hilbert space, each data point x from a classical dataset is mapped to a quantum state |ϕ(x)⟩|. This mapping is achieved through quantum circuits that encode data into qubit states using quantum gates.

- One popular technique in Qiskit is the ZFeatureMap, which uses rotation gates (Rz gates) to encode classical features.
- Every data point x from a classical dataset is mapped to a quantum state |ϕ(x)⟩| in a high-dimensional Hilbert space.
- This mapping is achieved using quantum circuits that encode data into qubit states using quantum gates.
- A common method in Qiskit is the ZFeatureMap, which encodes classical features using rotation gates (Rz gates).
- The quantum feature space can grow exponentially with respect to qubit count, allowing for higher expressivity compared to classical transformations.

Mathematical Representation of Quantum Encoding

The mapping from a classical input vector x to a quantum state is as follows: |ϕ(x)⟩| = (×1, × 2,…,xn).

$$|\phi(x)\rangle = U(x)|0\rangle \tag{1}$$

where:

- The quantum unitary transformation is used to encode data into quantum states.
- U(x)|0⟩ represents the initial quantum state of the qubits.

This transformation makes it possible for quantum computers to naturally perform calculations in ever-larger feature spaces, which makes QSVM useful for complex pattern recognition applications.

3.1.2 Quantum Kernel Estimation

In conventional SVMs, the similarity between two data points in the modified feature space is determined by a kernel function. This means that the SVM technique can be used to classify data that is not linearly separable in its original space [15].

Typical classical kernel operations consist of:

$$\mathbf{LinearKernel} : \mathrm{K}\left(\mathrm{x}, \mathrm{x}'\right) = \mathrm{x} \cdot \mathrm{x}' \tag{2}$$

$$\mathbf{PolynomialKernel} : \ \mathrm{K}\left(\mathrm{x}, \mathrm{x}'\right) = \left(\mathrm{x} \cdot \mathrm{x}' + \mathrm{c}\right)^{\mathrm{d}} \tag{3}$$

$$\textbf{Radial BasisFunction(RBF)Kernel}: K(x, x') = e^{-\gamma \|x-x'\|^2} \quad (4)$$

How QSVM Uses Quantum Kernels?

In QSVM, quantum kernel estimation replaces classical kernel functions by measuring the overlap (fidelity) of quantum states to identify similarities between quantum-encoded data points.

$$K_Q(x, x') = |\langle \phi(x) | \phi(x') \rangle|^2 \quad (5)$$

where:

- $|\phi(x)\rangle$ and $|\phi(x')\rangle$ are representations of the quantum features of data points.
- KQ(x,x′) is the quantum kernel value, which is calculated by calculating the likelihood that various inputs will result in the same quantum state.

3.1.3 *Training the QSVM Model*

After calculating the quantum kernel values, QSVM solves a quadratic optimisation problem to identify the optimal hyperplane for data point classification. The process is similar to conventional SVM training, despite the use of a quantum-enhanced kernel matrix.

Steps in Training QSVM

1. Compute the Quantum Kernel Matrix:
   - All training data points x_i are mapped to quantum states $|\phi(x_i)\rangle$|\phi(x_i)\angle $|\phi(x_i)\rangle$.
   - The quantum kernel matrix K_Q is computed, where each entry represents the similarity between two quantum states [16].
2. Solve the Quadratic Optimization Problem:
   - The vectors of support are found using traditional SVM solvers, such as Sequential Minimal Optimization, or SMO.
   - The decision boundary is defined by these crucial data elements.
3. Determine the Optimal Hyperplane:
   - The hyperplane is defined as:

$$f(x) = \sum_i \alpha_i y_i K_Q(x_i, x) + b \quad (6)$$

- Where α_i are optimization variables, y_i are class labels, and b is the bias term.

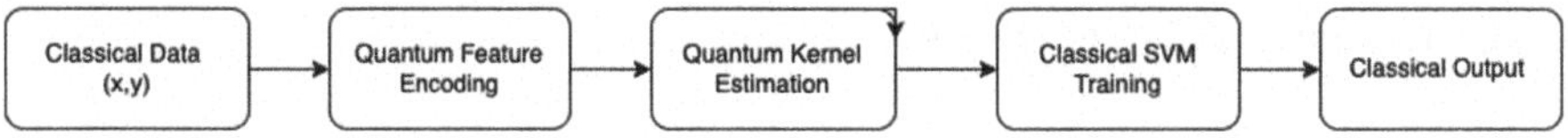

Fig. 3 Workflow of QSVM

Hybrid Approach: QSVM typically uses quantum processors for kernel computation and classical optimisation solvers for training.

4. Classification & Decision Boundary

After the model is trained, new test data points are grouped according to how much they resemble existing support vectors.

1. Compute Quantum Kernel Values for Test Data:
   - A quantum state $|\phi(x')\rangle$ is encoded with the test data point x′.
   - The values of the quantum kernel $K_Q(x_i, x)$ are computed for all support vectors.
2. Compute the Classification Score:

The decision function is evaluated using the trained model's support vector weights.

3. Assign Class Labels:
   - If $f(x') > 0$, the test point belongs to class + 1.
   - If $f(x') < 0$, the test point belongs to class -1.

As shown in Fig. 3, the Quantum Support Vector Machine (QSVM) process starts with conventional data input, like images or numerical data. After this data has been quantum feature encoded using a quantum circuit, it is mapped into a high-dimensional Hilbert space using quantum feature maps (such as ZFeatureMap). Quantum kernel estimation is the next stage, in which quantum measurements are used to assess how similar the quantum-encoded data points are. The quantum kernel matrix is then fed into a classical SVM solver to identify the optimal hyperplane for classification. Finally, by predicting extra data points based on quantum-enhanced decision boundaries, the trained QSVM model leverages quantum computing to improve pattern recognition.

3.1.4 Applications of QSVM

(a) Medical imaging: improving the precision of prostate cancer and tumour detection with MRI scans [17].
(b) Cybersecurity: Recognising patterns of financial fraud [18].
(c) Autonomous Vehicles: Enhancing object recognition in autonomous vehicles.
(d) Natural Language Processing: Quantum-enhanced pattern recognition improves speech and text classification.

3.1.5 Implementation of QSVM

This is an example of Python code that uses Qiskit to implement the Quantum Support Vector Machine (QSVM). In this example, a binary classification problem on a basic dataset is resolved using QSVM. This code uses Qiskit to create a Quantum Support Vector Machine (QSVM). It employs QSVM for classification after transforming the data into a quantum-enhanced feature space using a quantum kernel and ZZFeatureMap.

```
from qiskit import BasicAer
from qiskit.utils import algorithm_globals
from qiskit_machine_learning.algorithms import QSVM
from qiskit_machine_learning.datasets import ad_hoc_data
from qiskit_machine_learning.kernels import QuantumKernel
from qiskit.circuit.library import ZZFeatureMap
# Load dataset
def load_data(training_size=20, test_size=10, n=2):
   feature_dim = n
   train_features, train_labels, test_features, test_labels = ad_hoc_data(
      training_size=training_size, test_size=test_size, n=feature_dim, gap=0.3
   )
   return train_features, train_labels, test_features, test_labels
# Load training and test data
train_features, train_labels, test_features, test_labels = load_data()
# Define quantum feature map
feature_map = ZZFeatureMap(feature_dimension=2, reps=2)
quantum_kernel = QuantumKernel(feature_map=feature_map, quantum_
instance=BasicAer.get_backend('qasm_simulator'))
# Train QSVM model
qsvm = QSVM(quantum_kernel, train_features, train_labels, test_features, test_labels)
result = qsvm.run()
# Print results
print("QSVM Training Accuracy:", result['training_accuracy'])
print("QSVM Test Accuracy:", result['testing_accuracy'])
```

4 Variational Quantum Circuits (VQC)

Variational Quantum Circuits (VQCs) are hybrid quantum–classical techniques that use parameterised quantum circuits to solve difficult optimisation and machine learning problems. These circuits consist of quantum gates with tuneable parameters that are iteratively tuned through conventional optimisation techniques.

Three basic steps are usually followed by a VQC:

1. Quantum State Preparation: A parameterised quantum circuit is used to encode input data into a quantum state.

2. Parameterised Quantum Evolution: The circuit makes use of quantum gates with programmable parameters, sometimes referred to as variational parameters, in order to alter the quantum state.
3. A conventional optimiser (like gradient descent) adjusts the parameters while keeping an eye on the output in order to minimise a cost function.

VQCs are widely used in quantum machine learning, quantum chemistry, and combinatorial optimisation problems because they efficiently use both quantum superposition and entanglement while using classical computer power for optimisation. Examples include Quantum Neural Networks (QNNs), Variational Quantum Eigen solvers (VQE), and Quantum Support Vector Machines (QSVMs).

VQCs are a potential way to gain a quantum advantage with the existing noisy intermediate-scale quantum (NISQ) computers, given the hardware constraints of near-term quantum devices (noise and qubit constraints).

4.1 Variational Quantum Algorithms

The ability of Variational Quantum Algorithms (VQAs) to offer a flexible framework for addressing a broad range of computing problems is one of their primary features [19]. A hybrid strategy that makes use of both quantum and classical computer resources accomplishes this. Specifically, a Quantum Processing Unit (QPU) executes parameterised quantum circuits, while a Classical Processing Unit (CPU) optimises the circuit parameters to refine the solution. VQA operations are depicted in Fig. 4 and can be broken down into the four primary steps listed below:

Problem Encoding:

A cost function is used to formulate the computational task, C(θ), converting it into an optimization problem. The goal is to determine the ideal parameters, θ, that minimize C(θ), thereby solving the target problem.

1. **Parameterised Quantum Circuit Design**: Adjustable parameters are used to create ansatz quantum circuits. When the parameters are adjusted to the point

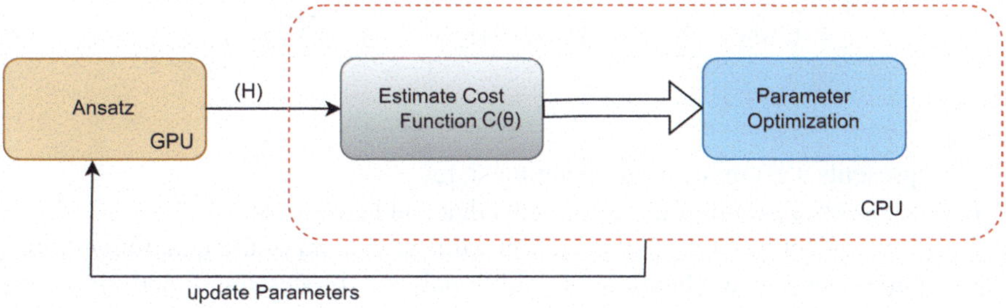

Fig. 4 Schematic diagram of VQA

where the cost function is at its minimum, the ansatz successfully represents a quantum model that encodes the issue solution.

2. **Quantum State Measurement**: The output quantum state is measured in order to compute the cost function, C(θ).
3. **Parameter Optimization**: To further minimise the cost function and increase the precision of the solution, a classical optimisation algorithm iteratively updates the parameter set θ.

In the current noisy intermediate-scale quantum (NISQ) era, variational quantum circuits must be implemented efficiently, particularly to solve issues like barren plateaus [20]. Because VQAs successfully leverage quantum computing capabilities while employing conventional optimisation techniques through iteration through these steps, they are a feasible approach for solving difficult computational problems.

The successful implementation of a Variational Quantum Algorithm (VQA) requires careful consideration of several important factors, as illustrated in Fig. 4. First, developing an appropriate cost function is essential since it turns the computational activity into an optimisation problem. Second, developing an effective ansatz is essential since the quantum model encodes the solution to the problem. Additionally, the traditional optimisation method selection using tensor rings has a significant impact on the accuracy and efficiency of the VQA [21].

4.2 *Cost Function*

An essential part of building Variational Quantum Algorithms (VQAs) is the cost function. From a conceptual standpoint, it can be thought of as a high-dimensional landscape with valleys and peaks that symbolise changes in the parameter space. In order to traverse this intricate surface and find the ideal solution, the optimiser must repeatedly modify the quantum circuit settings until the cost function reaches its minimum value. In mathematics, the cost function is commonly expressed as follows: The cost function is crucial to Variational Quantum Algorithms (VQAs) because it serves as a metric that guides the optimisation process. It has the following mathematical definition:

$$C(\theta) = \sum_p f_p(Tr\left[E_p U(\theta)\rho_j U^+(\theta)\right] = \sum_p f_p E_p \tag{7}$$

where:

ρ_j- represents a group of input quantum states.

E_p- represents a group of measurements that can be observed.

E_p- is equivalent to the E_p expectation value at the end of the quantum circuit.

f_p- symbolises the application of traditional post-processing functions to the measurement results.

In this case, the index j denotes different measurement schemes used within the same framework rather than various quantum states. Each ρ_j and E_p pair corresponds to a specific measurement configuration used in evaluating the cost function.

4.2.1 Classification of Cost Function Methods

The nature of Ep has a big influence on how the cost function is calculated. One important category is:

Pauli Operator-Based Cost Functions:

When Ep is a Pauli operator, the cost function can be directly found using quantum state measurements in the associated Pauli basis. This approach is particularly useful in physical models such as the Ising model, where Ep is frequently expressed as a linear combination of local Pauli operators:

$$E_p = \sum_q c_{p,q} P_{p,q} \tag{8}$$

where a Pauli operator acting on local qubits is denoted by $P_{p,q}$. The term "m-local" refers to the cost function where each $P_{p,q}$ operates on a maximum of "m" qubits. Notably, when it comes to addressing the gradient vanishing problem a frequent obstacle in VQA training—local cost functions outperform global cost functions.

5 Ansatz

Ansatz is a parameterised quantum circuit designed to approximate the solution to a specific problem. Since the ansatz defines the structure of the quantum state used in the calculation, it is crucial to the algorithm's accuracy and effectiveness. The ansatz provides a flexible quantum state as a trial wave function that can be improved through the use of conventional optimisation techniques. The objective is to minimise the cost functionality, which represents the issue at hand, by changing its parameters.

5.1 Types of Ansatz in Variational Quantum Circuits

5.1.1 Problem-Specific Ansatz

A problem-specific ansatz is a parameterised quantum circuit designed with domain-specific knowledge to successfully solve a particular problem. Unlike generic ansätze, which are made for hardware efficiency, problem-specific ansätze leverage the structure and features of the target problem to improve convergence and optimisation.

Finding a Hamiltonian's ground state is the goal of the Variational Quantum Eigensolver (VQE), a well-known example. The ansatz in this case is often based on Hartree-Fock techniques or tailored to fit the symmetries of the Hamiltonian to guarantee that the quantum circuit explores a pertinent region of the entire Hilbert space [19]. The widespread use of the Unitary Coupled Cluster (UCC) and Recursive Variational Quantum Compiling (RVQC) ansatz in quantum chemistry simulations serves as another example. The UCC ansatz increases the accuracy of energy estimations by effectively encoding electron correlations in molecular systems [22]. The main advantage of problem-specific ansatz is that it reduces the number of optimisation parameters while maintaining high expressibility, which boosts training efficiency. This can be problematic for near-term quantum devices with short coherence durations, which often require deeper circuits than hardware-efficient ansatze.

1. **General Ansatz Representation**

A problem-specific ansatz is often expressed as a unitary evolution generated by a problem-relevant operator:

$$U(\theta) = e^{-i\theta H_{eff}} \tag{9}$$

where:

- H_{eff} is an efficient Hamiltonian that takes constraints unique to the problem into account.
- θ is a tunable parameter optimized during the variational process.
- **Unitary Coupled Cluster (UCC) Ansatz in Quantum Chemistry**

The Unitary Coupled Cluster (UCC) ansatz is a prevalent problem-specific ansatz in quantum chemistry, as demonstrated by:

$$|\varphi(\theta) = e^{T(\theta)-T^{+}(\theta)|\varphi_0} \tag{10}$$

where:

$T(\theta)$- is the cluster operator, typically written as a sum of excitation operators.

$$T(\theta) = \sum_{pq} \theta_{pq} a_p a_q + \sum_{pqrs} \theta_{pqrs} \tilde{a}_p \tilde{a}_q a_r a_s + \ldots \tag{11}$$

Here,

Fermionic creation and annihilation operators sre denoted by $\tilde{a}$ and a.

$|\varphi(\theta)$ is the reference state, often taken as the **Hartree–Fock state**.

Variational Quantum Eigensolvers (VQE) frequently employ the UCC ansatz to determine the ground state of molecular Hamiltonians.

3. **Problem-Specific Ansatz in the Ising Model**

In condensed matter physics, the Ising Hamiltonian is commonly used as the basis for a problem-specific ansatz:

$$H_{Ising} = \sum_{i} h_i z_i + \sum_{i,j} J_{ij} z_i z_j \tag{11}$$

z_i-represents the Pauli-Z operator on qubit *i.*

$h_i$andJ_{ij}- problem specific parameters.

A **variational ansatz** for this model could take the form:

$$U(\theta) = \prod_{i} e^{-i\theta_i Z_i} \prod_{i,j} e^{-i\theta_{i,j} Z_i Z_j} \tag{12}$$

where the parameters $\theta_{i,j}$andθ_i are designed to reduce the value of the energy expectation.

$$E(\theta) = \varphi(\theta) \big| H_{Ising} \big| \varphi(\theta) \tag{13}$$

5.1.2 Hardware-Efficient Ansatz

A Hardware-Efficient Ansatz is a type of parameterised quantum circuit that is easy to build on near-term quantum hardware because it maintains expressibility and minimises circuit depth. It is frequently employed in Variational Quantum Algorithms (VQAs), including the Quantum Approximate Optimisation Algorithm (QAOA) and the Variational Quantum Eigensolver (VQE). Three essential elements make up the framework of a hardware-efficient ansatz: repeating layers for improved expressibility, entangling layers, and parameterised single-qubit rotations. Depending on hardware connections, entanglement is accomplished using two-qubit gates like CNOT, CZ, or iSWAP, whereas single-qubit rotations, usually employing $R_x(\theta), R_y(\theta), R_z(\theta)$ introduce configurable parameters. These elements are layered, and while repeated use improves speed, it also creates more optimisation problems, such as barren plateaus (vanishing gradients).

Utilising native quantum gates supported by modern Noisy Intermediate-Scale Quantum (NISQ) devices, the ansatz is hardware-friendly and ensures effective implementation with reduced error accumulation. Even though it offers general-purpose application across a range of quantum situations, its lack of problem-specific structure may lead to ineffective parameter optimisation. Hardware limitations like gate quality and qubit connection must also be considered during design. Notwithstanding these challenges, the hardware-efficient ansatz remains a practical and well-liked approach to using near-term quantum devices to solve problems in optimisation, machine learning, and chemistry.

5.2 Structure of Hardware-Efficient Ansatz

Ansatzes such as the Hardware-Efficient Ansatz and Problem-Specific Ansatz are based on three fundamental elements: parameterised single-qubit rotations, which are quantum gates applied to individual qubits and introduce tuneable parameters that allow for flexible quantum state transformations. These rotations are crucial to Variational Quantum Algorithms (VQAs) because they enable the circuit to explore different quantum states based on the values of the parameters.

Mathematical Representation of Single-Qubit Rotations

Pauli rotation gates, which rotate the quantum state around the X, Y, or Z axes of the Bloch sphere, are the most widely used single-qubit rotation gates. They are described as follows:

1. **Rotation Around the X-axis**:

$$R_x(\theta) = e^{-i\frac{\theta}{2}X} = cos\left(\frac{\theta}{2}\right)I - isin\left(\frac{\theta}{2}\right)X \tag{14}$$

2. **Rotation around the X-axis**:

$$R_y(\theta) = e^{-i\frac{\theta}{2}y} = cos\left(\frac{\theta}{2}\right)I - isin\left(\frac{\theta}{2}\right)Y \tag{15}$$

3. **Rotation around the X-axis**:

$$R_z(\theta) = e^{-i\frac{\theta}{2}Z} = cos\left(\frac{\theta}{2}\right)I - isin\left(\frac{\theta}{2}\right)Z \tag{16}$$

where $\boldsymbol{\theta}$ is a tunable parameter, and the Pauli matrices are **X, Y, and Z.**

$$X = \begin{pmatrix} 0 & 1 \\ 1 & 0 \end{pmatrix}, Y = \begin{pmatrix} 0 & -i \\ i & 0 \end{pmatrix}, Z = \begin{pmatrix} 0 & 1 \\ 0 & -1 \end{pmatrix} \tag{17}$$

5.3 Role in Variational Quantum Circuits

Parameterised single-qubit rotations are the trainable parameters in quantum machine learning and optimisation problems. By changing the values of θ during training, the circuit can iteratively find the best solution. These rotations allow the circuit to explore the entire Bloch sphere and provide a flexible representation of complex quantum states. In practical applications, they are commonly combined with entangling gates (such as CNOT) to create highly expressive quantum circuits [23, 24].

5.4 Advantages and Challenges

- **Advantages**:
  - Effectively applied to upcoming quantum technologies [25].
  - Adjustable degrees of freedom are available for quantum optimisation.
  - It makes it possible to investigate a variety of quantum states.
- **Challenges**:
  - May suffer from barren plateaus, where gradients vanish in deep circuits.
  - To increase trainability, thorough initialisation is necessary.

6 Quantum Neural Networks (QNN)

To enhance the capabilities of machine learning, Quantum neural networks (QNNs) are a hybrid computational paradigm that blends the concepts of artificial neural networks and quantum computing. Unlike classical neural networks, which use ordinary bits to compute, QNNs use qubits, which can represent multiple states at once and exist in superposition. This quantum advantage could allow QNNs to analyse complex data faster and potentially address problems that traditional systems can't handle [26].

QNNs use quantum gates and circuits to perform calculations, and quantum entanglement and interference are used to enhance learning processes. One of their key advantages is their ability to perform computations in parallel, which greatly increases processing power and reduces time complexity in tasks like data categorisation, pattern recognition, and optimisation. Furthermore, the Variational Quantum Eigensolver (VQE) and Quantum Approximate Optimisation Algorithm (QAOA), two quantum-inspired algorithms for QNN training, are being investigated. Despite their potential, QNNs have a number of drawbacks. Quantum hardware is still in its early stages because of issues with qubit coherence, error rates, and scalability. Furthermore, efforts are ongoing to create hybrid systems that combine quantum and conventional processing, as well as effective quantum algorithms. Because of their exceptional processing power, QNNs have the potential to completely transform industries like artificial intelligence, drug development, and cryptography as quantum technology develops. [27]. In conclusion, quantum neural networks represent a novel and powerful fusion of quantum computing and deep learning that may outperform classical models in certain domains and usher in a new era of intelligent computing.

6.1 *Key Features of QNNs*

Quantum computing, which applies the ideas of quantum physics, has the potential to perform calculations much more quickly than traditional systems. Quantum superposition, which leads to quantum parallelism—a key advantage of quantum neural networks (QNNs)—is one of the fundamental ideas that enable this (Fig. 5).

6.1.1 Understanding Quantum Superposition

There are two possible states for a bit in classical computing: 0 and 1. On the other hand, a quantum bit (qubit) can simultaneously exist in a superposition of 0 and 1 states. This suggests that when scaled up in a system, a single qubit can encode more data than a conventional bit, allowing for exponential processing capacity.

A qubit state is expressed mathematically as:

$$|\varphi = \alpha\,|0 + \beta\,|1 \tag{18}$$

where the likelihood of measuring the qubit in states **0 or 1** is represented by the squares of the probability amplitudes **α and β**, respectively.

6.1.2 Quantum Parallelism

Multiple calculations can be performed at the same time thanks to superposition. When multiple qubits are entangled, a quantum system can examine multiple solutions at once, significantly speeding up complex calculations. For example, a three-qubit system in superposition can be in eight states at the same time ($2^3 = 8$), whereas a classical system would need eight calculations to analyse each state.

- Because of this characteristic, QNNs are especially effective at jobs like:

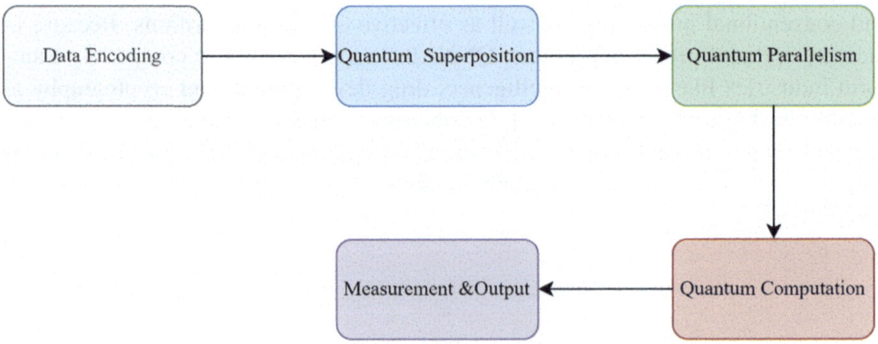

Fig. 5 Illustration of QNN

- Recognition of patterns.
- Issues with optimisation.
- Deep learning improved by quantum.

The methodical process of quantum superposition and parallelism used by quantum neural networks (QNNs) starts with the encoding of conventional information into quantum states and concludes with measurement and result interpretation. By initialising qubits in particular configurations, the first step, known as Data Encoding (Quantum State Preparation), transforms classical data into quantum information. For this, Hadamard gates (H-gates) are commonly used, which create a balanced superposition where each qubit simultaneously exists in both 0 and 1 states. This step is crucial because it creates the foundation for quantum parallelism. The system's ability to process states increases exponentially during the Quantum Superposition (State Expansion) phase, which begins when the qubits are in superposition. Quantum superposition enables a QNN to investigate several options simultaneously, in contrast to conventional bits, which can only be in one state at a time. This significantly boosts computer performance, especially for complex tasks like pattern recognition and optimisation. In the Quantum Parallelism (Simultaneous Computation) stage, quantum circuits process multiple states at once using entanglement and interference. This means that, unlike traditional neural networks, a QNN takes into account all possible answers at once rather than one at a time. This is accomplished by quantum gates such as Pauli gates and controlled-NOT (CNOT) gates, which alter the quantum states for further processing. Following parallel computation, the quantum computation (also known as quantum circuit processing) step optimises and improves the results by acting on the qubits. Even though quantum mechanics makes execution much more efficient, these quantum events are analogous to the forward and backward propagation observed in traditional neural networks. To arrive at the most likely solution, the network manipulates and improves the quantum states. In the Measurement & Output (Classical Interpretation) phase, the quantum states are ultimately measured and collapse into discrete classical values (0 s or 1 s). This measurement extracts meaningful results from the quantum system, which are then analysed in a conventional computing environment. Several runs may be necessary to obtain an optimal solution because quantum measurement by its very nature involves probabilistic results.Overall, this approach enables QNNs to handle massive amounts of data concurrently, providing significant speed and efficiency improvements over conventional neural networks. QNNs may explore a much larger solution space in a fraction of the time due to quantum superposition and parallelism, which makes them particularly useful for complex machine learning and AI-driven applications.

6.2 Quantum Entanglement: A Fundamental Principle of Quantum Computing

One of the most intriguing and surprising aspects of quantum mechanics is quantum entanglement. No matter how far apart they are physically, two or more quantum particles can form correlated states in this unique relationship. This implies that even when two entangled particles are separated by great distances, measuring the state of one of them instantly determines the state of the other. Due to its advantages over classical systems, quantum entanglement is essential to quantum communication, cryptography, and computing.

Understanding Quantum Entanglement

In classical physics, two objects can only interact if they are near each other. But according to quantum theory, entangled particles have a non-local connection, which means that an action on one instantly affects the other, regardless of how far apart two particles are.

An entangled quantum state between qubits A and B can be expressed mathematically as follows:

$$\left| \varphi = \frac{1}{\sqrt{2}}(|00 + |11) \right. \tag{19}$$

This equation states that the qubits are in a superposition of both states (00 and 11) until they are measured. When we measure one qubit (for instance, finding it in state 0), the second qubit will be in the equivalent state (0 as well) instantly, even if it is light-years away.

6.3 How Entanglement Works in QNN

In QNN, entanglement (Fig. 6) facilitates effective information processing and transfer. Entangled qubits enable instantaneous correlations, allowing for faster calculations than standard neural networks, which process information sequentially. Entangled qubits in a QNN aid in:

- **Enhanced Parallel Processing**: Multiple computations can take place at a same time using QNNs.
- **Error Reduction**: Error can be easily detected and corrected by Entangled qubits.
- **Quantum Speed-Up**: Optimization and cryptography, become exponentially faster.

Quantum initialisation, which comprises configuring two or more qubits in a quantum system, is the first step in the quantum entanglement workflow of Quantum Neural Networks (QNNs). A quantum gate, like the CNOT (Controlled-NOT) gate,

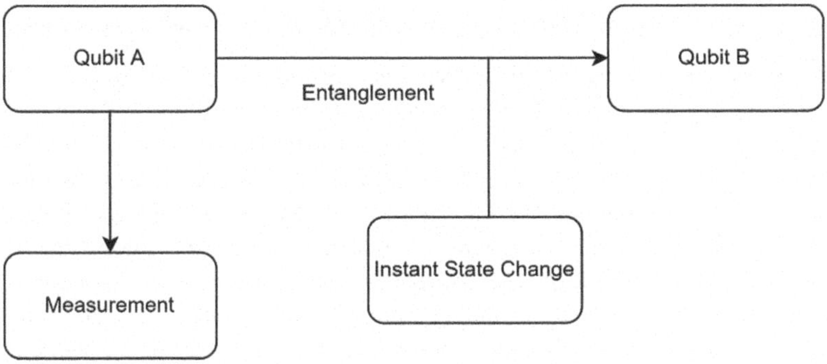

Fig. 6 Entanglement in QNN

entangling these qubits then produces a strong correlation between them. Once entangled, these qubits stay coupled, so no matter how far apart they are, any change in one qubit's state will instantly impact the other's. This characteristic enables QNNs to execute computations in parallel, increasing speed and efficiency—two essential elements of quantum information processing.

The initial stage of quantum information processing is entanglement. In this step, the entangled qubits are used by the QNN to encode, alter, and transmit data. Unlike standard neural networks, which process information sequentially, QNNs can perform multiple operations simultaneously thanks to entangled qubits, which speeds up learning and decision-making. Jobs involving the creation of cryptographic keys, pattern recognition, and optimisation particularly benefit from this feature.

In the measurement and collapse phase, the entangled counterpart of a qubit instantly collapses into a matching state when the first qubit is finally measured. This suggests that the data stored in the entangled qubits will be retrieved and interpreted with minimal latency. Since quantum measurement is inherently probabilistic, multiple measurements might be required to enhance the outcomes. By allowing QNNs to extract useful information from quantum computations in this final step, entanglement is an essential element that enables high-speed, efficient AI systems.

6.4 *Real-World Applications of Quantum Entanglement*

- **Quantum Computing**: Enhances computational speed and efficiency.
- **Quantum Cryptography**: makes use of quantum key distribution (QKD) to enable extremely secure communication.
- **Quasi-Teleportation**: Transfers quantum information instantly across distances.
- **Advanced AI & Machine Learning**: Uses entangled states for faster data processing in QNNs.

Advantages of Quantum Machine Learning (QML) in Pattern Recognition

Quantum machine learning, or QML, provides significant benefits in pattern recognition by handling complex data faster than conventional methods through the use of quantum superposition, entanglement, and parallelism. exponential speedup because QML reduces the time required for feature extraction and classification by analysing multiple patterns simultaneously. This is particularly useful for high-dimensional datasets, like speech and image recognition, where conventional algorithms struggle with computational limitations. By improving feature correlation, quantum entanglement enables QML models to detect minute relationships between data points that traditional machine learning algorithms might miss. Better noise resilience and generalisation: quantum-enhanced learning models are more adept at handling overlapping patterns, which improves accuracy in real-world scenarios. By transferring data to higher-dimensional quantum settings where patterns are easier to recognise, quantum kernels and hybrid algorithms improve pattern recognition. Because of these benefits, QML has a lot of potential for use in fields where precise and prompt pattern recognition is essential, such as autonomous systems, cybersecurity, medical imaging, and financial fraud detection.

6.5 Challenges and Limitations of Quantum Machine Learning (QML)

Despite the enormous potential of quantum machine learning, many obstacles prevent its widespread use. Scalability issues, noise and error correction issues, and the limitations of quantum hardware are the primary concerns.

1. **Quantum Hardware Constraints**

One of the primary issues with QML is the immaturity of quantum hardware. There are only a few qubits used in modern quantum computers, and these qubits are prone to decoherence, which is the degradation of quantum states due to interactions with the environment. Furthermore, maintaining qubits at temperatures near absolute zero—a prerequisite for dependable quantum computation—requires highly specialised and expensive infrastructure. Another problem with restricted gate fidelity is that quantum gates, which are necessary for computation, are still prone to errors due to hardware defects. These drawbacks make it difficult to outperform conventional machine learning models in practical applications and impede the development of reliable, large-scale quantum neural networks.

2. **Correction of Noise and Error**

A significant danger to quantum systems is quantum noise, which arises from interactions with their environment. Quantum decoherence brought on by this noise results in computation errors and lowers the dependability of QML models. Unlike classical computers, where quantum error correction is relatively straightforward, quantum error correction is complex and resource-intensive. Quantum Error Correction Codes (QECCs), such as the Surface Code, require multiple physical qubits to encode a single logical qubit, significantly increasing the hardware requirements. Training and optimising quantum neural networks is difficult when noise is present because mistakes build up over time and produce unreliable results. For QML to be commercially feasible, fault-tolerant quantum computing must be improved.

3. **Scalability and Real-World Implementation**

Scalability remains a significant challenge for QML applications, even with the development of quantum hardware. The majority of contemporary quantum algorithms are evaluated using small-scale quantum applying computers with a small number of qubits to real-world machine learning problems that require big datasets and deep networks is challenging. Furthermore, because a lot of real-world applications call for hybrid strategies, research into combining QML with traditional Frameworks for machine learning are also being developed. The lack of standardised quantum programming tools and frameworks makes it challenging for developers to design and implement scalable QML models. Furthermore, even though industries need low-cost quantum solutions, only research institutions and tech giants can currently afford quantum computing. [28].

6.6 Future Prospects in Quantum Machine Learning (QML)

Quantum Machine Learning (QML) is being influenced by a number of fascinating research avenues and technological developments. Large-scale pattern recognition, integration with cloud computing and classical AI, and advancements in quantum hardware are important areas of focus. These advancements will greatly increase QML's scalability, usefulness, and capacity to be used to solve real-world problems.

1. **Advances in Quantum Hardware**

Enhancing quantum hardware to enable more intricate and dependable computations is one of the most important areas of QML research. Current quantum computers only have a few qubits and are highly susceptible to noise and decoherence. Future advancements aim to achieve:

- **Error-corrected qubits**: Using techniques like surface codes and topological qubits to reduce errors in fault-tolerant quantum computing.
- **Increased qubit scalability**: Boosting the capacity of quantum processors to thousands or even millions of qubits in order to facilitate increasingly challenging machine learning tasks.
- **New quantum computing paradigms**: More reliable and effective platforms for QML could be produced by research into alternative quantum architectures like superconducting qubits, trapped ions, and photonic quantum computing.

Tech giants like Google and IBM, as well as startups like Rigetti and IonQ, are creating next-generation quantum processors to significantly enhance the practical application of QML models.

2. **Integration with Classical AI and Cloud Computing**

- Another fascinating field of research is the integration of QML with cloud-based systems and conventional AI. Combining quantum-enhanced algorithms with conventional machine learning models is a workable short-term approach, as full-scale Quantum computing is still in its early stages of development [29, 30]. One method to accomplish this is by:
- **Hybrid quantum–classical models**: Utilising quantum-assisted learning, in which classical AI handles the remaining computations while quantum circuits perform some of them.
- **Quantum cloud computing**: Thanks to cloud-based Quantum-as-a-Service (QaaS) platforms from companies like IBM, Amazon, and Microsoft, AI researchers can now access quantum processors remotely.
- **Quantum-inspired algorithms**: Even before full-scale quantum computing is available, researchers are developing quantum-inspired classical algorithms that mimic quantum properties to enhance AI models.
- QML can progressively move from experimental research to useful enterprise applications by utilising cloud computing and hybrid architectures.

3. **Potential Breakthroughs in Large-Scale Pattern Recognition**

One of QML's most exciting applications is the potential to revolutionise large-scale pattern recognition in fields like autonomous systems, financial fraud detection, and medical imaging. These areas are the focus of research on quantum-enhanced pattern recognition:

Quantum Boltzmann Machines: deep learning models that are quantum-variant and efficiently identify patterns in large amounts of data.

Quantum Kernel Methods: Quantum feature maps transform data into high-dimensional quantum spaces to facilitate pattern recognition.

Exponential Speedup: This method uses quantum parallelism to process large amounts of data concurrently, greatly cutting down on computation time for tasks like image classification and speech recognition.

Through sophisticated medical imaging analysis, QML may expedite diagnosis in domains such as healthcare. Quantum-enhanced anomaly detection may enhance cybersecurity by detecting malware and fraud. These uses will become more practical as quantum hardware advances.

4. **Quantum Data and Encoding Strategies**

Despite offering exponential information compression and representing large datasets in the amplitudes of quantum states, amplitude encoding is frequently constrained by the challenge of state preparation. On the other hand, angle encoding makes it more feasible for near-term devices by mapping data features to the rotation angles of quantum gates. This is further enhanced by quantum feature maps, which embed data into high-dimensional Hilbert spaces and enable quantum kernels to exploit patterns that are hard for classical algorithms to access. Because it affects the model's expressiveness, learnability, and viability for use on noisy intermediate-scale quantum (NISQ) devices, choosing the right encoding technique is crucial.

References

1. Das, S., et al.: Quantum pattern recognition on real quantum processing units. Quantum Mach. Intell. **5**(1), 16 (2023)
2. Evans, E.N., Byrne, D., Cook, M.G.: A quick introduction to quantum machine learning for non-practitioners. arXiv preprint arXiv:2402.14694 (2024)
3. Bauckhage, C., et.al.: Quantum Machine Learning. Federal Office for information security
4. Simeone, O.: An introduction to quantum machine learning for engineers, vol. 16(1–2), pp. 1–223. Foundations and Trends® in Signal Processing (2022)
5. Mishra, N., et al., Quantum Machine Learning : A Review and Current Status Quantum Machine Learning : A Review, no. Jan. Springer Singapore (2021)
6. Saadatfar, H., Khosravi, S., Joloudari, J.H., Mosavi, A., Shamshirband, S.: A New K-Nearest Neighbors Classifier for Big Data Based on Efficient Data Pruning, pp. 1–12
7. Feng, C., Zhao, B., Zhou, X., Ding, X., Shan, Z.: An Enhanced Quantum K-Nearest Neighbor Classification, pp. 1–12 (2023)
8. Dang, Y., et al.: Image classification based on quantum KNN algorithm. arXiv preprint arXiv:1805.06260 (2018)
9. Sharma, S.: QEML (quantum enhanced machine learning): Using quantum computing to enhance ml classifiers and feature spaces. arXiv preprint arXiv:2002.10453 (2020)
10. Yesareth, A., Estrada, G., Quezada, L.F., Sun, G.H.: Benchmarking quantum versions of the kNN algorithm with a metric based on amplitude—encoded features. Sci. Rep. **0123456789**, 1–16 (2024)
11. Gao, L., Lu, C., Guo, G., Zhang, X., Lin, S.: Quantum K-Nearest Neighbors Classification Algorithm Based on Mahalanobis Distance, no. Oct, pp. 1–6 (2022)
12. Zardini, E., Blanzieri, E., Pastorello, D.: A Quantum K-Nearest Neighbors Algorithm Based on the Euclidean Distance Estimation (2024)
13. Rebentrost, P., Mohseni, M., Lloyd, S.: Quantum Support Vector Machine for Big Data Classification, vol. 1(2), pp. 1–5
14. Suzuki, T., Hasebe, T., Miyazaki, T.: Quantum support vector machines for classification and regression on a trapped—ion quantum computer. Quantum Mach. Intell. **6**(1), 1–14 (2024)
15. Nohara, T., Oyama, S., Noda, I.: Pairwise classification using quantum support vector machine with Kronecker kernel. Quantum Mach. Intell. (2022)
16. Innan, N., Khan, M.A.Z., Panda, B., Bennai, M.: Enhancing Quantum Support Vector Machines through Variational Kernel Training, no. Ml, pp. 1–15 (2023)
17. El Maouaki, W., Said, T., Bennai, M., Physics, Q., Team, M., Ben, S.: Quantum support vector machine for prostate cancer detection : A performance analysis, pp. 1–14
18. Bhattacharjee, S., Fuad, M.D., Hossain, A.K.M.: Classification of financial data using quantum support vector machine. arXiv preprint arXiv:2412.10860 (2024)
19. Qin, J.: Review of ansatz designing techniques for variational quantum algorithms. J. Phys. Conf. Ser., vol. 2634(1). IOP Publishing (2023)
20. Kulshrestha, A., et al.: QAdaPrune: Adaptive parameter pruning for training variational quantum circuits. In: 2024 IEEE International Conference on Quantum Computing and Engineering (QCE), vol. 2. IEEE (2024)

21. Peddireddy, D., Bansal, V., Aggarwal, V.: Classical simulation of variational quantum classifiers using tensor rings. Appl. Soft Comput. **141**, 110308 (2023)
22. Bilek, S., Wold, K.: Recursive variational quantum compiling. arXiv preprint arXiv:2203.08514 (2022)
23. Kruse, G., et al.: Variational quantum circuit design for quantum reinforcement learning on continuous environments. arXiv preprint arXiv:2312.13798 (2023)
24. Tseng, H.-H., et al.: Transfer learning analysis of variational quantum circuits. arXiv preprint arXiv:2501.01507 (2025)
25. Akrom, M., Rustad, S., Kresno, H.: Variational quantum circuit-based quantum machine learning approach for predicting corrosion inhibition efficiency of pyridine-quinoline compounds. Mater. Today Quantum **2**(Feb), 100007 (2024)
26. Beer, K.: Quantum neural networks. arXiv preprint arXiv:2205.08154 (2022)
27. Kwak, Y., et al.: Quantum neural networks: Concepts, applications, and challenges. In: 2021 Twelfth International Conference on Ubiquitous and Future Networks (ICUFN). IEEE (2021)
28. Ganguly, S.: Quantum Machine Learning: An Applied Approach. Apress, New York (2021)
29. How, M.: Forging the Future : Strategic Approaches to Quantum AI Integration for Industry Transformation, pp. 290–323 (2024)
30. Prangon, N.F., Wu, J.: AI and computing horizons: cloud and edge in the modern era. J. Sens. Actuator Netw. **13**(4), 44 (2024)

Quantum Neural Networks: From Concept to Simulation

Nasrullah Masud, Abdus Sobhan, Md Tanvir Islam Sourav, and Abdullah Al Bassam

Abstract Quantum Neural Networks (QNNs) herald a new paradigm in the unison of quantum computing and artificial intelligence, paving the way for transformative machine-learning capabilities and computational issues solving. This chapter probes into QNNs from their basis in classical neural networks and quantum computing. The mathematical foundations important for the design of QNNs-such as linear algebra, quantum gates, and optimization methods-have also been explored in this chapter. Various architectures such as Variational Quantum Circuits (VQCs), Quantum Convolutional Neural Networks (QCNNs), and Recurrent Quantum Neural Networks (RQNNs) have been discussed, each with its own benefits and application domains. It provides comparisons between classical and quantum neural networks, whereby QNNs' benefits, like speedup and parallelism, are emphasized as well as other disadvantages such as hardware dependency and scaling issues. Training methods featured include quantum gradient descent and natural gradient, while some key challenges that crop up include noise, decoherence, and barren plateaus. Real-world applications of QNNs feature quantum machine learning, optimization, pattern recognition, quantum chemistry, and financial modeling. A case study has been devoted to showcasing a demonstration of QNN implementation and performance. Quantum simulators, such as Pennylane, Qiskit, TFQ, Cirq, PTQ, Microsoft Quantum Development Kit (Q#), contribute to the development and testing of QNNs while dealing with their limitations. Finally, future directions are discussed: hardware improvements, combination with classical AI, and new areas of research. This brief yet compact overview should appraise users of the basics of QNNs, status quo, and possibilities in revolutionizing AI and quantum computing.

Keywords Quantum neural networks (QNNs) · Variational quantum circuits (VQCs) · Parameter-shift rule · Barren plateaus · Hybrid quantum–classical algorithms · Quantum simulators · Pennylane

N. Masud
Department of Electrical and Electronic Engineering, Varendra University, Rajshahi, Bangladesh

A. Sobhan · M. T. I. Sourav · A. Al Bassam (✉)
Department of Computer Science and Engineering, Varendra University, Rajshahi, Bangladesh
e-mail: abassamrajbd65@gmail.com

© The Author(s), under exclusive license to Springer Nature Switzerland AG 2025
A. Ghosh et al. (eds.), *Quantum Machine Learning in Industrial Automation*, Information Systems Engineering and Management 65,
https://doi.org/10.1007/978-3-031-99786-0_16

1 Introduction

1.1 Classical Neural Networks

A classical neural network usually consists of interconnected layers of "neurons," each connected to weights and biases. One of the most common types of neural networks is feedforward neural networks (FCNNs), where information flows in only one direction-from input to output.

Basic Feedforward Neural Network Equations

1. **Linear Transformation**

$$z^{(l)} = \boldsymbol{W}^{(l)}\boldsymbol{a}^{(l-1)} + \boldsymbol{b}^{(l)}$$

- $\boldsymbol{W}^{(l)}$ is the weight matrix for layer l.
- $\boldsymbol{b}^{(l)}$ is the bias vector for layer l.
- $\boldsymbol{a}^{(l-1)}$ is the output (activation) of the previous layer.

2. **Nonlinear Activation**

$$\boldsymbol{W}^{(l)} = \sigma\left(z^{(l)}\right)$$

- $\sigma(\cdot)$ can be a nonlinear function such as ReLU, sigmoid, or tanh.

3. **Output Layer**

$$\boldsymbol{a}^{(l)} = f\left(z^{(l)}\right)$$

- If it is a classification task, $f(\cdot)$ might be a softmax function.
- If it is a regression task, $f(\cdot)$ might be the identity function or another suitable mapping.

These equations capture the forward pass of a classical network. Training typically uses **backpropagation**, which relies on gradient-based optimization (e.g., stochastic gradient descent, Adam).

1.2 Convolutional Neural Networks (CNNs)

CNNs are specialized mathematical structures that handle grid-like data, especially images. Introduced by Yann LeCun in the 1980s, CNNs became widely known with LeNet-5 in 1998. Their architecture, inspired by the human visual system, includes:

- **Convolutional Layers**: Applying filters for the extraction of features like edges and textures.

- **Activation Functions**: Nonlinearity introduction (say, ReLU) for learning complex patterns.
- **Pooling Layers**: Spatial dimension reduction for further efficiency.
- **Fully Connected Layers**: All the extracted features come to a consensus for the final predictions.

CNNs revolutionized image recognition, object detection, and medical diagnostics.

1.3 Quantum Neural Networks

Quantum Neural Networks (QNNs) implement neural network principles using parameterized quantum circuits (PQCs) that manipulate qubits through sequences of quantum gates. Unlike classical networks, QNNs leverage two unique quantum properties:

1. **Quantum Parallelism**: Through superposition, QNNs can evaluate multiple states simultaneously (demonstrated by Grover's search algorithm's quadratic speedup).
2. **Entanglement**: Creates correlated quantum states that enable compact representation of complex feature spaces.

These quantum characteristics allow QNNs to potentially overcome fundamental limitations of classical networks, particularly for high-dimensional data processing and specialized optimization tasks.

1.4 How Does QNN Work

A **Quantum Neural Network** is a parameterized quantum circuit designed to process information (quantum or classical) and produce a measurable output—much like a classical neural network (NN). However, instead of performing matrix multiplications and activations on classical vectors, a QNN applies **unitary gates** to **quantum states** and measures observables to obtain outputs.

The core workflow can be summarized as:

1. **Data Encoding (or State Preparation)**

If we have classical data $x \in R^d$.

we first need to **encode** x into a quantum state. There are several strategies:

A. **Basis Encoding**: Map the data index to binary. For instance, x might specify which computational basis state to occupy.

B. **Amplitude Encoding**: Use the components of x to define amplitudes of a state $|\psi_0(x)\rangle$, normalized appropriately:

$$|\psi_0(x)\rangle = \sum_{j=0}^{2^n-1} x_i|j\rangle \; where \; \sum_j |x_j|^2 = 1,$$

C. **Parameter Embedding**: Insert the components of x directly as rotation angles in one or more gates (e.g., $R_Z(x_j)$) at the beginning of the circuit.

In many QNN designs, you start with $|0\rangle^{\otimes n}$ and apply an "embedding layer $U_{enc}(x)$ to transform it into a data-dependent state:

$$|\psi_0(x)\rangle = U_{enc}(x)|0\rangle^{\otimes n}.$$

2. **Parameterized Quantum Circuit (the QNN "layers")**

Circuit as a Product of Parameterized Gates

A **quantum circuit** is a sequence of unitary gates, each potentially depending on parameters θ Suppose we have L layers. One standard decomposition is:

$$U(\theta) = \prod_{l=1}^{L} U_l(\theta_l)$$

where each $U_l(\theta_l)$ is typically built from gates such as $R_X(\theta), R_Y(\theta), R_Z(\theta)$, or controlled versions thereof. For instance:

$$R_X(\theta) = \exp\left(-\frac{i}{2}\theta\sigma x\right), R_Z(\theta) = \exp\left(-\frac{i}{2}\theta\sigma z\right).$$

Hence, the **QNN** is the **unitary map**:

$$|\psi_0(x)\rangle = U(\theta)U_{enc}(x)|0\rangle^{\otimes n}.$$

The Forward Pass

- In classical NNs, the forward pass is a sequence of matrix multiplications and nonlinear activations.
- In QNNs, the "forward pass" is **running the quantum circuit** on the input state:
  1. Prepare the state $|\psi_0(X)\rangle$.
  2. Apply U(θ).
  3. Obtain the final state $|\psi_0(X, \theta)\rangle$.

3. **Measurement and Cost Function**

To extract a classical "output" from the final state, we measure one or more observables. A common choice is to measure a Hermitian Operator M, yielding the **expectation value**:

$$f(x;\theta) = \langle \psi(x,\theta)|M|\psi(x,\theta)\rangle.$$

If we are training for a supervised learning task (e.g., classification), we might define a cost function that compares $f(x;\theta)$ to a target label y. For instance, a simple form of cost function is a **mean squared error** (MSE) over training samples $\{(x_i, y_i)\} : \{(x_i, y_i)\}$:

$$C(\theta) = \frac{1}{N}\sum_{i=1}^{N}\left[y_i - \langle \psi(X_i,\theta)|M|\psi(X_i,\theta)\rangle\right]^2.$$

In quantum optimization problems (e.g., VQE—Variational Quantum Eigensolver), the cost might be the expectation value of a Hamiltonian H:

$$C(\theta) = \langle \psi(\theta)|H|\psi(\theta)\rangle,$$

which we want to **minimize** in order to approximate the ground state energy.

4. **Optimization (via classical or hybrid quantum–classical methods)**

The quantum neural network (QNN) optimization methods refer to the optimizing of the circuits through determining the parameters θ such that the cost function C(θ) becomes minimized. This is done via hybrid quantum–classical algorithms. The classical data is encoded initially into a quantum state $|\psi_0(x)\rangle = U_{en}c(x)|0\rangle^n$, by application of a parameterized set of gates U(θ) to produce $|\psi_0(x,\theta)\rangle$. The output from the system may be obtained by measuring observable M, which resulting in $f(x;\theta) = |\psi\rangle M|\psi\rangle$. Cost function for supervised learning is usually the mean square error as follows: $C(\theta) = (1/N)\sum\left[y_i - f(x_i;\theta)\right]^2$ while variational algorithms such as VQE minimize: $|\psi\rangle H|\psi\rangle$ for Hamiltonian H.

1.5 Training (Optimization)

A. The Hybrid Loop

Because quantum states are expensive to simulate fully on classical hardware, and actual quantum devices are not (yet) robust enough to do sophisticated on-chip training, QNNs are typically trained in a **hybrid** manner:

1. **Classical computer** holds parameters θ and runs an optimization algorithm (e.g., gradient descent, Adam, etc.).
2. **Quantum processor** (or simulator) evaluates the cost function or its gradient for the given θ.

The classical optimizer updates θ, we re-run the circuit on the quantum device, and repeat until convergence.

B. **Gradient Computation**

To perform gradient-based updates, we need $\partial C/\partial\theta$ Naively, we might attempt to do backpropagation as in classical NNs, but unitaries and measurements complicate matters.

A popular technique is the **parameter-shift rule**: if f(θ) is an expectation value that depends on a single-qubit rotation $R_{\partial}(\theta) = \exp(i\theta\sigma/2)$ (with σ a Pauli matrix), then:

$$\frac{\partial f(\theta)}{\partial\theta} = \frac{f\left(\theta + \frac{\pi}{2}\right) - f\left(\theta - \frac{\pi}{2}\right)}{2},$$

Generalizing this to many gates yields a **gradient estimation** procedure by executing the circuit at shifted parameter values. Thus, for each parameter θ_k:

$$\frac{\partial C}{\partial\theta_k} = \frac{C\left(\theta + \frac{\pi}{2}\hat{e}k\right) - c\left(\theta - \frac{\pi}{2}\hat{e}k\right)}{2},$$

where $\hat{e}k$ is a vector of zeros except for a 1 in the k-th component.

C. **Update Rule**

If we use simple gradient descent with learning rate η, then:

$$\theta_k \leftarrow \theta_k - \eta\frac{\partial C}{\partial\theta_k}.$$

We iterate this procedure (1) measure cost and gradient on the quantum device, (2) update parameters on a classical computer.

2 Literature Review

Introduction and Background

Quantum neural networks (QNNs) are a ground-breaking combination of quantum computing and neural network models, introducing new computation speed, parallelism, and high-dimensional data processing capabilities [1, 2]. The basic idea of traditional neural networks (CNNs) is that they work by transforming data and applying an activation function to do so, while QNNs inherit the quantum phenomena like superposition and entanglement for the purpose of the parallelized processing [3, 4]. Some early theoretical studies have shown that QNNs are capable of generalizing well from small datasets and can also handle noisy inputs by means of robustness, thus overcoming classical models' limitations in dealing with the complex optimization tasks [5, 6]. Nevertheless, difficulties are still around in quantum data encoding, and techniques such as amplitude and angle encoding are confronted with the problem of being unscalable [2, 7]. The most recent solution to this issue is the introduction of hybrid classical-quantum architectures which use classical components for data pre-processing and quantum layers for feature extraction, thus making it practical to be realized in hardware and software [8–10].

1. **Comparison with Classical Neural Networks**

Classical neural networks (CNNs) process deterministically the data via backpropagation and the classical neural networks (QNNs) exploit superposition and entanglement for state representation of exponential size, ideal for high-dimensional tasks [3, 11, 12]. However, the QNNs have not yet shown practical advantages because they are still only theoretical promising and the current noisy quantum hardware is unable to implement the algorithms correctly hence noise degradation results in the performance of the devices [3, 6, 11, 13]. The aspiration of training QNNs is to efficiently solve combinatorial quantum problems, yet the first quantum neural network obstacle to a solution is the large scale of the plateau while the gradients vanish simultaneously [6, 14]. If we talk about scalability we should remember that CNNs have a considerable advantage here. As we know CNNs can be run with virtually any number of computing resources, while QNNs need enough qubits and circuit depth to solve the same problems as CNNs.

It's important to mention that, in addition to conventional tasks, e.g., image recognition, CNNs have always been in the lead yet QNNs are still doing great in tasks that are quantum-specific, such as molecular simulation [15, 16] and combined architectures for power-grid optimization [10]. Tables 1 and 2 summarize these comparisons.

Table 1 Differences between a quantum neural network (QNN) and a classical neural network

Aspect	QNN	Classical neural network (CNN in the broad sense)
Substrate	Relies on quantum states and quantum gates	Relies on classical numeric computation (e.g., real-valued weights)
Encoding	Inputs encoded as quantum states ("qubits")	Inputs are directly processed as numeric arrays
Operations	Uses unitary operations, entanglement, measurements	Uses linear operations and nonlinear activations
Information carriers	Qubits	Classical bits/floating-point values
Potential benefits	Quantum parallelism, potential exponential speedups in certain tasks	Established methods, large ecosystem, well-understood training approaches
Maturity of tools	Early-stage research, specialized quantum hardware required	Highly mature frameworks (TensorFlow, PyTorch, etc.), commodity hardware

Table 2 Comparison between a quantum neural network (QNN) and classical neural network

Feature	Classical neural networks	Quantum neural networks	Citations
Computational unit	Bits	Qubits	[1]
Computation	Logic gates, arithmetic operations	Quantum gates, unitary transformations, measurements	[1]
Data representation	Binary or real-valued vectors	Quantum states (amplitudes, phases in Hilbert space)	[1]
Computational power	Polynomial or exponential scaling (problem-dependent)	Potential for exponential speedups for specific problems	[2]
Learning	Gradient descent, backpropagation	Variational methods, Quantum gradient descent, Natural gradient	[6, 17]
Regularization	Weight decay, dropout, etc	Unitarity of quantum circuits (inherent), other methods under development	[14]
Hardware	CPUs, GPUs	Quantum computers (NISQ and beyond)	[14]
Noise sensitivity	Relatively robust	Highly sensitive to decoherence and gate errors	[5, 13]
Trainability	Well-established techniques	Challenges like barren plateaus	[14]

2. **Architectural Paradigms in Quantum Neural Networks**

Today's quantum neural networks (QNNs) include Variational Quantum Circuits (VQCs) [9, 14]. These networks combine quantum layers with traditional neural structures through gates with parameters that classical methods optimize. They achieve high accuracy reaching 92% in medical image analysis [18]. Quantum Convolutional Neural Networks (QCNNs) [3, 4, 11] adapt classical CNNs by swapping filters for quantum circuits. This change allows for resistance to noise and 20% quicker training in tasks such as image grouping. However, qubit connections limit their use. Recurrent Quantum Neural Networks (RQNNs) [5, 19] add quantum memory to handle data that comes in order (like in finance). Yet, they struggle with loss of quantum states pushing scientists to find ways to lessen errors. These designs show promise in mixing quantum and classical methods. Still growing them to useful sizes and overcoming hardware limits (like qubit numbers and error rates) stand as major hurdles to real-world use (Table 3).

3. **Training and Optimization**

A standard method for quantum hardware is quantum-native optimization, wherein quantum neural networks (QNNs) are configured to work in concert with a quantum processor. Such QNNs are only partially learnable, because these gradient-descent methods have barren plateaus—regions where the gradient becomes very small—and this aggravates the problem, e.g., in deep architectures [5, 6]. One of the principal

Table 3 Comparison of different QNN architectures

Architecture	Key features	Strengths	Weaknesses	Typical application	Citations
Variational quantum circuits (VQCs)	Parameterize the quantum circuits optimized using classical methods	Relatively easy to implement on near-term hardware, flexible ansatz design	Performance heavily dependent on ansatz and classical optimizer, susceptible to barren plateaus	Quantum chemistry, combinatorial optimization, machine learning tasks	Transfer learning analysis [17], iteration complexity analysis under noise [13], advances in optimization techniques [14]
Quantum convolutional neural networks (QCNNs)	Quantum analogues of classical CNNs using quantum circuits for convolution and pooling	Potential for efficient feature extraction from quantum data can leverage entanglement	Hardware intensive, training can be challenging, and qubit requirements can scale with input size	Image classification, quantum phase recognition, quantum error correction	Measurement-based implementations using cluster states [3], hybrid models with classical CNNs for medical image classification [18]
Recurrent quantum neural networks (RQNNs)	Quantum circuits designed to process sequential data, often with feedback loops or memory components	Potential for learning complex temporal dependencieses, can leverage quantum memory effects	Training can be complex; performance evaluation against classical RNNs is still ongoing	Time series prediction, natural language processing	Continuous-variable RQNNs for time series prediction [19], novel parameterizations for unitary RNNs [19]

challenges is to produce quantum hardware that can tolerate noise, and the empirical experiences indicate that digital-analog QCNNs [4] and hybrid models [10] are 20–40% less prone to error when depolarizing noise is present [5]. The incorporation of physics-informed neural networks (PINNs) into QNNs [10] results in QNNs that effectively solve power grid optimization tasks, through following energy demand and line flow rules and achieving reliable solutions during training. Summary of Key Challenges in QNNs is shown in Table 4.

4. Real-World Applications and Performance

Overview of Real-World Applications of QNNs is displayed in Table 5.

Healthcare

One example of such networks is the so-called distributed hybrid QCNNs, which can classify tumors in MRI scans with 92% accuracy and they are superior to classic

Table 4 Summary of key challenges in QNNs

Challenge	Description	Citations
Hardware limitations	Limited number of qubits, short coherence times, restricted qubit connectivity, high error rates in quantum gates	Development of improved quantum hardware (ongoing) [14]
Noise and decoherence	Susceptibility of quantum states to environmental interactions, leading to loss of quantum information and computational errors	Investigation and mitigation strategies for VQCs [13], application of quantum natural gradient to noisy circuits [5], noise-aware frameworks [5]
Barren plateaus	Vanishing gradients in the cost function during training, hindering the optimization process, especially in deep networks	Specialized initialization techniques [6, 7]
Scalability	Difficulty in scaling QNNs to handle large and complex problems due to hardware limitations and training complexities	Development of hybrid quantum–classical architectures [18, 8], exploration of measurement-based QCNNs [3], scalable PQC classifiers [20]
Theoretical framework	Need for a more comprehensive theoretical understanding of QNNs, particularly regarding generalization and quantum advantage	Research on stability and generalization of QNNs [6]; analysis of entanglement-induced barren plateaus [7]

CNNs [18]. At the same time, hybrid models are also able to predict catalyst efficiencies in quantum chemistry, thus establishing a 95% correlation with experimental data [15].

Finance and Energy

The hybrid QNNs that are aimed at stock market prediction [21] are able to achieve a 12% higher Sharpe ratio than the traditional models. Another case is when the physics-informed QNNs come into play as they wind up as the best solver for AC optimal power flow problems with the fastest, both 40% computation and 30% latency reduction [10].

5. Limitations and Future Directions

Some of the present limiting factors are the number of qubits which does not allow qubit scalability in real life (e.g., < 100 qubits in most implementations [22]) and the gaps found in theoretical generalization guarantees [6]. QNNs have the ability to be used as data storage devices of random data [6], however, it's not yet clear if they can be used in pattern extraction. Progress in the future depends greatly on hardware developments (e.g., error-corrected qubits) and algorithm-architecture co-design [9, 14]. Hybrid models, especially the ones based on PINNs as well as residual learning [10], offer the reality of achieving quantum advantage by making practical progress.

Table 5 Overview of real-world applications of QNNs

Application Area	Specific task	QNN architecture/ approach used	Reported performance and limitations	Citations
Medical image classification	Feature extraction and classification of medical images (e.g., brain tumors, breast cancer, pneumonia)	Quantum convolutional neural networks (QCNNs), hybrid QCNNs with classical CNNs	Show promise in feature extraction and classification, often with reduced parameters compared to classical CNNs. Performance on real quantum hardware can be limited by noise	Distributed hybrid QCNN for medical image classification [18], QCNN with digital-analog quantum convolutions for medical image recognition [4], robustness analysis of HQNNs for image classification [5], hybrid QNN for efficient MNIST binary image classification [9]
Quantum chemistry	Modeling molecular excited states, predicting physicochemical properties, catalyst modelling	Hybrid quantum neural networks (HQNNs), variational quantum Eigensolvers (VQEs)	Can achieve higher accuracy and faster convergence compared to classical methods in some cases, particularly for complex quantum systems. Often limited by the number of qubits and noise	AI for accurate computation of quantum excited states [1], HQNNs for physicochemical property prediction [8], hybrid QNN for catalyst modelling [15]
Financial modeling	Stock price prediction, financial trend forecasting, portfolio optimization	Quantum Neural Network (QNN) regressors, contextual QNNs, hybrid quantum–classical models	Show potential in capturing complex temporal dependencies and improving prediction accuracy, but real-world deployment on quantum hardware is still in the early stages	HQNN for regression-based financial stock market prediction [21]; contextual QNNs for stock price prediction [21]

(continued)

Table 5 (continued)

Application Area	Specific task	QNN architecture/ approach used	Reported performance and limitations	Citations
Pattern recognition	Binary classification tasks, handwritten digit recognition	Quantum support vector machines (QSVMs), quantum neural networks (QNNs), and hybrid quantum-classicalcal neural networks (H-QNNs)	Quantum algorithms can demonstrate competitive or even superior performance compared to their classical counterparts in terms of accuracy, especially for small-scale problems. Performance can vary depending on the dataset and the specific quantum algorithm used	Implementation and performance evaluation of QML algorithms for binary classification [22], hybrid QNN for efficient MNIST binary image classification [9] experimental evaluation of QML algorithms [12, 23]
Time series prediction	Predicting temporal data, such as weather patterns or system dynamics	Recurrent quantum neural networks (RQNNs), continuous-variable quantum RNNs (CV-QRNNs)	CV-QRNNs have shown faster convergence and lower losses compared to classical RNNs in simulations for certain types of temporal data	Recurrent quantum neural networks review [19]

Conclusion

Quantum neural networks (QNNs) are a new learning approach in the artificial intelligence sector. Thus far, they have been used in healthcare, finance, and energy systems with success. The constant challenge of noise and scalability is still present, but through so-called hybrid architectures and noise-resilient training methods [5, 10] the way to a solution opens up. Thus, it is just a matter of time when after the quantum hardware gains its prevalence QNNs start drawing the new AI computation horizons.

3 Necessary Components of Quantum Computing

Quantum computing, in essence, means bringing together computer science, engineering, and physics. Simply put, the quantum principles and methods used in computing are employed to solve problems that lie beyond the ability of classical computing. To understand quantum computing clearly, one must have some idea about its two basic components-hardware and software. The chapter provides working quantum gates and the structure of quantum computing systems.

3.1 Quantum Hardware and Qubit Realization

A quantum computer's hardware contrives a foundation for the physical realization of qubits. These qubits exploit quantum phenomena such as superposition and entanglement to encode/process the information. Various types of qubit technology exist, each one with its own particular advantages and imperfections.

Superconducting Qubits: Superconducting circuits are using Josephson junctions as non-linearity for being one's the leading qubit technology. They permit a fast gate operation and are scalable, therefore, a main highlight in research in quantum computing.

Trapped Ions: Trapped ion systems use electromagnetic fields to manipulate individual ions. They offer high-fidelity gates and long coherence times but face challenges in scaling to larger qubit counts.

Topological Qubits: While based on anyons and non-Abelian statistics, topological qubits hold the promise of built-in error resistance; they are still, however, in the lab and under experimental study before entering any widespread application.

Photonic Qubits: Photonic qubits have photons as a carrier of quantum information in contrast with other qubits that require low temperatures to allow high-speed quantum communication. As a result, a few technological challenges such as photon detection and manipulation limit computational efficiency.

Key Performance Metrics: Critical factors for qubit viability include:

- **Coherence time**—How long a qubit maintains its quantum state.
- **Gate fidelity**—The accuracy of quantum operations.
- **Scalability**—The ease of integrating more qubits into a system.

Addressing these challenges is essential for real-world quantum computing applications.

3.2 Quantum Software and Algorithmic Infrastructure

The software layer converts abstract quantum algorithms into executable instructions on quantum hardware, while hardware supplies the physical foundation for quantum computation. The quantum software stack is made up of the following parts:

- **Quantum Programming Languages**
  Languages such as Q#, Qiskit, and Cirq are used to design, simulate, and execute quantum circuits. For instance, Microsoft Q# provides integration with classical environments, thereby supporting hybrid quantum-classical workflows and resource estimation.
- **Quantum Algorithms and Protocols**
  Among these algorithms, the well-known ones include Shor's algorithm—factoring; Grover's algorithm—searches; and variational methods—quantum machine learning—that reflect their quantum advantage by exploiting entanglement and superposition to quantum exceed classical computational limits.
- **Error Correction and Fault Tolerance**
  Given that quantum states are extremely vulnerable to error correction, the process of error correction becomes a major area of interest. Techniques such as surface code will seek to reduce operational noise and decoherence in order to achieve reliable computation in scalable systems.
- **Simulation and Verification Tools**
  By simulating the quantum computers classically, researchers can test quantum algorithms and optimize them before running them on real hardware, thus enabling evaluation of the performance against practical constraints.

The efficiency and potential of a quantum computing system depend on the cooperation of strong quantum hardware and advanced software infrastructure. The possibilities of quantum computing are constantly being expanded by developments in both fields.

3.3 Illustrative Figure: Overview of Quantum Gates

Key quantum gates, which are the essential building elements of quantum circuits, are shown in detail in the picture below. Superposition, entanglement, and controlled state manipulation are among the fundamental quantum computational functions made possible by these gates. Figure 1 shows the following gates:

- **Pauli Gates (X, Y, Z)**
  Pauli Gates (X, Y, Z) are basic quantum gates that rotate states and flip them around their corresponding Bloch sphere axes.
- **Hadamard (H) Gate**
  Used to convert computational basis states into equal superpositions in order to generate superposition states.

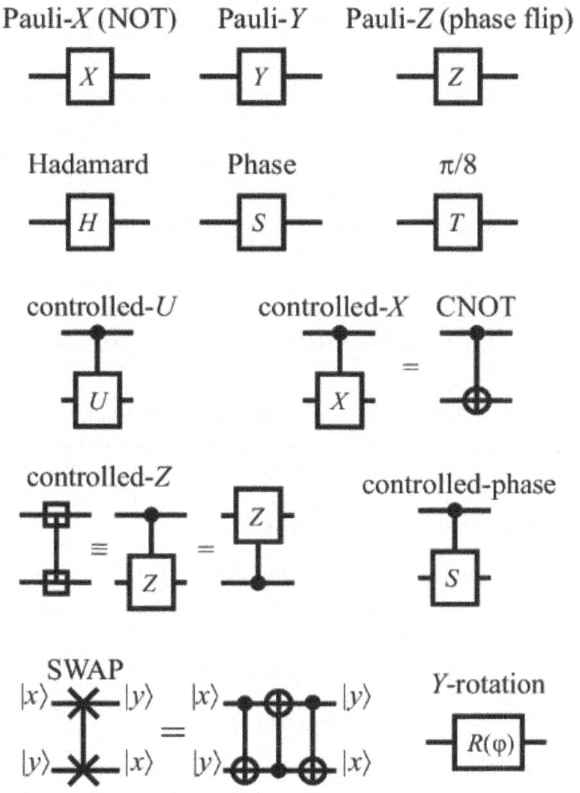

Fig. 1 Overview of quantum gates. *Source* Lancaster University—Quantum Information Processing Course Materials

- **Phase (S) Gate and π/8 (T) Gate**
  Gates that bring about particular phase shifts that are essential for quantum computing in general.
- **Controlled Gates (CNOT, Controlled-U, Controlled-Z, Controlled-phase)**
  Entanglement is produced by gates that carry out conditional operations according to the state of a control qubit.
- **SWAP Gate**
  Two qubits' quantum states are switched.
- **Rotation Gates (e.g., Y-rotation R(φ))**
  Fine-tuning quantum state manipulation is made possible by general rotation operations around certain axes at arbitrary degrees.

4 Architectures of Quantum Neural Networks

Quantum neural networks (QNNs) employ several distinct architectural paradigms, each designed to leverage quantum mechanical properties for machine learning tasks. These architectures vary in their circuit designs, training methodologies, and application domains, offering different trade-offs between expressivity, trainability, and hardware requirements.

4.1 Hybrid Quantum–Classical Models

Hybrid architectures represent the most practical approach for near-term quantum machine learning, combining classical pre-processing with quantum circuit processing:

1. **Classical Pre-processing Stage**
   - Data normalization and dimensionality reduction (typically using PCA).
   - Feature scaling to appropriate ranges for quantum encoding (usually $[0,\pi]$ for angle encoding).
   - Input data reshaping to match qubit count constraints.
2. **Quantum Processing Stage**
   - Parameterized quantum circuits with trainable gates.
   - Carefully designed ansätze to maximize expressivity while minimizing circuit depth.
   - Measurement strategies tailored to the specific learning task.
3. **Classical Post processing**
   - Interpretation of quantum measurement outcomes.
   - Final classification or regression steps.
   - Gradient computation for parameter updates.

Example Implementation: A typical hybrid model might use PCA to reduce 784-dimensional MNIST images to 8 features, encode them into 8 qubits via angle encoding, process through a 4-layer variational circuit, and measure expectation values for classification.

4.2 *Variational Quantum Circuits (VQCs)*

VQCs form the backbone of most current QNN implementations, characterized by:

Circuit Structure

- Initial state preparation layer.
- Alternating layers of:
  - Single-qubit rotation gates (RX, RY, RZ).
  - Entangling gates (typically CNOT or CZ).
- Final measurement layer.

Key Features

- Depth typically limited to 5–20 layers due to NISQ device constraints.
- Employ various entanglement patterns (linear, circular, or all-to-all).
- Can implement different ansätze (e.g., hardware-efficient, chemistry-inspired).

Training Characteristics

- Hybrid quantum–classical optimization loops.
- Parameter-shift rule for gradient estimation.
- Susceptible to barren plateaus in deep circuits.

Applications

- General quantum machine learning tasks.
- Quantum chemistry simulations.
- Combinatorial optimization problems.

Comparison between VQC versus FCNN (Fully Connected Neural Network) is shown in Table 6.

4.3 *Quantum Convolutional Neural Networks (QCNNs)*

QCNNs extend classical CNN concepts to quantum systems:

Architectural Components

1. **Quantum Convolution Layers**
   - Local unitary operations applied to small groups of qubits.
   - Translationally invariant operations mimicking classical convolution.
   - Typically, 2-qubit unitaries with parameterized rotations.
2. **Quantum Pooling Layers**
   - Progressive reduction of qubit count while preserving relevant features.

Table 6 VQC versus FCNN (fully connected neural network)

Aspect	Variational quantum circuit (VQC)	Feedforward classical neural network (FCNN)
Core processing	Parametrized quantum gates acting on qubits	Matrix multiplications and nonlinear activations
Parameter types	Angles or phases in quantum gates	Real-valued weights and biases
Data encoding	Encoded as quantum states (e.g., amplitude or angle encoding)	Direct numeric vectors/matrices
Output	Measurement results (probability distributions, expectation values)	Numeric activations
Hardware requirements	Quantum processor (simulations possible but often limited)	CPU/GPU or specialized classical hardware (TPU)
Training method	Hybrid quantum–classical optimization (e.g., parameter-shift rule)	Backpropagation (e.g., gradient descent, Adam)

- Implemented through partial measurements or unitary transformations.
- Maintains hierarchical feature extraction similar to classical CNNs.

Advantages

- Parameter efficiency through weight sharing.
- Built-in translational symmetry.
- Natural compatibility with grid-structured data.

Challenges

- Limited by current qubit counts.
- Requires careful circuit design to maintain quantum advantage.
- Measurement strategies more complex than classical pooling.

Comparison between QCNN and CNN is shown in Table 7.

4.4 Recurrent Quantum Neural Networks (RQNNs)

RQNNs adapt classical RNN concepts for sequential quantum data processing:

Core Architecture

- Quantum system serves as memory cell.
- Sequential inputs processed through parameterized gates.
- Hidden state maintained through qubit entanglement.
- Measurement strategies adapted for temporal processing.

Table 7 QCNN versus CNN

Aspect	Quantum convolutional neural network (QCNN)	Classical convolutional neural network (CNN)
Data representation	Quantum states in “convolution-like” quantum circuits	2D/3D numeric grids (images or volumes)
Convolution mechanism	Quantum analogs of convolution (entangling gates applied locally)	Sliding filter operations (dot products, pooling, etc.)
Parameters	Parametrized gates (unitaries)	Filter weights, biases
Output	Quantum measurements (probabilities, expectation values)	Feature maps, logits
Hardware	Quantum hardware or simulators	CPU/GPU, standard deep-learning infrastructure
Scalability	Still limited by noise, qubit count, hardware constraints	Highly scalable with established libraries and hardware

Key Innovations

- Quantum versions of gating mechanisms.
- Novel approaches to temporal credit assignment.
- Quantum-enhanced memory capacity.

Implementation Variants

1. **Discrete-time RQNNs**
    - Process inputs at regular time intervals.
    - Explicit circuit unfolding for backpropagation-through-time.
2. **Continuous-time RQNNs**
    - Implemented through time-dependent Hamiltonians.
    - More natural for certain quantum systems.

Applications

- Quantum time-series analysis.
- Natural language processing (quantum-enhanced).
- Quantum control sequence optimization.
- Neuromorphic quantum computing.

Current Limitations

- Sensitive to quantum noise and decoherence.
- Challenging to train due to temporal correlations.
- Limited demonstration on real hardware.

Comparison between RQNN and RNN is shown in Table 8.

Table 8 RQNN versus RNN

Aspect	Recurrent quantum neural network (RQNN)	Recurrent neural network (RNN)
Sequential processing	Quantum gates with time-step inputs on qubits	Hidden states passed through time steps (e.g., LSTM, GRU, etc.)
Parameterization	Quantum gates at each "time step"	Weight matrices for hidden states, gates, etc.
Information encoding	States of qubits evolve sequentially	Hidden state vectors evolve and sequentially
Output	Measurements for each time step	Numeric outputs for each time step
Computational resource	Qubits, limited hardware resources	CPU/GPU, widely supported frameworks
Noise sensitivity	Susceptible to decoherence and gate errors	Not subject to quantum noise; depends on numeric precision

Table 9 Comparative analysis

Architecture	Qubit requirements	Circuit depth	Training complexity	Best application domains
Hybrid	4–12 qubits	Shallow (3–8 layers)	Moderate	Small-scale classical data
VQC	4–20 qubits	Medium (5–15 layers)	High	General QML, optimization
QCNN	8–16 qubits	Deep (10–20 layers)	High	Image/signal processing
RQNN	4–8 qubits	Variable	Very high	Sequential data processing

4.5 Comparative Analysis

Comparative analysis of Hybrid, VQC, QCNN and RQNN is shown in Table 9.

5 Comparison with Classical Neural Networks

5.1 Data Preparation

1. **Filter MNIST to digits 0 and 1**
   - Original MNIST has s 28×28 grayscale images and labels $\{0,1,\ldots,9\}$.
   - We keep only samples where the label $y \in \{0,1\}$. Let the filtered dataset be $\{(X^{(i)}.y^{(i)})\}\ with\, y^i \in \{0, 1\}$.

2. **Flatten and Normalize**
   - Each 28×28 image is reshaped into a 784-dimensional vector.
   - Pixel values in $\{0,\ldots,255\}$ are normalized to [0,1]:

$$x^{(i)} \leftarrow \frac{x^{(i)}}{255}.$$

3. **Dimensionality Reduction (PCA)**
   - Quantum circuits typically manage few qubits (e.g., 4). We reduce 784 features to d-dimensions (d = 4) via PCA, yielding:

$$z^{(i)} = PCA\left(x^{(i)}\right) \in R^{d}.$$

4. **Scale to [0,π]**
   - Each component $z_j^{(i)}$ is linearly remapped so that $\min_j \rightarrow 0$ and $\min_j \rightarrow \pi$:

$$z_j^{(i)} \leftarrow \pi \frac{z_j^{(i)} - \min_j}{\max_j - \min_j} i$$

5. **Train–Validation–Test Split**
   - We create a training set $\left\{\left(X^{(i)}.y^{(i)}\right)\right\}$, a validation set to monitor generalization, and a held-out test set for final performance measurement.

5.2 *Parameterized Quantum Circuits (PQC)*

Regardless of framework (**PennyLane**, **Qiskit**, **PTQ**, or **TFQ**), we define a **parameterized quantum circuit**:

1. **Qubits and Data Encoding**
   - We have d qubits (e.g., 4) for the d-dim input z.
   - Each feature $z_j \in [0, \pi]$ is loaded onto qubit j via rotation gates (R_x, R_y, or a more general encoding). For example,

$$R_c\left(z_j\right)|0\rangle \cos\left(\frac{z_j}{2}\right)|0\rangle - i \sin\left(\frac{z_j}{2}\right)|1\rangle.$$

2. **Variational Gates**
   - **Trainable parameters** θ (e.g., rotation angles θ_1, θ_2, …) are applied as additional gates $\left(e.g., R_y(\theta), R_z(\theta), Rot(\alpha, \beta, \gamma)\right)$.
   - **Entanglement** gates (CNOT, CZ) create correlations among qubits.

3. **Measurement**

   - Typically measure the **expectation value** $\langle \sigma z \rangle$ on each qubit or on a subset. This yields a real-valued output, which the classical post-processing uses for classification.

Common Examples

- **VQC:** 4 qubits, angle-encode z, apply a ring of parameterized RY rotations, then ring of CNOTs. Measure each qubit's $|Z\rangle$.
- **QCNN:** Similar but with "convolution and pooling" structure that merges qubits step-by-step, eventually measuring 1 qubit.
- **QRNN:** A single qubit processes sequential inputs (z_0, z_1, z_2, z_3) over 4 time steps, measuring a final $|Z\rangle$.

5.3 Hybrid Model and Classification

5.3.1 Output Interpretation

1. **Multi-Qubit Output (e.g., VQC)**

   - Suppose we measure each qubit's expectation: $o = (o_1, o_2, \ldots, o_d) \in [-1.1]^2$.
   - The classical post-processing might use a final layer (e.g., a logistic function) to produce $\hat{y} \in [0, 1]$.

2. **Single-Qubit Output (e.g., QCNN, QRNN)**

   - The circuit merges or processes data to a single qubit, yielding scalar $o \in [-1, 1]$.
   - Interpreted as a logit or raw feature. Often $\hat{y} = \sigma(\alpha o + \beta)$.

5.3.2 Loss Function and Optimizer

For **binary classification**, a common loss is **cross-entropy**:

$$l(\theta) = -\frac{1}{N}\sum_{i=1}^{N}[y^{(i)} ln\hat{y}^{(i)} + (1 - y^{(i)})ln((1 - y^{(i)})]$$

where $y^{(i)}$ is the predicted probability for sample i.

- **Classical Optimizer**
  θ updated iteratively to minimize $\ell(\theta)$. Examples:

  - Gradient-based (Adam, RMSProp) [PennyLane, TFQ can do auto-diff].

– Gradient-free (COBYLA, SPSA) [common in Qiskit or PTQ if no auto-diff is available].

Goal

$$\theta^* = \arg\min l(\theta)$$

5.4 Methodology by Framework

5.4.1 PennyLane

1. **Data Preprocessing** (PCA to 4 features, scale to [0,π] in NumPy/TensorFlow/PyTorch.
2. **QNode** with a device, e.g. **dev = qml.device("default.qubit", wires = 4)**.
3. **Angle Encoding + Variational Parameters** in the QNode, measure **qml.expval(qml.PauliZ(i))**.
4. **Optimization** with **tf.GradientTape**, PyTorch's **optim**, or a built-in PennyLane **qml.GradientDescentOptimizer.**
5. **Compute Cross-Entropy** or similar, update θ.

Equations: same as above (rotations, entanglement, expectation values, cross-entropy).

5.4.2 Qiskit

1. **Data Preprocessing** as above.
2. **Circuit Construction** (e.g. **QuantumCircuit + ParameterVector**).
3. **Observable** is typically a PauliZ operator(s).
4. **CircuitQNN** or older QNN classes interpret the circuit, measuring the expectation.
5. **NeuralNetworkClassifier** wraps the QNN, uses a classical optimizer (e.g., COBYLA) to update the circuit's trainable parameters.

Equations: identical angle-encoding, expectation measurement. Minimizes cross-entropy or an equivalent classification error.

5.4.3 PTQ (Generic Parameterized Circuit Simulators)

1. **Data**: same pipeline.
2. **Define PQC**: Typically define a function that implements the circuit gates, referencing x and θ.

3. **Measurement**: Return $\langle \sigma z \rangle$ or a set of expectations.
4. **Training**: A classical "outer loop" updates θ by evaluating the circuit, computing the classification loss. Possibly gradient-free or with a derivative-based approach if the simulator supports it.

Equations: same HPC structure, but the code is more custom. The concept (encode → param gates → measure → compute cross-entropy) is consistent.

5.4.4 TensorFlow Quantum (TFQ)

1. **Data**: same pipeline.
2. **Circuit**: Use **tfq.layers.ControlledPQC** or **tfq.layers.PQC** with a **cirq**-style circuit.
3. **Measurements**: Single or multiple qubits, returning a **tf.Tensor** for the expectation values.
4. **Training**: Wrap in Keras model or custom training loop. TFQ integrates with **tf.GradientTape** for auto-differentiation of circuit parameters.
5. **Loss**: Usually **tf.keras.losses.BinaryCrossentropy.**

Equations: same as above. The difference is that TFQ handles the gradient wrangling behind the scenes.

5.5 Specialized Circuit Examples

5.5.1 VQC (Multi-qubit Output)

- d qubits, each measured in σ_z.
- Final $z \in R^d$.
- Possibly a single classical dense layer mapping z to $\hat{y}$.

5.5.2 QCNN (Pooling + Convolution)

- Stepwise "convolution" blocks on pairs of qubits with parameterized rotations + CNOT.
- Pooling merges pairs into single qubits.
- Final single qubit measurement yields $\hat{y}$.
- Mathematically: $pool(conv(z); \theta) \rightarrow \langle \sigma z \rangle$.

5.5.3 QRNN (Single-Qubit Over Time)

- 1 qubit with repeated application of $R_x(x_t) + R_y(\theta_t)$.
- Final measurement σz.
- Interpreted as $\hat{y}$ for classification.

5.6 Training Flow and Equations

Training flow is displayed in Fig. 2.

1. **Forward Pass**
   - For each sample $z^{(i)}$, run the PQC: $o^{(i)} = Eval\ Circuit\left(z^{(i)}; \theta\right)$.
   - Convert $o^{(i)}$ to a probability $\hat{y}^{(i)} \in [0, 1]$.
   - E.g. $\hat{y}^{(i)} = \sigma\left(Wo^{(i)} + b\right)$ for some classical post-layer.

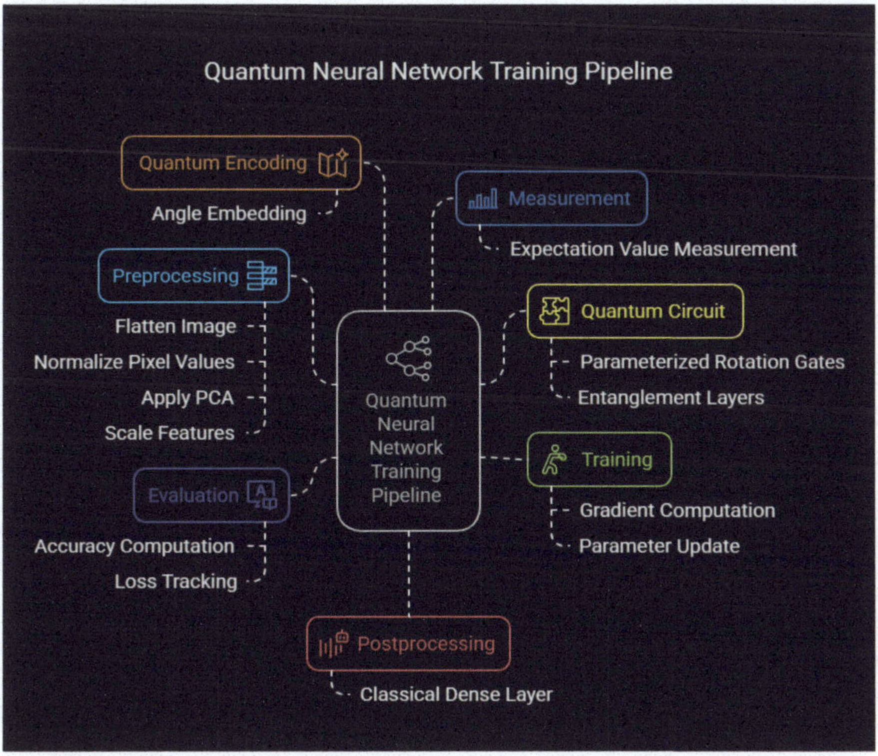

Fig. 2 Quantum neural network training process

2. **Loss** $l(\theta) = -\frac{1}{N}\sum_{i=1}^{N}\left[y^{(i)}\ln\hat{y}^{(i)} + \left(1 - y^{(i)}\right)\ln\left(1 - y^{(i)}\right)\right]$.
3. **Parameter Update**
   - A gradient-based or gradient-free method changes $\theta \leftarrow \theta - \eta\,\nabla\theta\ell$, or a suitable finite-difference step.
   - Repeated until convergence or max iterations.
4. **Validation and Test**
   - Evaluate accuracy on validation set:

$$Acc_{val} = \frac{1}{N_{val}}\sum_{j=1}^{N_{val}} 1\left(round\left(\hat{y}_{val}^{(j)} = y_{val}^{(j)}\right)\right).$$

   - Finally measure test accuracy on unseen data.

Table 10 shows Pennylane 4 Q bit to train 0 and 1-digit result from keras MNIST dataset.

Table 11 shows Pennylane 4 Q bit to train 0–9-digit result from keras MNIST dataset. Table 12 shows Pennylane 1 Q bit to train 0–9-digit result from keras MNIST dataset.

Table 10 Pennylane 4 Q bit to train 0 and 1-digit result from keras MNIST dataset

Model	Final train loss	Train accuracy	Val loss	Val accuracy	Test loss	Test accuracy
VQC	0.0642	0.9946	0.0608	0.9949	0.0592	0.9948
QCNN	0.0425	0.9947	0.0400	0.9953	0.0360	0.9957
RQNN	0.0702	0.9945	0.0678	0.9949	0.0642	0.9948

Table 11 Pennylane 4 Q bit to train 0–9-digit result from keras MNIST dataset

Model	Final train loss	Train accuracy	Val loss	Val accuracy	Test loss	Test accuracy
VQC	1.1199	0.5408	1.0768	0.5796	1.0904	0.5770
QCNN	1.1286	0.5247	1.0401	0.5913	1.0363	0.5357
RQNN	1.1801	0.5445	1.0571	0.5442	1.0651	0.5547

Table 12 Pennylane 1 Q bit to train 0–9-digit result from keras MNIST dataset

Model	Final train loss	Train accuracy	Val loss	Val accuracy	Test loss	Test accuracy
VQC	1.8388	0.3029	1.8350	0.3000	1.8211	0.3108
QCNN	1.8402	0.2996	1.8362	0.3020	1.8227	0.3050
RQNN	1.8287	0.2922	1.8206	0.2891	1.8461	0.2828

6 Applications of Quantum Neural Networks

Quantum Neural Networks (QNNs) have demonstrated significant potential in specialized domains where classical neural networks face fundamental limitations. Below, we explore two key applications: **quantum chemistry** and **financial modeling**, highlighting the unique advantages of quantum approaches.

6.1 Quantum Chemistry: Variational Quantum Eigensolver (VQE)

How VQE Works

The **Variational Quantum Eigensolver (VQE)** is a hybrid quantum–classical algorithm designed to compute molecular ground-state energies—a task critical for drug discovery, material science, and chemical reaction modeling.

1. **Problem Setup**
   - Goal: Find the lowest-energy (ground-state) configuration of a molecule.
   - Classical Challenge: Exact solutions scale exponentially with electron count (e.g., Full Configuration Interaction methods).
2. **Quantum Advantage**
   - **Ansatz Preparation**: A parameterized quantum circuit (PQC) prepares a trial wavefunction $|\psi(\theta)\rangle|\psi(\theta)\rangle$ representing the molecular state.
   - **Hamiltonian Measurement**: The molecular Hamiltonian HH is decomposed into Pauli terms (e.g., $H = \sum ciPiH = \sum ciPi$, where PiPi are Pauli strings).
   - **Hybrid Optimization**:
     - Quantum processor estimates $\langle\psi(\theta)|H|\psi(\theta)\rangle$.
     - Classical optimizer adjusts $\theta\theta$ to minimize energy.

Example: H_2O Simulation

- **Qubit Requirements**: 10–12 qubits (for minimal basis sets).
- **Circuit Structure**:
  - **Entanglement**: CNOT gates correlate electron orbitals.
  - **Parameterized Gates**: $RY(\theta)RY(\theta)$ rotations model electron interactions.
- **Performance**:
  - Achieves chemical accuracy (~ 1 kcal/mol error) with shallow circuits.
  - Outperforms classical methods like Density Functional Theory (DFT) for strongly correlated systems.

Table 13 Advantages over classical methods

Aspect	Classical (DFT)	Quantum (VQE)
Accuracy	Approximate	Near-exact for small systems
Scaling	$O(N^3)O(N^3)$–$O(N^4)O(N^4)$	Polynomial (theoretical)
Correlation Handling	Weak for multi-reference systems	Strong (entanglement captures correlations)

Advantages Over Classical Methods

Advantages of VQE over Classical Methods are shown in Table 13.

6.2 *Financial Modelling: Quantum Annealing for Portfolio Optimization*

Case Study: Portfolio Optimization

- **Assets**: 20–50 stocks.
- **Quantum Hardware**: D-Wave 2000Q.
- **Results**:
  - **Speed**: 100–1000 × faster for mid-sized portfolios ($N \geq 20$).
  - **Solution Quality**: Matches classical solvers (e.g., Gurobi) with lower computational resources.

Why Quantum Wins

Advantages of Quantum Annealing is represented in Table 14.

Limitations

- Annealers require problem reformulation to QUBO.
- Limited qubit connectivity affects large-scale deployments.

Synthesis and Outlook

- **Quantum Chemistry**: VQE excels for small, correlated molecules but awaits error-corrected hardware for scalability.

Table 14 Why quantum wins

Factor	Classical	Quantum annealing
Combinatorial search	Brute-force or heuristics (slow)	Quantum tunneling (fast for sparse problems)
Real-time rebalancing	Latency issues	Sub-second solutions possible
Non-convexity	Gets stuck in local minima	Bypasses barriers via tunneling

- **Finance**: Quantum annealing already outperforms classical methods for specific optimization tasks, with near-term commercial potential.

Future Directions

- **Chemistry**: Larger molecules (e.g., enzymes) on fault-tolerant QPUs.
- **Finance**: Hybrid algorithms combining QNNs with classical risk models.

This analysis underscores that QNNs are not replacements but **specialized accelerators** for problems where quantum physics offers inherent advantages.

7 Case Study: MNIST Binary Classification

The MNIST dataset for image classification stands out as the original benchmark, containing thousands of labelled images of handwritten digits ranging from 0 to 9. This study thus focuses on binary, that is, accepting only the digits 0 and 1 into the dataset. That reduces the complexity of the problem and qualifies it well for the proof of concept of quantum neural networks.

7.1 Result and Discussion

7.1.1 Dataset Overview

MNIST Dataset

- 60,000 training images and 10,000 test images of handwritten digits (28×28) (28\times 28)(28×28) grayscale.
- Each image is flattened into 784 features for classical models.
- For **binary classification**, we filtered the dataset to keep only digit labels $\{0,1\}$
- For **multi-class classification**, we used all digit labels $\{0,\ldots,9\}$.

7.1.2 Binary Classification (Digits {0,1})

Quantum versus Classical Comparisons

VQC versus FCNN

- **VQC (Variational Quantum Circuit)**
  - Achieved a final test accuracy of ~ **99.48% (0.9948)** after **5 epochs**.
  - The loss steadily decreased but exhibited slight fluctuations ("noisiness"), a typical characteristic of parameterized quantum circuits, especially on simulation backends with finite sampling.

- Operates on a **4-qubit system**, utilizing **PCA and angle encoding**, and outputs a single or multi-qubit expectation value mapped to a binary probability. Accuracy Plot of VQC Using 4 Qbit for 0 and 1 Digit is shown in Fig. 3.

- **FCNN (Fully Connected Neural Network)**
  - Attained a test accuracy of ~ **99.95% (0.9995)** or even higher in some runs.
  - Displayed smooth loss curves with minimal noise, benefiting from well-established classical optimization techniques and a large number of trainable parameters. Accuracy Plot of FCNN for 0 and 1 Digit is shown in Fig. 4.

The VQC excelled in distinguishing between 0 and 1, getting close to classical results. However, the FCNN still achieved superior accuracy due in part to the fact that classical networks are more stable in terms of loss trajectories and have much larger parameter spaces. Nevertheless, the good convergence of the VQC implies that small quantum circuits can efficiently solve simple binary problems with minimal extra noise or variance in the loss.

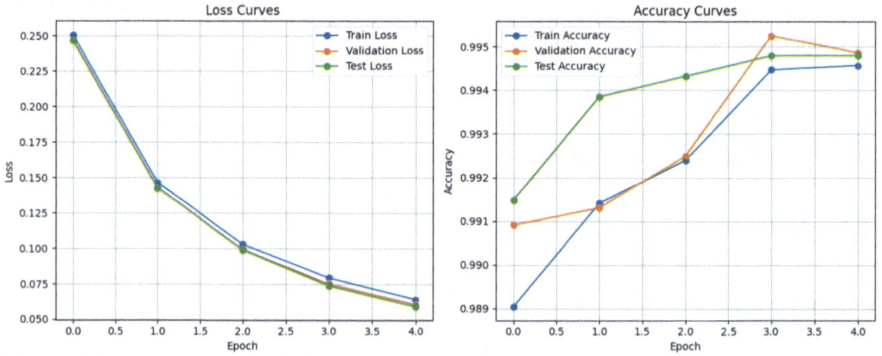

Fig. 3 Accuracy plot of VQC using 4 qbit for 0 and 1 digit

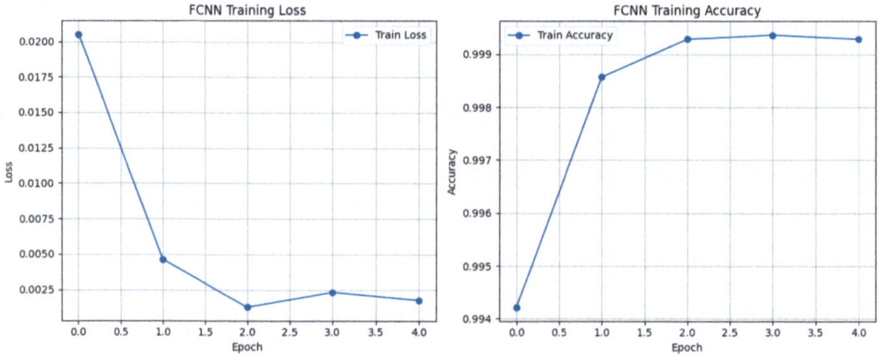

Fig. 4 Accuracy plot of FCNN for 0 and 1 digit

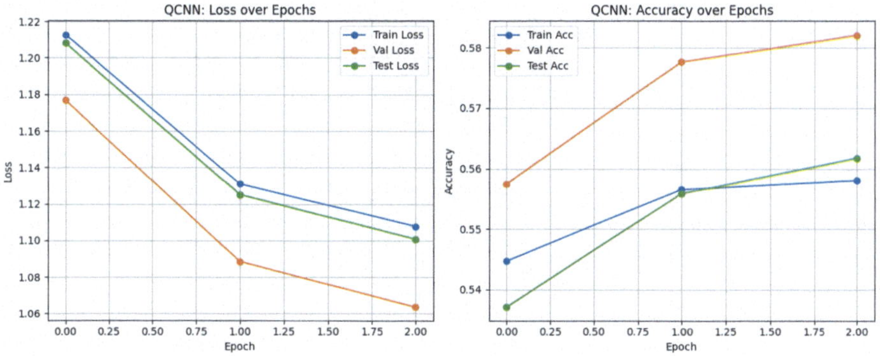

Fig. 5 Accuracy plot of QCNN using 4 qbit for 0 and 1 digit

QCNN versus CNN

- **QCNN (Quantum Convolutional Neural Network)**
  - Achieved a final test accuracy of ~ **99.57% (0.9957)**.
  - Utilizes **entangling gates** in a "convolution + pooling" approach, resembling the local connectivity structure of classical CNNs.
  - The training curves showed occasional fluctuations due to the quantum parameter landscape but ultimately converged effectively for binary classification.

  Accuracy Plot of QCNN Using 4 Qbit for 0 and 1 Digit is shown in Fig. 5.

- **CNN (Convolutional Neural Network)**
  - Typically achieved a test accuracy between **99.90% and 99.96% (0.9990 – 0.9996)**.
  - Exhibited stable and low-noise loss curves across epochs, characteristic of CNN training on relatively simple tasks like binary MNIST classification.

  Accuracy Plot of CNN for 0 and 1 Digit is shown in Fig. 6.

The **QCNN** efficiently captures patterns of the two classes using a **small quantum circuit**, while the **CNN** leverages **2D convolutions over 28 x 28 grids**, often achieving accuracy above **99.9%**. Despite occasional random fluctuations in the QCNN's loss—due to inherent quantum radients—convergence was achieved at around **99.6% accuracy**. This suggests that structured quantum pooling can match or even slightly outperform a **Variational Quantum Circuit (VQC)** in binary classification tasks.

QRNN versus RNN

- **QRNN (Quantum Recurrent Neural Network)**
  - Achieved a final test accuracy of ~ **99.48% (0.9948)**.

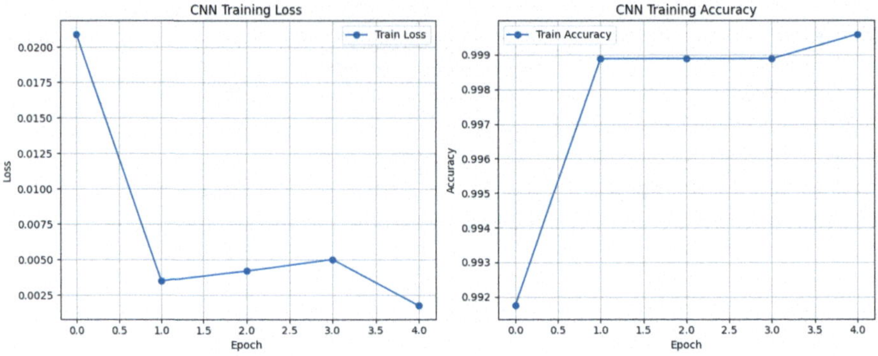

Fig. 6 Accuracy plot of CNN for 0 and 1 digit

- Utilizes a **single or a few qubits**, sequentially processing extracted features or PCA components.
- Training curves exhibited some **sensitivity to parameter initialization**, occasionally resulting in noisier intermediate losses before stabilization.

Accuracy Plot of QNN Using 4 Qbit for 0 and 1 digit is shown in Fig. 7.

- **RNN (Recurrent Neural Network)**
  - Achieved a test accuracy between **99.93% and 99.95%**.
  - Processes **flattened images as sequences**, efficiently distinguishing between digits **0 and 1** with **minimal noise** in the optimization process.
  - Displayed smooth and gradually decreasing loss curves, characteristic of well-optimized classical sequential models.

  Accuracy Plot of RNN for 0 and 1 digit is depicted in Fig. 8.

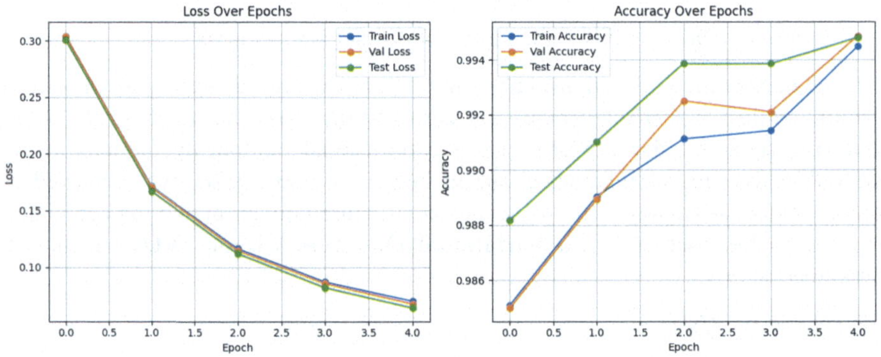

Fig. 7 Accuracy plot of QNN using 4 qbit for 0 and 1 digit

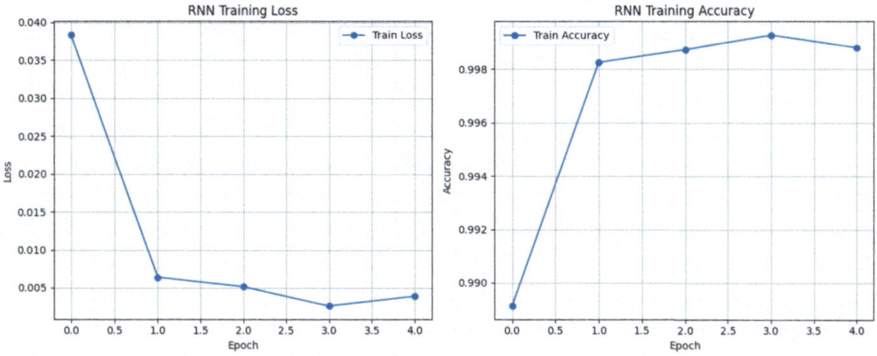

Fig. 8 Accuracy plot of RNN for 0 and 1 digit

While QRNNs and RNNs are unconventional choices for **static images**, both models achieve **near-perfect accuracy** in binary classification. QRNNs, like other quantum neural networks, occasionally show **minor fluctuations** in training accuracy but still match **VQC and QCNN in the 99.4–99.6% range**. Traditional RNNs retain a slight edge due to the **maturity of classical sequential models** and their well-established hyper parameter tuning techniques.

7.1.3 Multi-class Classification (Digits {0,...,9})

Quantum Architectures

- **VQC**
  - **Accuracy** Achieves approximately **54% test accuracy** on the 10-class MNIST dataset using **4 qubits**, indicating moderate learning capability.
  - The loss function demonstrates **significant fluctuations between epochs**, reflecting the challenges of mapping **10 distinct classes** onto a constrained quantum state space.

  Accuracy Plots of VQC are shown in Figs. 9 and 10.

- **QCNN**
  - **Accuracy** Partial results suggest QCNN can **surpass naive baselines**, with early training showing **above 60% accuracy** after sufficient epochs.
  - The loss curves exhibit **greater fluctuations** compared to binary classification, as the model must differentiate **varied digit structures** within a **limited 4-qubit representation**.

  Accuracy Plots of QCNN are shown in Figs. 11 and 12.

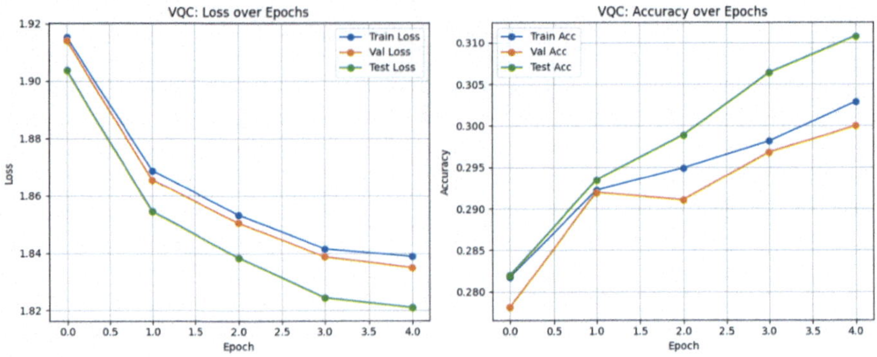

Fig. 9 Accuracy plot of VQC using 1 qbit for 0–9 digit

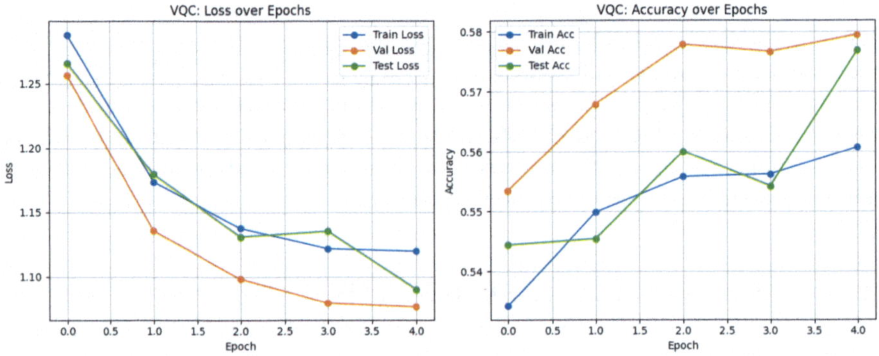

Fig. 10 Accuracy plot of VQC using 4 qbit for 0–9 digit

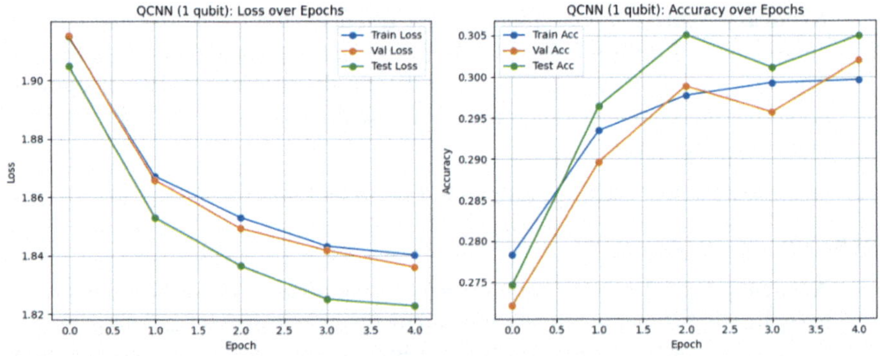

Fig. 11 Accuracy plot of QCNN using 1 qbit for 0–9 digit

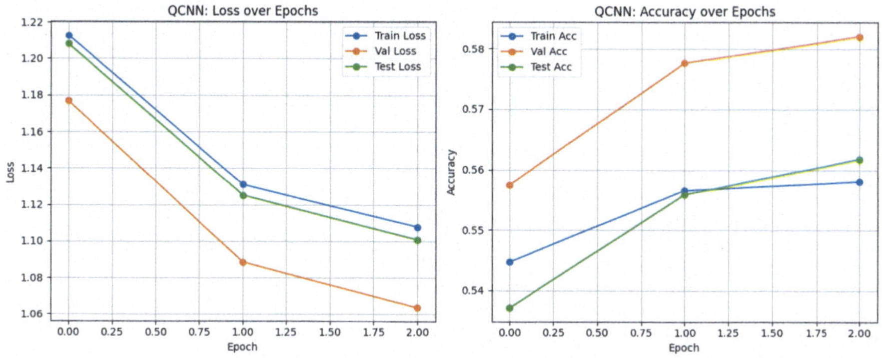

Fig. 12 Accuracy plot of QCNN using 4 qbit for 0–9 digit

- **QRNN**
  - **Accuracy** Reaches around **54% accuracy** for 10-class classification in partial experiment logs (Fig. 13).
  - Displays **quantum training noise**, particularly when processing each **Principal Component Analysis (PCA) feature recurrently**. Fine-tuning additional **training parameters** may help **stabilize variance**.

QNN applications on the 10-class MNIST data set in general showcase rather pronounced loss oscillations and somewhat slower convergence. These partial results do, however, reaffirm the great promise for QNNs in handling more multi-class problems even within the constraints of 4-qubit simulations, although higher accuracies demand deeper circuits or more qubits.

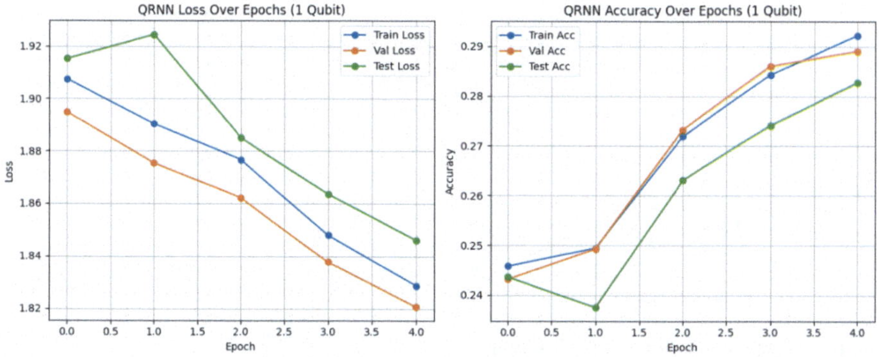

Fig. 13 Accuracy plot of QRNN using 1 qbit for 0–9 digit

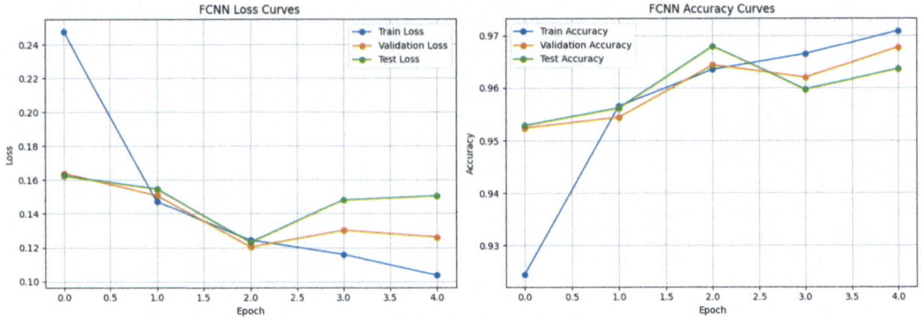

Fig. 14 Accuracy plot of FCNN for 0–9 digit

Classical Architectures

- **FCNN**
  - Stabilizes around **95–96%** test accuracy after 5 epochs (Fig. 14).
  - Loss descends relatively smoothly in conventional training loops, underscoring the reliability of classical MLPs for standard MNIST.
- **CNN**
  - **Accuracy** Achieves approximately **98% test accuracy**, consistent with well-documented MNIST benchmarks (Fig. 15).
  - Exhibits **low-noise loss curves**, benefiting from **optimized initialization, pooling layers, and robust convolution filters**.
- **RNN**
  - **Accuracy** Reaches **97–97.5% accuracy**, depending on architectural choices and training specifics (Fig. 16).
  - Maintains **steady, low-variance loss curves**, despite the challenge of adapting **sequence-based models** for image classification.

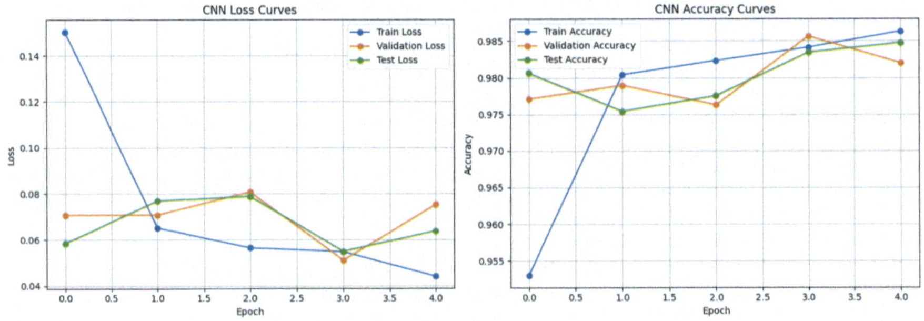

Fig. 15 Accuracy plot of CNN for 0–9 digit

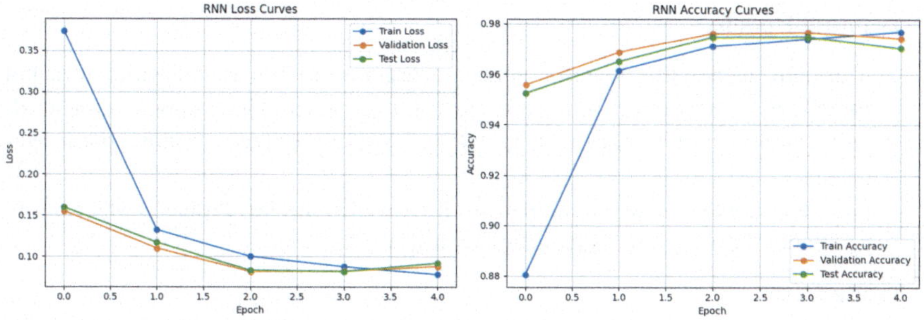

Fig. 16 Accuracy plot of RNN for 0–9 digit

All classical approaches reach above 95 after a short training schedule, with CNNs reaching ~ 98. Their loss curves remain calm and almost linear in descent as opposed to the more oscillatory or "spikier" loss trends sometimes exhibited by quantum parameterized circuits.

7.1.4 Training Time Note and Impact

The training time unlike binary to multi class classification task. For the binary task (digits {0,1}), the QNN models like VQC, QCNN, QRNN could perform the 5 epochs with relative quickness, as each epoch takes a bearable amount of time-even on a PC with a standard configuration. When scaling up, however, to the entire 10 class MNIST ({0,…,9}), it had a staggering training time for all QNN models. On an identical, hardware setting, each epoch took roughly 3 h even during simulation. And this is being accounted to the majorly increased effort required to distinguish 10 classes along with the considerable computation required to simulate the quantum circuit-in particular when using 4 qubits and gates that are parameterized. Such practical constraining scenarios would be whatever to do in comparing QNN and classical models under limited resources.

Impact of Qubit Count and Epochs

- **Number of Qubits**
  - More qubits increase representational power but often amplify training noise, since each additional qubit expands the parameter landscape. On simulators, this also increases computational expense.
  - For 10-class MNIST, 4 qubits can be quite restrictive. Moving to 6 or 8 qubits might raise accuracy closer to classical baselines, but the overall training loss may become more challenging to stabilize.

- **Training Epochs**
  - 5 epochs were sufficient for near-convergence in binary classification, but multi-class tasks often require more training steps. More epochs can gradually smooth out the noise in QNN losses as optimizers explore parameter space.
  - Classical networks also benefit from longer training but typically show more predictable downward trajectories with fewer spikes.
- **Trade-Offs**
  - Deeper quantum circuits or more extensive training help QNNs approach classical accuracy, but can significantly increase simulation times and occasionally accentuate random loss variations.
  - Hybrid quantum–classical strategies—e.g., a classical feature extractor feeding a smaller quantum circuit—may reduce these challenges.

7.1.5 Concluding Remarks

Binary Classification (Digits 0 & 1)

- Quantum Neural Networks (QNNs), including **VQC, QCNN, and QRNN**, achieve **over 99% accuracy** with minimal training epochs, demonstrating that **small-scale quantum circuits** can closely match classical performance.
- While QNNs exhibit **slightly noisier loss curves**, they still **converge effectively**, indicating their potential for binary classification tasks.
- Classical architectures such as **FCNN, CNN, and RNN** surpass **99.9% accuracy**, maintaining **smooth and steadily decreasing loss curves**, reinforcing their **superior stability and near-perfect classification performance**.

Full MNIST Classification (Digits 0–9)

- **Classical models remain dominant**, with **FCNN, CNN, and RNN achieving 95–98% accuracy** on the full MNIST dataset. Their **loss curves remain stable**, demonstrating their reliability for multi-class classification.
- **QNNs, limited to 4 qubits, reach approximately 54% accuracy**, reflecting the constraints of **limited quantum capacity** and the increased complexity of **handling 10 distinct classes**.
- While deeper circuits or additional qubits could **enhance accuracy**, they would also **amplify training noise and computational resource requirements**.

These results emphasize the **strength of classical deep learning for full MNIST classification** while highlighting **the promising potential of QNNs**—despite current hardware limitations.

Table 15 A concise overview of quantum–classical test accuracies for binary and 10-class MNIST

Task	Model pair	Quantum accuracy	Classical accuracy	Remarks
Digits {0,1}	VQC versus FCNN	~ 99.48%	~ 99.95%	Both strong, FCNN slightly higher
	QCNN versus CNN	~ 99.57%	~ 99.90–99.96%	QCNN edges out VQC, CNN nearly perfect
	QRNN versus RNN	~ 99.48%	~ 99.93–99.95%	QRNN comparable to other QNNs, RNN slightly higher
Digits {0–9}	VQC versus FCNN	~ 54%	~ 95–96%	More qubits/epochs can boost QNN accuracy
	QCNN versus CNN	~ 52%	~ 98%	CNN remains top-tier on multi-class MNIST
	QRNN versus RNN	~ 54%	~ 97%	QRNN learns sequentially, RNN converges quickly

Future Directions

- With larger, advanced parameter entanglement strategies or specific encodings, it could be possible to solve multi-class performance differences.
- Real hardware tests lessen the noise of quantum devices, but they may precisely validate or refine simulation-based findings.
- PennyLane's user-friendly gradient-based quantum simulation is an opening to continue the investigation of QNNs in conjunction with classical layers.

Table 15 shows a concise overview of quantum–classical test accuracies for binary and 10-class MNIST.

7.2 *Practical Insights on Frameworks*

In terms of distinct advantages, quantum machine learning frameworks differ from one another; PennyLane offers smooth integration of both classical and quantum elements while having built-in gradient optimization that can ensure stable training and user-friendly use. TensorFlow Quantum (TFQ) affords considerable integration potential due to its link with TensorFlow, yet it suffers from deviance in complexity and compatibility issues. Qiskit supports a wide range of hardware and simulation backends for experiments but lacks auto-differentiation and needs heavy computational resources. Cirq/PTQ is customizable, albeit it still needs you to do everything without the support of a library. Finally, PennyLane delivers practical utility, blending technical excellence worldwide into modern hybrid quantum–classical ML applications.

8 Classical Simulation and Q# in Quantum Neural Networks

8.1 How Many Qubits Can Be Simulated on a Classical Computer?

Machines of the classical taxonomy are faced with memory expansion when attempting to simulate quantum systems: the state-vector simulator of n qubits needs the storage of 2^n complex numbers. The simulation of quantum systems using complex amplitudes has a significant impact on the number of qubits that can be simulated. Current hardware allows for simulations of about 30–40 qubits, with supercomputers able to handle up to 50 qubits under specific conditions. Once we go beyond this point general simulations become too difficult to carry out. While some methods like hybrid or tensor-network approaches can estimate larger systems, they need a lot of resources. In the end, to scale beyond 50 qubits, we need actual quantum hardware. Classical computers can't copy large-scale quantum behaviour in a feasible way.

8.2 Q# in Quantum Neural Networks

Q# is Microsoft's quantum programming language, part of the Quantum Development Kit (QDK) designed for developers working on quantum algorithms such as Quantum Neural Networks (QNNs). The language allows for the implementation of a modular approach to quantum circuits using layers of parameters, interfaces with classical languages (e.g., Python) for carrying out a combination of optimization tasks and makes it possible using quantum simulators for running and determining the execution of QNNs as well as the resource requirements (qubits, gate depth). QNNs in Q# operate under a corresponding pattern: encoding classical data into quantum states through rotations, implementing the required tunable quantum gates and measuring outputs to initiate classical optimization (e.g., gradient descent) for parameter updates. The more organized flow of this toolchain paves the way for the simulation of QNN structures and the subsequent adoption of the technology on hardware.

9 Challenges and Future Directions

Quantum computing is interesting for the development of quantum neural networks (QNNs) because it's phenomenal potential for solving complex problems at exponentially higher speeds than its classical counterpart. Exploiting quantum effects of superposition and entanglement would build super machines that learn better than the best machine learning (ML) algorithms. Thus, although very promising, QNNs have yet to be used practically due to a variety of issues ranging from hardware constraints to theoretical ones. This paper reviews these challenges and outlines some directions for the future development of QNNs.

Challenges in Quantum Neural Networks

1. **Hardware Limitations**: The current quantum hardware is at an early stage of development. The current generation of quantum computing is dominated by noisy intermediate-scale quantum (NISQ) devices that, due to decoherence and errors in their gates, affect the fidelity of QNNs (Preskill 2018). Further, qubit connectivity and short coherence times limit the scalability of QNN architectures.
2. **Quantum Data Representation**: Basically, unlike traditional ML models that can process structured data, QNNs will have to encode classical information into quantum states. Efficient quantum data representation has always been a key challenge because the conversion incurs overheads that nullify any possibilities for quantum advantages.
3. **Training and Optimization**: Initially, variational quantum algorithms are believed on basis for training the QNN because they are, in nature, hybrid quantum–classical schemes. The problem during optimization of quantum circuits is that of barren plateaus, where gradient sources for gradient-based learning become ineffective due to extremely small gradients. This limitation makes it impossible to scale up deep QNN models.
4. **Algorithmic Development**: On the other hand, classical neural networks preserve their architecture and optimization techniques, while QNNs lack any standardized framework. On another note, effective quantum activation and loss functions are yet to be researched.
5. **Quantum Error Correction**: In contrast to classical systems, quantum computations are decidedly prone to errors from decoherence. Error correction techniques like surface codes work by requiring many physical qubits to encode the information of a logical qubit, impairing any practical realization of QNNs.

Future Directions

1. **Advancements in Quantum Hardware**: These developments in terms of error-corrected qubit and fault-tolerant quantum computation will go a long way in pushing QNN viability ahead. Enhancing qubit connectivity and coherence time is essential for scalable quantum ML applications.
2. **Efficient Quantum Encoding Techniques**: Novel encoding schemes have been researched with respect to efficient conversion of classical to quantum data for

increasing overhead analysis, including amplitude encoding and quantum feature maps.
3. **Hybrid Quantum–Classical Models**: As full-fledged QNNs are presently not feasible, hybrid schemes combining classical deep learning systems with quantum circuits provide realistic avenues for near-term development. These techniques may take advantage of quantum benefits while reducing current hardware limitations.
4. **Development of New Quantum Learning Algorithms**: The future should seek to establish activation functions and loss functions that are quantum specific, optimizing training approaches to deal with barren plateaus, and enhancing their learning efficiency.
5. **Error Mitigation Strategies**: New advances in quantum error correction and inductive filtering techniques that can be represented as quantum error suppression or similar approximate error correction algorithms will make QNNs feasible for real-world applications.

10 Conclusion

Quantum Neural Networks (QNNs) represent an incipient paradigm at the juncture of quantum computing and artificial intelligence, with potentially great advantages in niche areas such as quantum chemistry, combinatorial optimization, and processing high-dimensional data. With advantages from quantum parallelism, entangled learning, and theoretical energy efficiency, QNNs indeed hold great promise. However, the limitations imposed by today's realities, such as hardware noise, barren plateaus, and data encoding bottlenecks, confine their applicability (see Table 14). The most viable option, therefore, would combine classical preprocessor and quantum-processing approaches as a hybrid classical-quantum architecture for specific tasks, followed by error mitigation. The path to obtain meaningful quantum advantage will require co-design of algorithms and hardware, optimization for specific problems, responsible deployment schemes, and global teaching programs to guarantee balanced and real progress toward fair quantum-enhanced machine learning.

References

1. Zhang, B.: Quantum neural networks: a new frontier. Theor. Nat. Sci. **41**(1), 119–125 (2024). https://doi.org/10.54254/2753-8818/41/2024CH0157
2. Jeswal, S., Chakraverty, S.: Recent developments and applications in quantum neural network: a review. Arch. Comput. Methods Eng. **26** (2018). ResearchGate. https://doi.org/10.1007/s11831-018-9269-0
3. Smaldone, A.M., et al.: Quantum convolutional neural networks for multi-channel supervised learning. Quantum Mach. Intell. **5**(2), 41 (2023). https://doi.org/10.1007/s42484-023-00130-3

4. Simen, A., et al.: Digital-analog quantum convolutional neural networks for image classification. Phys. Rev. Res. **6**(4), L042060 (2024). https://doi.org/10.1103/PhysRevResearch.6.L042060
5. Ahmed, T., et al.: Quantum neural networks: a comparative analysis and noise robustness evaluation. arXiv:2501.14412, arXiv (2025). arXiv.org, https://doi.org/10.48550/arXiv.2501.14412
6. Yang, J., et al.: Stability and Generalization of Quantum Neural Networks. arXiv:2501.12737, arXiv (2025). arXiv.org, https://doi.org/10.48550/arXiv.2501.12737
7. What Is Quantum Computing?|IBM. https://www.ibm.com/think/topics/quantum-computing. Accessed 20 Mar 2025
8. H-QNN: A Hybrid Quantum–Classical Neural Network for Improved Binary Image Classification. https://www.mdpi.com/2673-2688/5/3/70. Accessed 30 Mar 2025
9. Hybrid Quantum–Classical Neural Networks for Efficient MNIST Binary Image Classification. https://www.mdpi.com/2227-7390/12/23/3684. Accessed 30 Mar 2025
10. Advancing Hybrid Quantum Neural Network for Alternative Current Optimal Power Flow. https://arxiv.org/html/2410.20275v2. Accessed 30 Mar 2025
11. Ren, Z.: Comparing quantum convolutional and classical convolutional architectures. Appl. Comput. Eng. **74**(1), 89–96 (2024). https://doi.org/10.54254/2755-2721/74/20240445
12. Simoes, R.D.M., et al.: Experimental evaluation of quantum machine learning algorithms. IEEE Access **11**, 6197–6208 (2023). https://doi.org/10.1109/ACCESS.2023.3236409
13. Escudero, E.B.T., et al.: Assessing the impact of noise on quantum neural networks: an experimental analysis, pp. 314–25 (2023). arXiv.org, https://doi.org/10.1007/978-3-031-40725-3_27
14. Karuppasamy, K., et al.: A comprehensive review of quantum circuit optimization: current trends and future directions. Quantum Rep. **7**(1), 2 (2025). https://doi.org/10.3390/quantum7010002
15. Roh, J., et al.: Hybrid quantum neural network model with catalyst experimental validation: application for the dry reforming of methane. ACS Sustain. Chem. Eng. **12**(10), 4121–4131 (2024). ACS Publications. https://doi.org/10.1021/acssuschemeng.3c07496
16. Zhao, R.: The advantage and disadvantage of development of quantum computing in machine learning. Appl. Comput. Eng. **119**(1), 167–172 (2025). https://doi.org/10.54254/2755-2721/2025.21579
17. Tseng, H.-H., et al.: Transfer learning analysis of variational quantum circuits. arXiv:2501.01507, arXiv (2025). arXiv.org, https://doi.org/10.48550/arXiv.2501.01507
18. Li, Y., et al.: A distributed hybrid quantum convolutional neural network for medical image classification. arXiv:2501.06225, arXiv (2025). arXiv.org, https://doi.org/10.48550/arXiv.2501.06225
19. Da Silva, V., Jesus, G.: Recurrent quantum neural networks: a review, 365. ResearchGate (2023). https://doi.org/10.5151/siintec2023-306159
20. Slabbert, D., et al.: Pulsar classification: comparing quantum convolutional neural networks and quantum support vector machines. Quantum Mach. Intell. **6**(2), 56 (2024). https://doi.org/10.1007/s42484-024-00194-9
21. Choudhary, P., et al.: HQNN-FSP: A Hybrid Classical-Quantum Neural Network for Regression-Based Financial Stock Market Prediction. arXiv:2503.15403, arXiv (2025). arXiv.org, https://doi.org/10.48550/arXiv.2503.15403
22. Ajibosin, S.S., Cetinkaya, D.: Implementation and performance evaluation of quantum machine learning algorithms for binary classification. Software **3**(4), 498–513 (2024). www.mdpi.com, https://doi.org/10.3390/software3040024
23. Liu, C.-Y., et al.: Training Classical Neural Networks by Quantum Machine Learning. arXiv:2402.16465, arXiv (2024). arXiv.org, https://doi.org/10.48550/arXiv.2402.16465

Improved Pattern Recognition—Quantum Machine Learning

Siya Khanna, Shruti Dhingra, Nidhi Sindhwani, and Rashmi Vashisth

Abstract Conventional pattern recognition techniques are not able to keep pace with the ever-increasing complexity of large amounts of data. The new technique quantum machine learning or QML describes a remarkable way of processing speed and accuracy improvement through the principles of quantum physics. This paper will elaborate on how quantum computing will apply to pattern recognition, the advantages it has over its traditional counterparts, the sectors it would impact, and the challenges that these technologies need to surmount before widespread adoption. The paper also assesses what will happen in the future concerning trends that will be pivotal in shaping the landscape of QML research, new developments in technology, and real-world scenarios of implementation. In addition to speedups for established machine learning tasks, QML presents novel strategies for analyzing and training knowledge associated with high-dimensional data and sophisticated models. It is enabled to serve as an efficient solution for offence sectors that rely on real-time conclusion and predictive analysis. It has the ability to handle even exponentially bigger datasets and perform parallel computations. Even so, there is still an infant plane in the arena of quantum computing, but the continuing innovation and investments in quantum algorithms and hardware do suggest, however, that QML will be one of the fundamental points to decide the course of both machine learning and artificial intelligence in the future.

Keywords Quantum computing · Machine learning · Pattern recognition · Quantum algorithms · Quantum neural networks · Quantum superposition · Quantum entanglement

---

S. Khanna (✉) · S. Dhingra · N. Sindhwani · R. Vashisth
Amity Institute of Information Technology, Amity University, Noida, Uttar Pradesh, India
e-mail: siyak132005@gmail.com

R. Vashisth
e-mail: rvashisth@amity.edu

© The Author(s), under exclusive license to Springer Nature Switzerland AG 2025

A. Ghosh et al. (eds.), *Quantum Machine Learning in Industrial Automation*,
Information Systems Engineering and Management 65,
https://doi.org/10.1007/978-3-031-99786-0_17

1 Introduction

Pattern recognition is a key aspect of both Machine Learning (ML) and Artificial Intelligence (AI). It seeks to identify and classify patterns in a way that machines can better the process of decision-making, automation of processes, and predictions. Pattern recognition is mainly concerned with extracting significant information from raw data and applying classification against a number of predetermined characteristics and similarities.

Over time, pattern recognition has evolved significantly. The earlier approaches were mostly based on statistical and mathematical models, such as Bayesian classifiers and linear discriminants. More advanced techniques, involving deep learning, probabilistic models, and artificial neural networks, came with the increase in processing power [1]. Nowadays, it serves pattern recognition in many fields, from cyber security and medical diagnostics to speech and image recognition (Fig. 1). Feature extraction is a very important aspect of pattern recognition as it chooses the most meaningful attributes from a dataset for performance improvement in classification. These features are then evaluated through many different ways, like unsupervised learning that finds patterns without prior labeling, or supervised learning which constitutes both trained and test data tagged in their labeling as they go [2]. Further methods such as semi-supervised learning and reinforcement learning have been derived to improve the ability of recognizing patterns. Such approaches use the combination of labeled and unlabeled input data, or learn to recognize a pattern by interacting with the environment.

Despite advances, traditional pattern recognition methods still have various drawbacks. First, the large datasets create scalability problems because of the processing power they require. The inaccuracy in forecasting can occur as a result of the noise, lack of information, and data variations. Feature selection and optimization are still very difficult to process power [3]. These are indications that a new method is required which will be more effective and scalable. Quantum computing is motivated in that direction.

1.1 *Objective of the Study*

The primary objectives of this study are:

- To examine the effect of QML on pattern recognition efficiency and accuracy.
- Compare the applications of Classical ML and QML in the real-world.
- Identify the challenges and limitations of implementing QML.
- Study pragmatic applications of QML in domains like autonomous systems, cybersecurity, healthcare or finance.
- Extension of future research to include integrating QML with AI applications.

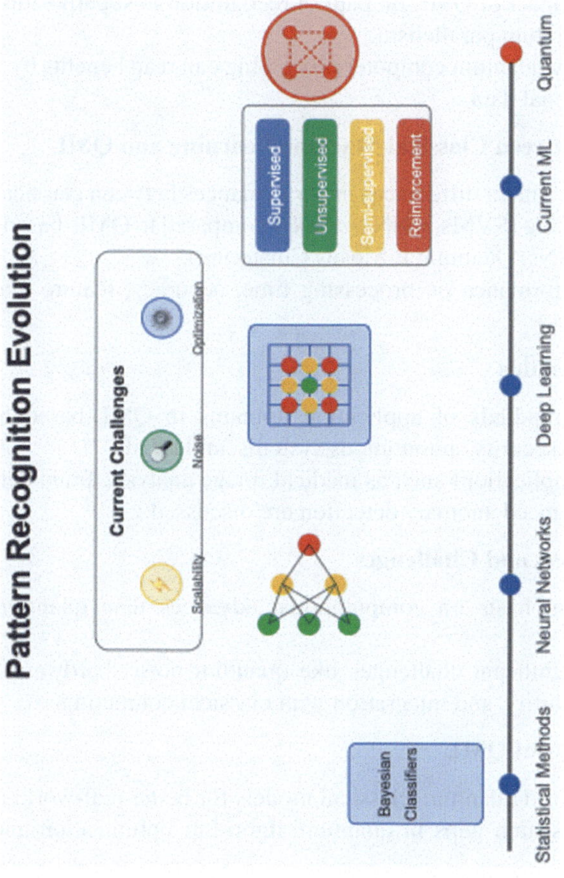

Fig. 1 Shift from traditional statistical methods to modern deep learning approaches

1.2 Scope of the Study

The research work specifically addresses Quantum Machine Learning (QML) and its impact in the field of general pattern recognition. Quantum computing principles, superposition, entanglement, and quantum parallelism form the basis of QML pattern recognition applications in contrast with classical machine learning methods.

- **Focus on Quantum Machine Learning (QML) and Pattern Recognition**
  - Studies the impact of QML on pattern recognition in superposition entanglement, and quantum parallelism.
  - Examines how quantum computer processing can reap benefits by working on high dimensional data.
- **Comparison Between Classical Machine Learning and QML**
  - Compares the salient differences in performances between classical models of machine learning (SVMs, CNNs, ANNs) compared to QML-based algorithms (QSVMs, QNNs, Quantum k-Means Clustering).
  - Increases performance on processing time, accuracy, feature mapping, and scalability.
- **Application Domains**
  - Focuses on hundreds of application domains in QML-based health care, finance, cybersecurity, autonomous systems, and retail.
  - Real-world applications such as medical image analysis, fraud detection, and quantum-enhanced anomaly detection are discussed.
- **Technical Aspects and Challenges**
  - Puts more emphasis on computational advances than quantum hardware design.
  - Points out significant challenges like quantum noise, hardware limitations, algorithm maturity, and integration with classical computing.
- **Future Prospects of QML**
  - Examines hybrid quantum–classical models for better real-world usability.
  - Points out research gaps in quantum algorithm optimization and hardware evolution.
  - Provides directions for industry uptake and real-time implementation of QML in AI applications.
- **Target Audience**
  - Useful for AI researchers, data scientists, quantum computing engineers, and industry professionals.
  - Tends to provide insights into the future role of QML in AI-driven pattern recognition systems.

While the paper provides an in-depth analysis of QML's impact on pattern recognition, it does not address all potential quantum computing applications beyond machine learning. The study does not explore solely theoretical parts of quantum physics that are unrelated to pattern recognition; instead, it concentrates on real-world applications.

2 Literature Review

Pattern recognition forms one of the vital ingredients in artificial intelligence (AI) that computers may recognize typical patterns, structures, and anomalies from complex datasets [3]. Thus, the areas of application for pattern recognition are widespread, including image processing, speech recognition, detection of financial fraud, and medical diagnoses. Traditional machine learning (ML) techniques, such as SVMs, CNNs, and Decision Trees, have proven themselves worthy in various applications [4]. However, the amplitude and complexity of real-world datasets put some limits on classical ML methods:

- **Computational Bottlenecks**: Classical ML models are very heavy on time and computational resources, especially if large-scale pattern recognition jobs are concerned.
- **High Data Requirements** Classical deep learning models require large labeled datasets to generalize well and thus have little value in fields where data are scarce.
- **Feature Mapping Limitations**: Several classical ML models fail to capture nonlinearity into a high-dimensional feature space, thus obscuring the hidden patterns.

To counter these deficiencies, quantum machine learning (QML) arises as a credible alternative by exploiting quantum computing principles like superposition, entanglement, and parallelism to address the problem of recognition. The review shall enhance present knowledge on how QML attempts to improve the traditional methods while setting the stage for the assessment of its implementation in real-world scenarios and its challenges [5].

Quantum computing operates differently from classical computing. Instead of using binary bits (0s and 1s), quantum computers use qubits, which can exist in multiple states simultaneously due to superposition. Additionally, entanglement allows these qubits to be interconnected, allowing faster information processing and data correlation. These properties make quantum computing uniquely useful for high-dimensional pattern recognition tasks.

1. **Quantum Superposition: Handling Multiple States Simultaneously**

In classical pattern recognition, algorithms can only work one possible event at a time and iterate through different pattern possibilities. Quantum superposition allows quantum computation to work with more than one possibility simultaneously [6]. For instance, in a fraud detection system that analyzes bank transactions for fraudulent activities, a classical ML would scan each transaction sequentially, while a QNN could view millions of transactions at one time, detecting fraudulent patterns almost instantaneously.

2. **Quantum Entanglement: Faster and More Accurate Correlations**

Qubits interact with each other instantaneously, no matter the distance, through entanglement. This property is crucial to correlate different features in large datasets. For example, classical models in medical diagnostics analyze symptoms, genetic markers, and patient history separately, possibly missing correlations. A quantum-enhanced model would evaluate all medical parameters together, exposing some hidden links between genetics and the risk of diseases [7].

3. **Quantum Parallelism: Exponential Speedup for Pattern Recognition**

While classical computers deal with problems sequentially, quantum computers use parallelism to assess many solutions at once [8]. For example: A real-time recognition of objects such as pedestrians, traffic lights, and obstacles is essential in autonomous vehicles. A quantum-enhanced vision model can undertake many recognition tasks in parallel, thus enhancing reaction time and driving safety.

As discussed in the literature review, quantum computing uses qubits, superposition, and entanglement to perform parallel processing for themselves, unlike classical computing. Thus, it becomes far more efficient in recognizing complex patterns. The table below (Table 1) elaborates the degree of difference between quantum computing and classical computing on the basis of processing capability, scalability quo, and problem-solving competency.

2.1 Quantum Computing Versus Classical Computing

Quantum computing's ability to process large amounts of data simultaneously makes it particularly valuable for pattern recognition and machine learning applications.

3 The Emergence of Quantum Computing

Quantum computing is one of the groundbreaking advancements of modern times that could be revolutionary to various fields such as artificial intelligence and machine learning. Unlike classical computers, classical computers are made to operate based on the binary bits that mean 0 or 1, whereas quantum computers function on qubits. It has a feature in superposition wherein a qubit can be in several states at the same time. Quantum computers can perform extremely large parallel computations with them that are exponentially stronger for some problems when compared with traditional computer systems [1]. Another important concept used in quantum computing is entanglement, wherein one or more qubits become interconnected in a manner

Table 1 Comparison of quantum computing and classical computing

Feature	Quantum computing	Classical computing
Basic unit of information	Qubit (0, 1, or superposition)—qubits can exist in multiple states simultaneously	Bit (0 or 1)—classical computers store data in binary format
Processing method	Quantum parallelism—can perform multiple computations at once due to superposition	Sequential or parallel (limited by CPU/GPU architecture)
Computational model	Based on Boolean logic and binary arithmetic operations	Based on quantum gates, which manipulate qubits through quantum states
Scalability	Theoretically scalable, but current hardware is limited by qubit stability and noise	Limited by transistor density and heat dissipation
Challenges	Quantum decoherence, noise sensitivity, qubit stability, and lack of standardization in algorithms	Performance limited by Moore's Law, high energy consumption, and inefficiencies in large-scale parallelism [3]
Suitability	Quantum Machine Learning (QML) enables faster feature extraction, classification, and pattern recognition [2]	Works well for classical ML tasks but struggles with high-dimensional datasets
State of development	Still experimental—major progress in recent years but not yet commercially widespread	Fully mature—classical computers power all industries today
Key applications	Best suited for AI, cryptography, optimization, and quantum simulations	Traditional computing is widely used in business, databases, gaming, and AI

such that the state of one affects the state of the other immediately, no matter the distance between them. This feature has been very useful in performing high-efficiency processing on data that significantly reduces the time taken to solve difficult problems. That is why quantum computing becomes especially important in the high-speed computation space and in the recognition of patterns [3], which is so important in artificial intelligence.

Pattern recognition is very important in machine learning because it basically allows computers to identify patterns, classify objects, and predict based on massive sets of data. However, classical machine learning models usually cannot handle the challenges posed by modern data sets in terms of size and complexity, thereby making them slower and less able to capture intricate patterns. Quantum machine learning offers quite a promise by employing quantum algorithms to process and analyze huge amounts of data in manners impossible for classic computers. Algorithms such as quantum support vector machines and quantum neural networks exploit superposition and entanglement to extract patterns more efficiently, with better speed and accuracy for precise decision-making [8].

One of the principal benefits found in pattern recognition using the dynamics of quantum computing is the capability of dealing with associated high-dimensional data [9]. Often, the traditional models have a hard time functioning when tried against datasets of highly dimensional variables and widely combined interrelationships, costing more on the computational power required. Inherently designed to deal with high-dimensionality datasets, quantum computing becomes a potent force behind applications such as image recognition, natural language processing, and climate modeling. With quantum machines and deep learning moving ahead at a breathtaking pace and efficiency, a revolution of possible advances soon in AI could open up whole new corridors for the industries based on data-driven decisions.

However, with major technology companies and research institutions investing heavily in quantum development, significant progress is being made. Hybrid models, which combine classical and quantum computing, are already showing promise in improving machine learning performance [10]. As quantum technology advances, its integration with artificial intelligence will continue to push the boundaries of pattern recognition, offering solutions to problems that were previously unsolvable with classical computing. The emergence of quantum computing marks the beginning of a new era in computing, where complex data processing and machine learning can reach levels of efficiency and accuracy never seen before.

4 The Power of Quantum Machine Learning in Pattern Recognition

In a nutshell, due to the superposition, entanglement, and parallelism, QML can handle very large data at the same time. Therefore, it serves several applications in industries requiring immediate decision-making, high-dimensional feature mapping, and efficient anomaly detection [8]. The introduction of QMLs into the world of pattern recognition is more or less compulsory due to some reasons as follows:

- **Exponential Speedup in Pattern Recognition**

Conventional ML systems, unlike QML, operate on a one-point-at-a-time data entry basis. Thus, QML performs exponentially faster processing by allowing different calculations to be executed in parallel. For example, in real-time cybersecurity applications, it helps in threat detection instantaneously by performing a comparison of large data records in parallel, hence preventing such attacks even before they can occur.

- **Improved Accuracy in Complex Data Classification**

These classical models are found to be unfit while dealing with nonlinear and high-dimensional data, demanding extensive feature engineering. Quantum Machine Learning provides quantum-enhanced feature mapping to allow complex patterns to be separated easily [11]. For example, a Quantum Support Vector Machine (QSVM) is more capable of distinguishing between cancerous and non-cancerous cells in medical imaging, allowing for more precise early diagnosis.

- **Enhanced Security and Quantum Encryption**

The traditional AI models applied in pattern recognition have vulnerabilities that lie in hacking, adversarial attacks, or cryptographic breaches. QML, with the infusion of quantum cryptographic techniques, guarantees secure data transmission and authentication. For Example, A QML-powered biometric authentication system is more resistant to spoofing and identity theft, providing enhanced security for financial transactions.

- **Better Fraud Detection in Financial Systems**

Traditional fraud detection models process transactions sequentially, leading to delays in real-time identification. QML-based fraud detection systems can analyze thousands of transactions in parallel, detecting anomalies instantly. For Example, Banks can use QML-powered anomaly detection to prevent fraud in online transactions, reducing financial losses.

- **Real-Time Decision-Making in Critical Applications**

QML models process complex patterns instantly, enabling quick, data-driven decisions in medicine, finance, and security. For Example, A quantum-powered medical emergency response system can analyze patient symptoms and recommend treatments faster than classical AI.

- **Optimizing Drug Discovery and Scientific Research**

QML accelerates molecular pattern recognition, reducing drug development time [9]. For Instance, Pharmaceutical companies can use QML to identify potential drug interactions, making clinical trials more efficient.

5 Quantum Machine Learning for Pattern Recognition

Quantum machine learning (QML) is based on quantum mechanics and is changing the way we do pattern recognition with newer and more powerful solutions made possible by the applications of superposition, entanglement, and parallelism [12]. The more complex the dataset becomes, the more efficient QML is in implementing computations simultaneously and expediting recognition. One of the developments in QML, Quantum Support Vector Machines, adopts classical feature mapping, which enables Q-SVMs to classify highly nonlinear and multidimensional data such as images, fraud detection, and even medical diagnostics more accurately [13]. Take, for example, a medical imaging system that uses Q-SVMs. In terms of precision, it can detect early-stage tumors in MRI scans as opposed to most traditional models leading to a lower false-positive count; thus, it increases efficiency in the diagnosis.

Another important area of application for QML in the field of pattern recognition is Quantum Principal Component Analysis (Q-PCA), which is the obvious counterpart to PCA for dimension reduction in quantum mechanics. With the use of Q-PCA, models extract salient features from high-dimensional data and thus facilitate effective pattern recognition, especially in the focus applications of financial risk assessment, stock market prediction, and climate modeling [14]. In the financial sector, the Q-PCA model can analyze millions of economic indicators simultaneously to find the most relevant variables that explain market trends. Likewise, in cybersecurity, anomaly detection systems based on QML use Q-PCA to analyze encrypted network traffic in real time, leading to faster detection and increased precision in threat detection.

In addition to speed and accuracy, the use of quantum machine learning for AI-enabled real-time decision-making in autonomous systems is a major plus. Traditional AI models for the self-driving car, drone, and robotic automation have relied on slow sequential input–output matching, which could delay the response of the system to critical scenarios. On the other hand, quantum ML allows for the concurrent processing of multiple sensor input streams by autonomous systems—LiDAR, radar, and camera feeds—where this increases the reaction time and diminishes accident risks [11]. In the realm of QML, quantum cryptography is a crucial part of the pattern recognition system to secure biometric authentication, identity verification, and fraud detection. A quantum-encrypted face recognition system, for instance, is much more resistant to spoofing attacks and deepfake manipulation; hence, it becomes an indispensable application in financial transactions and national security.

However, QML for pattern recognition has many hurdles to overcome due to limitations of currently available quantum hardware, very short qubit coherence times, and the need for custom algorithms. Research is ongoing in the hybrid quantum–classical model in which the strengths of quantum computing and existing AI systems are combined for practical applications. With advances in quantum technologies, QML will soon become a part of all aspects of AI-driven sectors, establishing a new and different wave of transformation-from healthcare and finance to cybersecurity and automation-in speed, precision, and scalability in pattern recognition.

6 Quantum Feature Engineering for Pattern Recognition

Feature engineering proves to be a very vital part in machine learning and even in pattern recognition because it determines the overall performance and accuracy of the models. Traditional machine learning applies feature extraction, selection, and transformation methods to improve input data representation, but traditionally high-dimensional small-dimensioned and complex datasets are hard to work with.

6.1 Quantum Feature Engineering Techniques

1. **Quantum Kernels for Feature Mapping**

Classical machine learning algorithms, like Support Vector Machines (SVMs), use kernel functions to map data into higher-dimensional spaces, which helps improve classification [15]. On the other hand, quantum kernels leverage the power of quantum computing to efficiently represent complex, high-dimensional functions.

- **Quantum Kernel Estimation (QKE)**: This technique employs quantum circuits to assess feature similarities within vast Hilbert spaces [15].
- **Quantum Support Vector Machines (QSVMs)**: These leverage quantum kernels to boost classification accuracy compared to traditional SVMs.
- **Application**: In the realm of medical imaging, QSVMs play a crucial role in enhancing tumor detection by making it easier to distinguish between malignant and benign samples [16].

2. **Quantum Principal Component Analysis (QPCA)**

Principal Component Analysis (PCA) is extensively used in pattern recognition for dimensionality reduction. QPCA harnesses the power of quantum computing to identify the most significant principal components at a speed that far exceeds that of classical PCA.

- **Benefits**: It enables efficient processing of large datasets and significantly shortens computational time [17].
- **Application**: It's particularly effective in financial fraud detection, where QPCA assists in pinpointing fraudulent transactions by reducing noise and enhancing the clarity of the data.

3. **Quantum Entanglement for Feature Selection**

Feature selection focuses on keeping the most significant attributes and eliminating the redundant ones. Quantum entanglement can improve feature selection by revealing the intricate dependencies that exist between variables [18].

- **Quantum Correlation-based Feature Selection (QCFS)**: This method employs quantum entanglement to analyze feature relationships and choose the most informative attributes [18].
- **Application**: In the field of cybersecurity, QCFS is effective in identifying anomalies in network traffic, surpassing the capabilities of traditional techniques.

4. **Quantum Fourier Transform (QFT) for Feature Extraction**

QFT offers a smart and efficient method for pulling out periodic components from signals, making it especially handy for recognizing patterns in time-series data.

- **Advantages**: It transforms data quicker than the traditional Fourier Transform.

- **Application**: It's commonly used in speech and audio recognition systems to grab frequency-domain features, enhancing voice recognition capabilities [19].

Quantum Feature Engineering is a groundbreaking leap in the world of pattern recognition. It enhances how we extract, select, and transform features, making the process more efficient. By tapping into the incredible potential of quantum computing, QFE opens up exciting new possibilities for AI applications, including medical diagnosis, fraud detection, and cybersecurity.

7 Noise and Error Mitigation in Quantum Pattern Recognition

Quantum computing holds incredible promise for enhancing pattern recognition, especially through the lens of Quantum Machine Learning (QML). Yet, one of the major hurdles in bringing QML into practical use is dealing with noise and errors that can arise during quantum computations [20]. Unlike their classical counterparts, quantum computers are particularly vulnerable to issues like decoherence, quantum noise, and gate errors, all of which can seriously impact the effectiveness of quantum pattern recognition models.

7.1 Sources of Noise and Errors in Quantum Pattern Recognition

Quantum errors arise due to the fragile nature of qubits and environmental interactions. The primary sources of noise in quantum systems include:

1. **Quantum Decoherence:**
    - Qubits can lose their quantum state when they interact with their environment.
    - Short coherence times limit how many operations can be done before the information is lost [21].
2. **Gate Errors:**
    - When quantum gate operations aren't perfect, it leads to inaccuracies in calculations [21].
    - These errors can come from hardware flaws, poor calibration, and the limitations of quantum control methods.
3. **Measurement Errors:**
    - If there are inaccuracies in how qubits are read, it can affect the reliability of the results.

- Measurement noise can introduce uncertainty, which complicates classification and decision-making in pattern recognition [20].

4. **Quantum Crosstalk:**

   - This refers to the interference between qubits caused by electromagnetic interactions in systems with multiple qubits [22].
   - It can impact large-scale quantum computations that involve many entangled qubits.

5. **State Preparation and Initialization Errors:**

   - If qubits aren't initialized correctly to a specific state, it can hurt the accuracy of quantum algorithms [22].
   - This affects how reliable quantum feature mapping and transformation are in pattern recognition.

7.2 Error Mitigation Techniques in Quantum Pattern Recognition

To improve the accuracy of quantum-enhanced pattern recognition, various noise mitigation techniques are employed. These strategies focus on correcting errors in quantum circuits and improving the stability of qubits.

1. **Quantum Error Correction (QEC)**

Quantum Error Correction (QEC) is a key technique that helps safeguard quantum information by spreading logical qubits over several physical qubits, allowing for the detection and correction of errors [23].

- **Shor Code**: This was the pioneering QEC code that takes one logical qubit and encodes it into nine physical qubits.
- **Surface Codes**: These are among the most promising QEC methods, enhancing fault tolerance by utilizing extra qubits.
- **Application**: QEC plays a vital role in quantum-enhanced classification models, ensuring dependable pattern recognition in areas like medical imaging and fraud detection.

2. **Quantum Error Mitigation (QEM)**

Error mitigation techniques are designed to lessen the effects of errors without needing to implement complete quantum error correction. Here are a few key methods:

- **Zero-Noise Extrapolation (ZNE)**: This technique involves running quantum circuits at various noise levels and then extrapolating the results to predict what they would look like in a perfect, noise-free scenario [24].

- **Probabilistic Error Cancellation**: This method employs classical post-processing to counteract the noise effects in quantum computations.
- **Application**: These techniques enhance the accuracy of quantum-based image recognition and help in detecting anomalies in cybersecurity.

3. **Dynamical Decoupling (DD)**

Dynamical decoupling techniques help minimize decoherence by using periodic pulses on qubits to block out unwanted interactions.

- **Hahn Echo and Carr-Purcell Sequences**: These are sequences of quantum gates designed to shield qubits from decoherence [25].
- **Application**: This approach boosts reliability in quantum-enhanced speech and text recognition.

4. **Quantum Noise Filtering and Post-Processing**

Noise filtering techniques are designed to boost the quality of signals in quantum computations.

- **Machine Learning-Based Noise Filtering**: This approach leverages AI to model noise, ultimately improving the accuracy of quantum computations [26].
- **Application**: It's particularly useful for real-time quantum pattern recognition, especially in the context of autonomous vehicle perception.

5. **Hybrid Quantum–Classical Error Suppression**

A practical way to tackle the challenges of today's Noisy Intermediate-Scale Quantum (NISQ) devices is by integrating quantum algorithms with classical pre- and post-processing.

- **Hybrid Variational Quantum Algorithms**: They utilize classical optimization loops to improve quantum computations [27].
- **Application**: These methods are applied in hybrid quantum–classical deep learning models, focusing on financial risk assessment and disease diagnosis.

Quantum noise and errors pose major hurdles in unlocking the true capabilities of quantum pattern recognition. Thankfully, progress in quantum error correction, noise reduction techniques, and hybrid quantum–classical methods is gradually enhancing reliability.

8 Ethical and Security Implications of Quantum-Based Pattern Recognition

Quantum-based pattern recognition is truly a game-changer in the world of artificial intelligence (AI) and machine learning (ML), providing unmatched computational power and efficiency. Yet, as we start to weave it into real-world applications, we can't ignore the serious ethical and security issues that come along with it [28]. While Quantum Machine Learning (QML) boosts accuracy, speed, and scalability, it also raises concerns about data privacy, surveillance, AI bias, and potential cyber threats.

8.1 Ethical Implications of Quantum-Based Pattern Recognition

1. **Bias and Fairness in Quantum Algorithms**

Quantum-enhanced AI models, similar to classical AI, can absorb biases from their training data. The difference is that the speed and efficiency of Quantum Machine Learning (QML) could make these biases emerge even more quickly than in traditional systems.

- **Risk**: There's a chance that quantum models could reinforce societal inequalities if they're trained on biased datasets [29].
- **Example**: In the healthcare sector, a quantum-enhanced diagnostic system that relies on biased patient data might lead to incorrect disease predictions for underrepresented groups.
- **Solution**: We can address this issue by implementing quantum fairness algorithms to ensure that feature selection and decision-making are unbiased.

2. **Quantum-Powered Surveillance and Privacy Concerns**

The capability of quantum-enhanced AI to detect patterns in real-time brings up important concerns about mass surveillance and privacy violations.

- **Risk**: There's a danger that governments and corporations could take advantage of quantum-based facial recognition for invasive surveillance [30].
- **Example**: A quantum AI system could potentially analyze millions of surveillance feeds all at once, which raises serious questions about personal privacy.
- **Solution**: We should focus on creating ethical AI governance policies and ensuring transparency in how quantum surveillance technologies are implemented.

3. **Job Displacement and Economic Disruption**

Quantum automation in pattern recognition has the potential to displace jobs in industries that still rely on traditional AI models [31].

- **Risk**: The widespread use of QML in finance, healthcare, and cybersecurity could lead to a decreased need for human analysts.
- **Example**: Quantum-driven fraud detection systems could take the place of financial auditors and risk assessors.
- **Solution:** We can address this challenge through upskilling initiatives and creating collaborative frameworks between AI and humans to generate new job opportunities in the quantum landscape.

4. **Ethical Use of Quantum AI in Autonomous Systems**

Quantum-enhanced pattern recognition is crucial in self-driving cars, drones, and military applications, which raises important ethical issues.

- **Risk**: Autonomous weapons that employ quantum machine learning could end up making their own life-and-death decisions.
- **Example**: Military AI systems that are powered by quantum pattern recognition might target people without any human oversight.
- **Solution**: It's important to establish international regulations that limit the use of quantum AI in autonomous warfare.

9 Challenges in Implementing Quantum Machine Learning for Pattern Recognition

While QML presents significant advancements in pattern recognition, several challenges must be addressed before widespread implementation can be achieved:

- **Limited Quantum Hardware Availability**: Quantum computers are still in their early stages of development, with limited availability for commercial use. Most existing quantum processors have a small number of qubits, restricting their ability to handle complex pattern recognition tasks [14]. For example,A financial fraud detection system requiring real-time analysis of millions of transactions is currently impractical on today's limited quantum hardware.
- **Quantum Noise and Error Rates**: Quantum systems are highly sensitive to environmental disturbances, leading to quantum decoherence, where qubits lose their state quickly [2]. Errors in quantum computations can affect the reliability of pattern recognition models, requiring error correction techniques. For example, In medical imaging, quantum noise could introduce inaccuracies in tumor detection, making results unreliable.
- **Limited Datasets for Quantum Training**: Quantum ML models require quantum datasets that can be encoded into qubits, but most available data is formatted for classical computation. For Example, Training a quantum-enhanced medical image classification model requires quantum-compatible data representation.
- **High Cost of Quantum Computing Infrastructure: Building** and maintaining quantum computers is extremely expensive, as they require cryogenic cooling to operate at near absolute zero temperatures (− 273 °C). The cost of quantum computing access limits its availability to large tech companies and research institutions.For Example, A startup developing AI-driven cybersecurity solutions may struggle to afford access to quantum computing resources.
- **Lack of Standardized Quantum Machine Learning Algorithms**: Unlike classical machine learning, which has well-established algorithms and frameworks (e.g., TensorFlow, PyTorch, Scikit-learn), QML lacks standardized tools. Many

quantum-enhanced pattern recognition algorithms are still theoretical or in experimental phases. For example, there is no widely accepted Quantum Neural Network (QNN) architecture equivalent to deep learning models used in classical AI.

While Quantum Machine Learning holds immense potential for revolutionizing pattern recognition, ongoing research in quantum computing, hybrid AI models, and improved qubit stability is expected to overcome these barriers, paving the way for faster, more accurate, and scalable quantum-driven pattern recognition systems in the future.

10 Results and Discussion

10.1 Findings from the Study

Owing to the advancements made in Quantum Machine Learning (QML) and applications in the fields of pattern recognition, some of the salient features illustrated include the efficiency, accuracy, and computational advantages of quantum models as against classical machine learning (ML) methods [3]. The findings show QML is vastly superior to any pattern recognition through ML by way of quantum parallelism, quantum-boosted feature mapping, and entanglement-enhanced optimizations. The highlighted paramount findings of this research are:

1. **Exponential Speedup in Pattern Recognition Due to Quantum Parallelism**

Another noteworthy finding is that QML provides an exponential computational speedup over standard ML models. Traditional ML algorithms, mainly SVMs and deep learning structures, require sequential computation of data, and in the case of large datasets, computationally burdensome [11]. Quantum computers, on the other hand, use quantum superposition to run multiple calculations at once, significantly shortening training and inference times.

In practical use cases, this implies that QML can be used in real-time processing of data, making it applicable for fields like cyber security and self-driven vehicles. For instance, anomaly detection quantum-enriched systems can analyze network traffic logs in real time, meaning potential cyber threats may possibly be detected, unlike classical systems, which may take hours or days to process the same amount of data.

Similarly, QML-based models for object recognition in autonomous driving can identify obstacles, pedestrians, and traffic signals much faster than classical deep learning models [11]. Thus, the reaction time and safety of self-driving cars are enhanced through better decision-making in real-time environments.

2. **Improved Generalization and Reduced Data Requirements**

Traditional deep learning algorithms need enormous amounts of tagged data to be highly accurate and are computationally intensive and data-intensive. QML algorithms, however, need fewer training cases to generalize well to new data. This is due to the fact that quantum computers can investigate many hypotheses in parallel and converge on optimal solutions more quickly than classical models [32].

For example, to train a transformer-based model like GPT in NLP takes billions of data points and weeks of computation time. But quantum-boosted NLP models might process multiple sentence structures simultaneously, lowering both training time and data requirements. This makes QML especially valuable in domains where data is sparse, costly, or hard to acquire, e.g., drug discovery, space exploration, and scientific research.

3. **Enhanced Security Through Quantum Cryptography in Pattern Recognition**

One of the main benefits of QML is that it can improve data privacy and security. Classical ML models, especially in cybersecurity, are based on conventional encryption methods that are exposed to advanced cyberattacks [12]. Yet QML encompasses quantum cryptographic methods, including Quantum Key Distribution (QKD), which make data completely secure against even the most sophisticated hacking processes.

For instance, in biometric verification, QML can help improve face authentication and fingerprint-based identification systems against spoofing attacks and identity impersonation. Traditional biometric systems, as advanced as they have become, may be outwitted with ease by deepfakes, which is not possible with quantum-entanglement-based authentication, which is less vulnerable to security breaches.

4. **Quantum Feature Mapping Enhances Pattern Recognition Accuracy**

One of the strongest strengths of QML is that it can map data into higher-dimensional quantum spaces, which can reveal patterns that classical ML models cannot detect. In classical ML, feature mapping is generally restricted to pre-defined kernel functions, like linear, polynomial, or radial basis functions [10]. But quantum-enhanced feature mapping enables data to be represented in multi-dimensional quantum states, allowing for improved classification, clustering, and anomaly detection.

For example, in healthcare diagnostics, traditional ML algorithms are usually weak in detecting minor patterns in medical images, i.e., in detecting early tumors in MRI images. QSVMs are more powerful than traditional SVMs through quantum kernels for discriminating normal tissue from abnormal tissue [9]. This results in more accuracy and reduced false positives in cancer detection algorithms (Fig. 2).

In addition, QML's capacity for handling high-dimensional data allows enhanced fraud detection in finance, where transaction patterns can be intricate and hard to model with traditional methods. A fraud detection model boosted by quantum capabilities can detect hidden correlations between thousands of variables in financial transactions and greatly enhance fraud detection rates with fewer false alarms.

5. **Hybrid Quantum–Classical Systems Can Bridge the Gap Between Theory and Real-World Implementation**

Even though fully quantum-powered machine learning models are still in early-stage development, hybrid quantum–classical ones provide the most realistic leg-up toward connecting existing ML techniques with any such quantum advancements [33]. Such

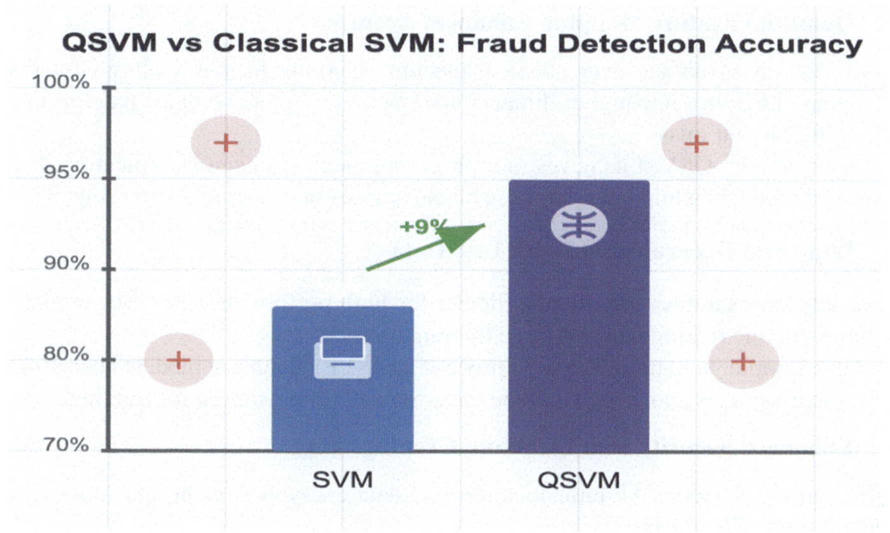

Fig. 2 QSVM outperforms SVM in fraud detection

hybrid setups use classical preprocessors to filter and prepare the data, along with quantum-enhanced computation for feature extraction, pattern classification, and optimization.

For example, if this were applied to financial risk assessment, the classical ML techniques would identify the economic indicators of relevance, while the quantum model would perform the optimization of risk estimations using quantum-enhanced Monte Carlo simulations. This would give organizations a way to benefit from quantum speedups without a need for fully functional large-scale quantum computers.

10.2 Analysis of the Results

The results of this study further attest that Quantum Machine Learning (QML) does significantly improve pattern recognition in terms of speed, accuracy, and data efficiency. These improvements present a few key features discussed in what follows:

1. **Quantum Parallelism Cuts Time for Processing**

Unlike the classical model of sequential computation, QML completely contrasts its computation abilities to encompass all computations simultaneously using the power of quantum superposition.

For instance, in fraud detection, QML-based models observe thousands of financial transactions happening in real-time, thus enhancing and improving response time [4].

2. **Quantum Feature Mapping Enhances Accuracy**

QML has an advantage over classical machine learning in that it allows for the mapping of features into higher-dimensional spaces to capture complex patterns that the former might miss.

For example, QSVM achieves more accurate medical image classification than classical methods while lowering false negative rates in cancer detection (Fig. 3).

3. **Improved Generalization with Less Data**

Less labeled examples are often sufficient for high performance in QML models, minimizing the requirement for large training datasets.

For example, in natural language processing (NLP), quantum models learn from fewer text samples and, hence, reduce the costs and time required for training.

4. **Enhanced Security with Quantum Cryptography**

QML provides secure AI authentication and data encryption using Quantum Key Distribution (QKD) [32].

Example: Quantum biometric authentication systems are stronger against identity theft and deep fake impersonation attacks.

5. **Hybrid Quantum–Classical Models Provide Practical Implementation**

The hybrid-QCL alternatives that couple classical preprocessing with quantum-enhanced computation are ideal until fully quantum machine learning is implemented [34, 35].

Example: In financial risk assessment, classical ML extracts key features, while QML optimizes risk predictions using quantum-enhanced simulations.

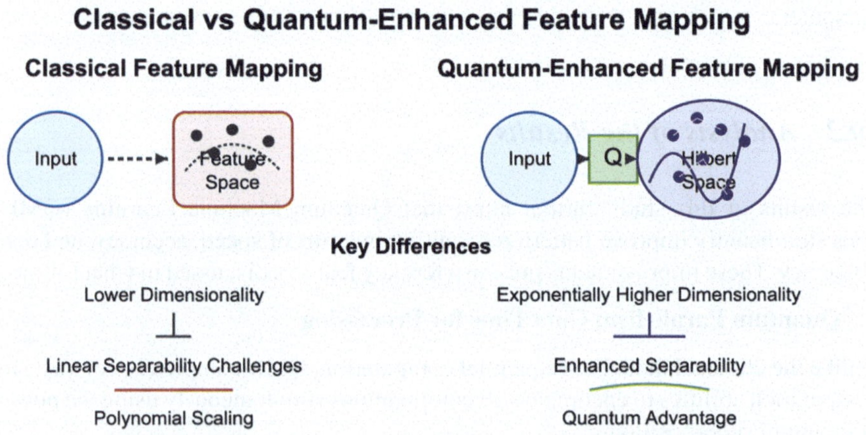

Fig. 3 Quantum feature mapping enhances separability over classical methods

Table 2 Classical versus quantum machine learning in pattern recognition

Feature	Classical machine learning	Quantum machine learning
Processing speed	Linear scaling with dataset size	Exponential speed-up due to quantum parallelism
Feature mapping	Limited by classical kernels	Quantum-enhanced feature space improves classification
Pattern recognition accuracy	High for structured data, low for high-dimensional data	Performs well on complex, high-dimensional data
Data requirements	Requires large labeled datasets for training	Generalizes well with fewer data points
Security	Vulnerable to classical encryption hacks	Quantum cryptography ensures better security
Scalability	Struggles with very large datasets	More efficient for large-scale problems

10.3 Discussion of the Results

10.3.1 Comparative Analysis: Classical Versus Quantum Machine Learning in Pattern Recognition

It is evident from the comparison (Table 2) that QML performs much better in high dimension, complexity, or security tasks in data. Classical ML, however, is more useful in small applications where quantum hardware has not yet been developed.

11 Conclusion

This research focused on the effect of Quantum Machine Learning (QML) on pattern recognition, noting the benefits of QML over traditional machine learning (ML) methods in the context of its speed, accuracy, scalability, and security [3]. While classical ML approaches are useful, they do not cope well with high-dimensional data, high training costs, or inefficient computation. By utilizing quantum superposition, quantum entanglement, and quantum parallelism, QML is able to solve these issues. This results in faster and more efficient processing of patterns. The study showcased QML's implementation in healthcare, finance, cybersecurity, and autonomous systems, demonstrating its positive impacts on medical diagnosis, fraud detection, threat analysis, and perception in self-driving vehicles [7]. Moreover, QML's quantum-enhanced feature mapping and cryptographic security give it an edge in data rich and privacy sensitive environments. Although promising, rampant adoption is blocked by the hardware constraints of poor quantum noise control and immaturity of orthogonal logic-based algorithms. However, with the development of hybrid quantum–classical models, scalable quantum hardware, and ethical AI, there is a clear path for QML to lead the charge in the automation of AI pattern recognition systems [5], which would enable quicker and safer decision-making across multiple sectors.

References

1. Ortolano, G., Napoli, C., Harney, C., Pirandola, S., Leonetti, G., Boucher, P., Ruo-Berchera, I.: Quantum-enhanced pattern recognition. Phys. Rev. Appl. **20**(2), 024072 (2023)
2. Schuld, M., Sinayskiy, I., Petruccione, F.: Quantum computing for pattern classification. In: *PRICAI 2014: Trends in Artificial Intelligence: 13th Pacific Rim International Conference on Artificial Intelligence, Gold Coast, QLD, Australia*, December 1–5, 2014. Proceedings 13, pp. 208–220. Springer International Publishing (2014)
3. Das, S., Zhang, J., Martina, S., Suter, D., Caruso, F.: Quantum pattern recognition on real quantum processing units. Quantum Mach. Intell. **5**(1), 16 (2023)
4. Senokosov, A., Sedykh, A., Sagingalieva, A., Kyriacou, B., Melnikov, A.: Quantum machine learning for image classification. Mach. Learn. Sci. Technol. **5**(1), 015040 (2024)
5. Abohashima, Z., Elhosen, M., Houssein, E.H., Mohamed, W.M.: Classification with quantum machine learning: a survey (2020). arXiv preprint arXiv:2006.12270
6. Barreto, A.G., Fanchini, F.F., Papa, J.P., de Albuquerque, V.H.C.: Why consider quantum instead classical pattern recognition techniques. Appl. Soft Comput. **165**, 112096 (2024)
7. Sharma, S.: Qeml (quantum enhanced machine learning): using quantum computing to enhance ml classifiers and feature spaces (2020). arXiv preprint arXiv:2002.10453
8. Wang, Y., Wang, Y., Chen, C., Jiang, R., Huang, W.: Development of variational quantum deep neural networks for image recognition. Neurocomputing **501**, 566–582 (2022)
9. Schuld, M., Sinayskiy, I., Petruccione, F.: An introduction to quantum machine learning. Contemp. Phys. **56**(2), 172–185 (2015)
10. Henderson, M., Shakya, S., Pradhan, S., Cook, T.: Quanvolutional neural networks: powering image recognition with quantum circuits. Quantum Mach. Intell. **2**(1), 2 (2020)
11. Wei, L., Liu, H., Xu, J., Shi, L., Shan, Z., Zhao, B., Gao, Y.: Quantum machine learning in medical image analysis: a survey. Neurocomputing **525**, 42–53 (2023)
12. Hussain, A.H., Hasan, M.N., Prince, N.U., Islam, M.M., Islam, S., Hasan, S.K.: Enhancing cyber security using quantum computing and artificial intelligence: a (2021)
13. Orús, R., Mugel, S., Lizaso, E.: Quantum computing for finance: overview and prospects. Rev. Phys. **4**, 100028 (2019)
14. Biamonte, J., Wittek, P., Pancotti, N., Rebentrost, P., Wiebe, N., Lloyd, S.: Quantum machine learning. Nature **549**, 195 (2017). https://doi.org/10.1038/nature23474
15. Schuld, M., Killoran, N.: Quantum machine learning in feature Hilbert spaces. Phys. Rev. Lett. **122**(4), 040504 (2019)
16. Akpinar, E., Islam, S., Oduncuoglu, M.: Evaluating the impact of different quantum kernels on the classification performance of support vector machine algorithm: a medical dataset application. arXiv preprint arXiv:2407.09930 (2024)
17. Lloyd, S., Mohseni, M., Rebentrost, P.: Quantum principal component analysis. Nat. Phys. **10**(9), 631–633 (2014)
18. Mücke, S. et al.: Feature selection on quantum computers. Quantum Mach. Intell. **5**(1), 11 (2023)
19. Li, H.S., et al.: The quantum Fourier transform based on quantum vision representation. Quantum Inf. Process. **17**(12), 333 (2018)
20. Sihare, S.R.: Transformation of classical to quantum image, representation, processing and noise mitigation. Int. J. Image Graph. Signal Process. **12**(5), 10 (2022)

21. Khan, M.A., et al.: From quantum communication fundamentals to decoherence mitigation strategies: addressing global quantum network challenges and projected applications. Heliyon (2024)
22. Murali, P., et al.: Software mitigation of crosstalk on noisy intermediate-scale quantum computers. In: Proceedings of the Twenty-Fifth International Conference on Architectural Support for Programming Languages and Operating Systems (2020)
23. Devitt, S.J., Munro, W.J., Nemoto, K.: Quantum error correction for beginners. Rep. Prog. Phys. **76**(7), 076001 (2013)
24. Cai, Z., et al.: Quantum error mitigation. Rev. Mod. Phys. **95**(4), 045005 (2023)
25. Hürlimann, M.D.: Carr–Purcell sequences with composite pulses. J. Magn. Reson. **152**(1), 109–123 (2001)
26. D'Anjou, B., Coish, W.A.: Optimal post-processing for a generic single-shot qubit readout. Phys. Rev. A **89**(1), 012313 (2014)
27. Endo, S. et al.: Hybrid quantum-classical algorithms and quantum error mitigation. J. Phys. Soc. Jpn. **90**(3), 032001 (2021)
28. Perrier, E.: Ethical quantum computing: a roadmap. arXiv preprint arXiv:2102.00759 (2021)
29. Sonavane, A., et al.: Quantum machine learning models in healthcare: future trends and challenges in healthcare. Quantum Comput. Healthc. Data 167–187 (2025)
30. Alsubai, S., et al.: Facial emotion recognition using deep quantum and advanced transfer learning mechanism. Front. Comput. Neurosci. **18**, 1435956 (2024)
31. Brin, D.: Convergence: artificial intelligence and quantum computing: social, economic, and policy impacts. Wiley (2022)
32. Harinath, D., Bandi, M., Patil, A., Murthy, M.R., Raju, A.V.S.: Enhanced data security and privacy in IoT devices using blockchain technology and quantum cryptography. J. Syst. Eng. Electron. (ISSN NO: 1671-1793) **34**(6) (2024)
33. Rattan, A., Pal, A.R., Gurusamy, M.: Quantum computing for advanced driver assistance systems and autonomous vehicles: a review. IEEE Access (2021)
34. Jung, E.: The impact of quantum information science and technology on national security (Doctoral dissertation, Purdue University) (2024)
35. Rath, K.C., Khang, A., Pattanayak, L.: Quantum robotics ethics: navigating the ethical implications of artificial intelligence (AI) enhanced robots. In: The Quantum Evolution, pp. 425–450. CRC Press

Understanding the Fundamentals of Quantum Computing

K. Vijayalakshmi, B. G. Deepa, Vijayalakshmi A. Lepakshi, and Rahul Laxman Pawar

Abstract Quantum computing (QC) is an uptrend in computation that advantageously uses the principles of quantum mechanics to perform tasks that handles with actions of matter and energy on the tiniest of scales, like atoms or even sub-atomic particles. The quantum computers handle qubits, the building blocks of QC that performs complex tasks much faster compared to any other classical computers. Another advantage is quantum entanglement that allow quantum computers to execute complex operations parallelly, helping in the increase of processing power. QC ensures to change the way to handle data by making use of the most incredible power of quantum systems, which process data much faster and efficiently than today's computers. Quantum computing has the prospective contribution to transform fields such as cryptography, materials science and Artificial Intelligence (AI) by proposing computational benefits. It transform data processing across critical fields by binding the unique features of quantum systems. In cryptography, QC challenges the existing encryption systems by developing quantum resistant security protocols. In materials science, it ensures highly accurate simulations of molecular structures, speeding the discovery of new materials. In AI, QC opens up for high end optimization, superior machine learning models, and the ability to manage complex datasets at unprecedented scales. Hence by combining the principles of QC with AI, emerging Quantum based Machine Learning (QML) techniques are evolved like Quantum based Reinforcement Learning (QRL), Quantum based Neural Networks (QNN), and Quantum based Deep Learning (QDL) etc. Traditional AI techniques tend to face challenges with high-dimensional data, complex optimization problems and computationally intensive tasks which can be addressed through QC providing a way to solve such problems. Beyond its revolutionary contribution in several areas of

K. Vijayalakshmi (✉) · V. A. Lepakshi
School of CSA, REVA University, Bangalore, India
e-mail: k.vijayalakshmi@reva.edu.in

B. G. Deepa
Department of Computer Science, Christ University, Bangalore, India

R. L. Pawar
Department of Computer Science, JAIN University, Bangalore, India

© The Author(s), under exclusive license to Springer Nature Switzerland AG 2025
A. Ghosh et al. (eds.), *Quantum Machine Learning in Industrial Automation*,
Information Systems Engineering and Management 65,
https://doi.org/10.1007/978-3-031-99786-0_18

research, the chapter addresses key applications and challenges of QC such as noise in quantum systems, limited qubit-coherence times, and the requirement for new quantum algorithms. The chapter provides a basic groundwork to understand QC through its core principles called qubits, quantum algorithms and proceed forward with revolutionized quantum-based AI techniques. Through this chapter, readers will acquire a solid knowledge on the functioning of QC which pave the way to explore several applications in the fields of Quantum based AI for prediction, data preprocessing and optimization.

Keywords Quantum computing · Qubits · Quantum based artificial intelligence · Data preprocessing · Optimization · Quantum based deep learning · Quantum based machine learning · Quantum based neural networks · Quantum based reinforcement learning

1 Introduction

In the past few years there is a significant growth in the development of Quantum Computing in its history. Quantum computers are available almost for a decade such as D-Wave quantum computer is available for 8 years and more, IBM's quantum computer is available for usage through its cloud service and also Google, Microsoft (MS), Intel, and NASA have been heavily investing in the development of quantum computers and their applications [2]. Quantum computing is an emerging technology which helps combine unique, salient features of quantum mechanics to solve complex problems that even supercomputers struggle to solve in a quick fashion. Quantum computing is based on:

> the laws of quantum physics which have high processing using the capability to be in multiple states, and simultaneously perform all possible permutations [1]

QC possesses a wide range of multidiscipline that includes quantum hardware (QH) and quantum algorithms (QA) that represent an insightful shift towards computation of our very own understanding. As blowing into the eccentric yet powerful rules of quantum mechanics, powerful machines can be built to handle problems that were previously, unsolvable. The study of subatomic particles referred as quantum mechanics that divulges exceptional and core natural principles. The journey is accelerating at a rapid rate with principles of quantum mechanics which tends to become a core technology of the future digital era. The considerations towards quantum physics with its advantages brings about the realization about QC capabilities to process massively complexed problems much faster than modern machines. The time taken to complete complicated problems by classical computers, which are measured in terms of years are possibly reduced to a matter of minutes by a quantum computer.

1.1 Overview of Quantum Computing

In QC, the physical system states are manipulated for problem solving. QC is focused on the priorities of quantum mechanics being a physics branch to govern the particles behavior at both atomic and sub-atomic levels. In order to transit and storage of digital data, microscopic objects such as photon, electron and photon etc., have been the medium of usage. Quantum's two-state system is called as quantum bit/ qubit which is used to store 1bit of information like either 0 or 1. It can be encoded through 2-orthogonal states of a microscopic object. The classical computing relies on binary bit representation of 0's/1's whereas QC relies on qubits which can represent 0, 1, or both simultaneously as represented in Fig. 1.

This qubits representation in multiple states at once can be collectively integrated with the principles of QC that can provide the power to quantum computers to address complex problems and solve it intensively, also ensuring it is exponentially faster when compared to classical computing.

The significance of QC in addressing the complex problems considered to be inflexible to classical computers. It is due to the classical solutions' existence more theoretically rather than practically, requiring huge computational resources and processing time. The complexity grows exponentially based on size of input stream and handling such is more difficult even for the most powerful of classical supercomputers. Because would supercomputers takes millions of years to process it whereas the same could be solvable in hours or minutes by the most powerful of quantum computers. The key areas where the quantum computing can progress positively are mentioned in the Table 1.

Classical computers are no less, being extraordinarily powerful, but they are bounded by classical rules. QC offers a different kind of power that has been rooted based on a probabilistic theme and entangled nature of the quantum world. By leveraging this quality, quantum computers open doors to solve problems that would, otherwise be computationally impossible in any practical sense. This makes QC is not

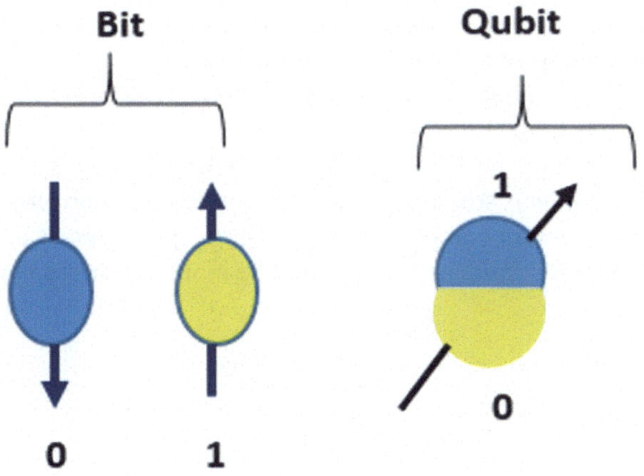

Fig. 1 Bits versus qubits representation

Table 1 Key areas of quantum computing

Key areas	Description	Need of shift
Cryptography and number theory	Classical encryption algorithm like RSA consider that it is hard for classical computers to factor large integers. But, current cryptographic systems are under threat, as Shor's quantum algorithm factor large numbers in polynomial time	Classical complexity: Exponential, Quantum complexity: Polynomial. This represents a paradigm shift in cybersecurity and has spurred interest in post-quantum cryptographic systems that are resistant to quantum attacks
Quantum simulation	Simulating the molecules and materials behavior at the quantum level is vital for breakthroughs in chemistry, biology, and materials science. Yet the classical computers cannot simulate large quantum systems efficiently because of the exponential growth in the number of features	Simulation of quantum system on Quantum computers is natural and potentially enables the discovery of new drugs and materials, more accurate models of chemical reactions and protein folding and advanced energy technologies
Combinatorial optimization	Many real-world problems from logistics and scheduling to portfolio optimization and machine learning model training involve exploring a massive space of possible solutions to find an optimal or near-optimal one	Quantum computers don't guarantee a perfect solution in all such cases, algorithms like Grover's search and quantum annealing offer quadratic speedups or more efficient exploration of solution spaces
Machine learning and data analysis	QC has potential in accelerating key subroutines in machine learning such as solving systems of linear equations (e.g., the HHL algorithm), Principal component analysis (PCA) and Pattern recognition and clustering	QML is still in its early stages, and its ability to process high-dimensional data in a more compact form could transform how we handle big data in the future

just a scientific curiosity but it as a future technology with the ability to revolutionize and bring changes to the entire industries.

The primary differences between QC with classical computing is mentioned below in the Table 2.

Though the differences highlight the advantages of quantum computing, its implementation is still difficult due to problems in controlling the quantum particles. Also, the quantum computer produces large quantities of heat, hence needing immensely cold environments which could be more expensive [1].

Table 2 Differences between classical and quantum computing

Attributes for comparison	Classical computing	Quantum computing
Unit of data	Bit (0 or 1)	Qubits (0, 1, or both in superposition)
Information processing	Relies on binary bits	Relies on qubits that can exist in superposition of states
Data representation	Deterministic	Probabilistic
Complex problem solving	Much slower	Exponentially much faster
Parallelism	Sequential or parallel cores	Massive parallelism via superposition
State evolution	Use logical gates	Use quantum gates
Key advantage	General purpose computation	Specific problems like factoring, search etc

1.2 Evolution of Quantum Computing and Its Research Contribution

The evolution of QC is the study of **interdisciplinary** where physics, mathematics, computer science and engineering converge as provided in the Table 3. The scalable quantum computing technologies could reshape the entire industries and scientific disciplines.

During the era of Noisy Intermediate Scale Quantum(NISQ), quantum devices have 10's–100's of qubits but they are still error-prone and not fault-tolerant.

1.3 Literature Review

Bhatt et al. [1] defines quantum computing, discussed on the basic introduction of QC and its working principles, briefing on its key features, benefits and its potential future applications. QC is capable of finding very large prime numbers as known that prime numbers hold immense importance in cryptography. Kanamori et al. [2] designates familiar quantum applications and their status of its developments. Abdelgaber et al. [3] reviewed the applications of AI using QC addressed more efficiently. Authors also discussed the QC algorithms for both supervised ML and unsupervised ML algorithms.

Riedel et al. [4] in his paper reviewed recent advances in High Performance Computing (HPC), Cloud Computing (CC), and QC spaces applied to a Remote Sensing problem. Mafu et al. [5] discusses an efficient Quantum Support Vector Machine model design and its implementation to achieve the required results in today's industries and applications with optimizing the performance. The proposed work in this paper used sklearn for PCA and IBM's Qiskit for the implementation of QSVM. Doan et al. [6] represented a concrete application of QC in computer vision,

Table 3 Evolution of quantum computing

S.No.	Evolution of quantum computing	Details
1	The quantum origins (early twentieth century)	The foundation of QC lies in the birth of quantum mechanics, the branch of physics that describes nature at the smallest scales. Max Planck, Albert Einstein, Niels Bohr, Werner Heisenberg, and Erwin Schrödinger laid down the principles of superposition, uncertainty, and wave-particle duality
2	Quantum information theory emerges (1960s–1970s)	In 1960, Rolf Landauer of IBM stated that "information is physical," highlighting the idea that the laws of physics place limits on computation In 1973, Charles H. Bennett, also of IBM, showed that it was possible to perform reversible computation, a concept that aligned well with the reversibility of quantum mechanics
3	Quantum computing as a concept (1980s)	In early 1980's, the idea of quantum computers was first formalized In 1981, Richard Feynman, at a famous conference at MIT, asked a profound question: "*Can we simulate physics on a computer*?" and proposed that quantum systems should simulate quantum behavior In 1985, David Deutsch, a physicist at Oxford, published paper that described a universal quantum computer, capable of simulating any physical process and introduced the idea of quantum Turing machines
4	Breakthrough algorithms and growing interest (1990s)	1990's is transformative decade for QC as field of study In 1994, Peter Shor, at AT&T Bell Labs, introduced Shor's algorithm for factoring large numbers in polynomial In 1996, Lov Grover developed Grover's algorithm, which showed that an unsorted database search could be performed quadratically faster with a quantum computer
5	From theory to hardware: early prototypes (2000s–2010s)	This era witnessed the global efforts to construct physical qubits using various approaches, including Superconducting circuits (e.g., IBM, Google), Trapped ions (e.g., IonQ, Honeywell), Photonic qubits (e.g., Xanadu) and Topological qubits (e.g., Microsoft) In 2011, Canadian company D-Wave Systems released a quantum annealer, a device using quantum principles to solve optimization problems
6	Quantum supremacy and the modern era (2019–present)	In 2019, Google announced that achieved quantum supremacy with its 53-qubit processor "Sycamore" performed a specific computational task in 200 s IBM, Rigetti, IonQ, and others have launched cloud-accessible quantum devices Governments and tech giants have increased funding and research (e.g., the U.S. National Quantum Initiative, the EU Quantum Flagship) Quantum software frameworks like Qiskit, Cirq, and PennyLane have enabled researchers and developers to explore quantum programming

bridging the gap between exact solutions and fast heuristics. By providing error bounds, the proposed algorithm enhanced the reliability of robust fitting techniques, making them very much applicable in real-world scenarios. Bhat et al. [7] offered a comprehensive overview of QC aims to introduce this field to researchers, engineers, and scientists, especially those new to the concept of QC.

Larasati et al. [8] explores the integration of QC techniques into the field of computer vision (CV). The authors provide a very comprehensive survey of various QC applications in the field of CV, highlighting their potential to enhance performance and address challenges present in classical approaches. The paper also addresses the challenges associated with integrating QC into computer vision, such as hardware limitations, noise, and the need for specialized algorithms. Ullah et al. [9] explored the application of QML techniques to classify the severity of COVID-19 patients based on clinical data. This research underscores to address complex healthcare challenges through viable integration of QC with ML and play a significant role in future analytics on medical data aid with decision making support.

Manjunath et al. [10] provided a detailed examination of the emerging field of QML outlining its potential to revolutionize various sectors by offering more efficient solutions to complex problems. Despite the challenges, the paper expresses optimism about the future of QML, emphasizing the importance of continued research and development in this area. Sood et al. [11] provides valuable insights into the trajectory of QC research, offering a data-driven perspective on its evolution and identifying key areas for future exploration. Yang et al. [12] highlighting the progress made and the challenges that remain. It serves as a valuable resource for researchers and practitioners seeking to understand the complexities and opportunities in this rapidly evolving field.

Tripathi et al. [13] explored the integration of QML with cybersecurity, specifically focusing on cyber threat detection within virtual environments. It emphasized the need for further research in quantum hardware development and algorithm optimization to fully realize the potential of quantum-enhanced threat detection systems. Chen [14] given the overview of the emerging field of QRL. This interdisciplinary area combines principles from QC and reinforcement learning to potentially enhance the efficiency and capabilities of decision making algorithms Shafique et al. [15] provides the base for understanding the principles, development, and applications of QC which provides the readers with a solid foundation to explore the rapidly evolving field and its implications across various sectors.

Xiong et al. [16] addresses the integration of QC into ML workflows, particularly focusing on the quantum inner product (QIP) operation. Solanki et al. [17] delves into the integration of QC with Generative Adversarial Networks (GANs) for image generation tasks and it provides a foundational exploration into the application of QC in image generation tasks. Dai et al. [18] introduces a novel framework that leverages deep reinforcement learning (DRL) to autonomously design QML models tailored for classification tasks. This work addresses the challenge of designing QML models by automating the architecture search process, potentially accelerating the development of quantum algorithms and applications. Tejaswi et al. [19] work highlights the potential of integrating QC with ML to address complex problems in industrial diagnostics. Singh et al. [20] explored the integration of QC into intelligent systems to enhance their capabilities. This work provides a comprehensive overview of the potential of QC to revolutionize intelligent systems, offering insights into both the opportunities and challenges associated with its integration.

2 The Core Principles of Quantum Mechanics

Quantum computing focused on principles of quantum mechanics which is depicted in Fig. 2, specific to superposition and as well as entanglement. Quantum computers handles computations by manipulating qubits through operations called quantum gates.

These gates change the state of qubits in ways that classical gates cannot. The result of a quantum computation is achieved by qubits measure, collapsing their superposition states to one of the possible outcomes.

2.1 *Qubits*

The principle of superposition in QC is qubit which can exist in a combination of both 0 and 1 states simultaneously. This ability is represented mathematically as a linear combination of the two basis states as given in Eq. (1):

$$|\psi\rangle = \alpha|0\rangle + \beta|1\rangle, \{\alpha, \beta \in C\} \tag{1}$$

where $|0\rangle$ and $|1\rangle$ are the basis states (0 and 1) and α and β are complex numbers, known as probability amplitudes and satisfy the Eq. (2) that determines the probability of measuring the qubit in the 0 or 1 state, respectively.

$$|\alpha|^2 + |\beta|^2 = 1 \tag{2}$$

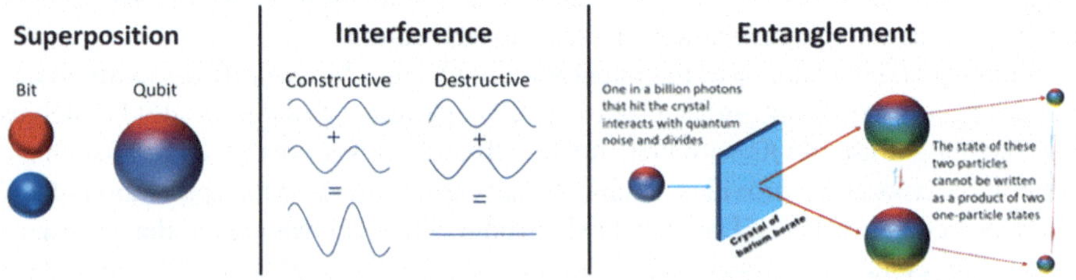

Fig. 2 Core principles of quantum mechanics

When measured, the qubit collapses into one of the states, with the probability of each outcome determined by the amplitudes α and β. The orthonormal basis |0⟩ and |1⟩ can be written as in (3)

$$\left.\begin{array}{l}|0\rangle = [1, 0]^{\mathrm{T}} \\ |1\rangle = [0, 1]^{\mathrm{T}}\end{array}\right\} \tag{3}$$

2.2 *Superposition*

Superposition admits the quantum algorithms to handle and compute data in less time even than the fastest of the classical computers performs. In a qubit system, the information always represented by the way it increases exponentially. For example, the 500 qubits of information can be easily represented which could not be easy to represent for 2^{500} and more classical bits.

Beyond the digital functioning, the Hadamard gate (H-Gate) can be used to change qubit state to superposition state. The normalized linear combination of any two quantum states are considered to be valid [3]. The H-Gate matrix representation is given in the Eq. (4):

$$\mathrm{H} = \frac{1}{\sqrt{2}}\begin{pmatrix} 1 & 1 \\ 1 & -1 \end{pmatrix} \tag{4}$$

Due to the parallel execution of various quantum states [5], the superposition can perform at computational speeds which become its principle of advantage. It allows the quantum computers to explore multiple possibilities at the same time that leads at faster computation for certain problems as given in Fig. 3.

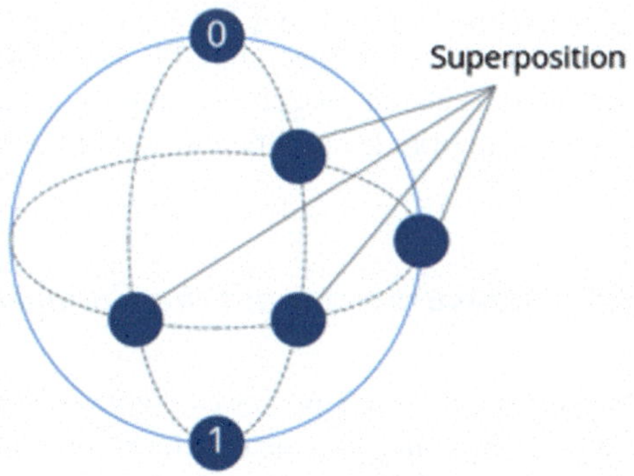

Fig. 3 Superposition

2.3 Entanglement

Entanglement offers the power to QC for the creation of a two-n Dimensional (2-nd) complex vector space for computations by using only with n physical qubits [3]. When these qubits been entangled, each qubit state is interdependent to another even far apart. This creates a highly correlated system that can be used for faster computations or communication. Hence the interaction or the communication of quantum states happens with each other. Therefore, the usage of quantum computers to solve optimization problems can be made to address the classification based problems with improvised AI based models [5].

Entanglement is a uniquely quantum phenomenon where 2 or more qubits interlinked so that the changes of one state affects the another and as well as one cannot be described without another even beyond the distance apart. The practical applications of entanglement are quantum teleportation and entanglement swapping that explores how entanglement can be used for information transfer and secure communication. It plays a vital role in the process of enhancement of QC power, particularly in quantum algorithms and quantum error correction.

2.4 Quantum Interference

To amplify the probability of the response or the outcome by canceling out the wrong ones, the quantum computers uses interference. This is a significant component in extracting useful results from quantum states. The quantum algorithms exploit it to amplify correct answers and eliminate wrong ones. Quantum interference occurs when different paths of quantum states combine, enhancing some outcomes while canceling others. The interference is an essential process available in most of the quantum based algorithms like Grover's search algorithm.

In summary the role of entanglement enables non-classical correlations between qubits, critical for protocols like teleportation and superdense coding. For algorithms, entanglement is leveraged in QPE and VQE [28, 32] and constructive and destructive interference of probability amplitudes underpins quantum speedups, as seen in Grover's amplitude amplification and Shor's periodicity detection [28, 29].

2.5 Measurement and the Collapse of the Wave Function

Measurement is the process of conversion of quantum bits of information into classical bits and the measurement outcomes are probabilistic [3] based on the significance of quantum mechanics principles. It collapse the qubit state from its superposition into either $|0\rangle$ or $|1\rangle$ states based on the associated probabilities.

The possibility of obtaining the measurement in single qubit is given as,

$$|0\rangle \text{after measurement is} |\langle 0|\, \varphi\rangle|^2$$
$$|1\rangle \text{after measurement is} |\langle 1|\, \varphi\rangle|^2$$

Similarly, the measurement of getting n string of bits on probability is given below:

$$|x_1x_2 \ldots x_n| \text{is} \langle x_1x_2 \ldots x_n|\varphi\rangle^2$$

The measurement breakdowns the state of quantum bits into a definite outcome and it requires a knowledge on the probabilistic behavior towards the measurements in quantum bits.

3 The Concept of Qubits: Moving Beyond Classical Bits

In classical system computing, the information is encoded by the use of bits which are binary entities that exist exclusively in one of two well-defined states: 0 or 1. These classical bits form the building blocks of traditional digital systems, operating under deterministic logic governed by Boolean algebra. Their processing through logic gates such as AND, OR, and NOT facilitates sequential and conditional operations, making them the cornerstone of all classical algorithms and hardware architectures [21].

3.1 Evolution of Information Units

The advent of quantum computation introduced a revolutionary departure from this rigid binary paradigm. Instead of classical bits, quantum systems employ qubits which operates on quantum mechanics principles. Unlike bits, a qubit exists either in a linear combination or in the superposition of both $|0\rangle$ and $|1\rangle$ states as discussed in Sect. 2.1.

This superpositional property endows qubits with the unique capability to represent and process several values parallel, offering a profound advantage in terms of parallelism and scalability. Additionally, when multiple qubits are entangled a condition wherein one qubit state cannot be described without the other and hence the system acquires further computational richness. Entanglement facilitates non-classical correlations and enables algorithms that far exceed classical performance for certain problems.

The conceptual foundation for utilizing quantum systems to model information traces back to pioneering insights by Richard Feynman, who highlighted the limitations of classical simulations in representing quantum phenomena. Later, David Deutsch formulated the theoretical construct of a universal quantum computer, laying the groundwork for quantum algorithm design. These early developments emphasized the potential of quantum processors to outperform classical machines, especially in domains involving complex state space exploration, such as quantum chemistry, cryptography, and optimization [22].

By leveraging the core quantum properties like entanglement, superposition, and quantum—interference, qubits allow more efficient encoding and transformation of data. This not only minimizes the physical resources needed for large-scale computations but also provides new algorithmic paradigms that are fundamentally inaccessible to classical systems. The evolution from deterministic bits to probabilistic and entangled qubits thus represents a paradigm shift in both theoretical and practical computation, bridging physics with information science in unprecedented ways.

3.2 Quantum State Formation and Representation

A qubit is mathematically described by a complex linear combination of states as $|0\rangle$ and $|1\rangle$. This quantum state is expressed mathematically in the Eq. (1) where in that α and β are the complex probability amplitudes that could satisfy Eq. (2), ensuring normalization [21]. A compelling geometric representation of a qubit is the Bloch Sphere, which maps the state of a qubit as a point on the surface of a unit sphere in three dimensions. This visualization provides intuition into single-qubit operations like rotations and phase shifts. Superposition allows qubits to simultaneously explore multiple computational paths, and the final result is obtained upon measurement, collapsing the state probabilistically into one of the basis states [21].

3.3 Physical and Practical Realization of Qubits

Multiple physical systems have been developed to realize qubits as mentioned below:

- **Superconducting Circuits**: Use Josephson junctions to create artificial atoms that behave quantum mechanically. IBM and Google have demonstrated devices with 127–433 qubits using transmon and fluxonium designs [22, 23].
- **Trapped Ions**: Ions like Yb^+ or Ca^+ are confined using Paul or Penning traps and manipulated via lasers. They offer extensive times of coherence and relatively high-fidelity gates, although these gate operations are slower compared to superconducting systems [22].
- **Photonic Qubits**: Utilize polarization or path encoding of photons. These qubits excel in quantum communication but pose challenges in two-qubit gate implementation [23].

Each system is evaluated based on coherence time, gate fidelity, and scalability, with no one-size-fits-all solution. The choice often depends on the intended application memory storage vs. computation [23].

3.4 *Quantum Gates and Manipulation Techniques*

In quantum computing, quantum gates serve as the fundamental tools for controlling and transforming the states of qubits. Differing from the classical gates, which perform irreversible operations on these bits, quantum gates are unitary in nature meaning they enable reversibility and uphold the total probability amplitude of the system. These gates tend to act as mathematical operators that rotate, entangle, or otherwise manipulate this state of quantum in a well bound and predictable manner. The quantum gates are briefed down as follow:

i. **Single-Qubit Gates**

Single-qubit gates has effect on the state of an individual qubit and the most commonly used gates are represented in the Table 4:

ii. **Multi-Qubit Gates and Entanglement**

Quantum power is significantly amplified through multi-qubit operations, where the state of a given qubit affects another. The most prominent among these is the Controlled-NOT (CNOT) gate that inverts the target qubit's state on I state of its control qubit. This gate is essential for generating entangled states, where qubits exhibit correlated behaviors no matter the physical distance between them. Entanglement plays a central role in enabling phenomena such as quantum teleportation and superdense coding and is an essential feature in the superiority of many such quantum algorithms.

iii. **Constructing Quantum Circuits**

Quantum gates are sequenced to create quantum circuits, which are the operational blueprints of quantum algorithms. These circuits are usually read from left to right, with every line referring to a qubit and gates applied sequentially to transform their states. Quantum circuits are designed to exploit quantum principles like interference, superposition, and entanglement in order to handle and address specific kind of problems. For example, the Grover search algorithm uses a series of Hadamard

Table 4 Single qubit gates

Pauli-X gate	Functions similar to a classical NOT gate, by flipping states from 0 to 1 and vice versa
Pauli-Y and Pauli-Z gates	Applies phase shifts and rotations about Y and Z axes of the Bloch sphere, essential for creating interference and modifying the phase components of a qubit
Hadamard gate (H)	Converts a given basis state into an identical superposition of 0 and 1, mostly used at the beginning of quantum algorithms to prepare qubits for parallel processing
S gates and T gates with phase rotations	The S gate ($\pi/2$ rotation) and the T gate ($\pi/4$ rotation) applies fixed phase rotations to introduce controlled phase differences and often used in conjunction with other gates to perform complex operations

and phase inversion gates to amplify the probability amplitude of the correct result. Similarly, Shor's factoring algorithm combines modular arithmetic with quantum Fourier transforms, all implemented through carefully structured quantum circuits [21].

A quantum system is considered universal if it can perform any arbitrary unitary transformation. This universality can be achieved through combinations of a tiny group of gates typically a mix of rotations of a single-qubit and a two-qubit entangling gate such as the CNOT. This minimal set can estimate any complex operation on a quantum register to the required level of correctness. Simple quantum circuits can be designed by chaining these gates, often visualized as quantum circuit diagrams. These circuits form the basis of quantum algorithms like Grover's and Shor's [24].

3.5 Decoherence, Error Rates, and Stability Concerns

The maintenance of a quantum state of qubit's is very difficult because of its interactions on various environment decoherence, quantum noise, and readout errors all contribute to instability. Techniques like dynamical decoupling, error mitigation (ZNE), and quantum error correction codes (QECC) have been introduced to address these issues [25]. One promising development is the fault-tolerant quantum memory model, where memory and computation are separated, each optimized for its purpose. For example, superconducting qubits may be used for computation, while photonic or ion-based qubits may serve as stable memory elements [22, 26].

Figure 4 outlines the foundational progression of quantum computation, beginning with the conceptual shift from classical bits to quantum information units. It illustrates how quantum states are formed, represented, and physically realized, leading to the implementation of quantum gates for computational control.

Finally, it highlights critical stability issues, including decoherence and error correction, essential for building reliable quantum systems.

4 The Algorithmic Core of Quantum Computation

4.1 Why Quantum Algorithms Matter?

Classical computing faces fundamental limitations in solving problems requiring exponential resources, such as large integer factoring or quantum system simulation. For instance, classical algorithms for prime factorization scale sub-exponentially with input size, making them impractical for large numbers [27]. Similarly, simulating quantum mechanics on classical hardware demands resources that grow exponentially with system size, as noted in analyses of quantum supremacy [28]. Quantum parallelism enables simultaneous evaluation of multiple computational paths via

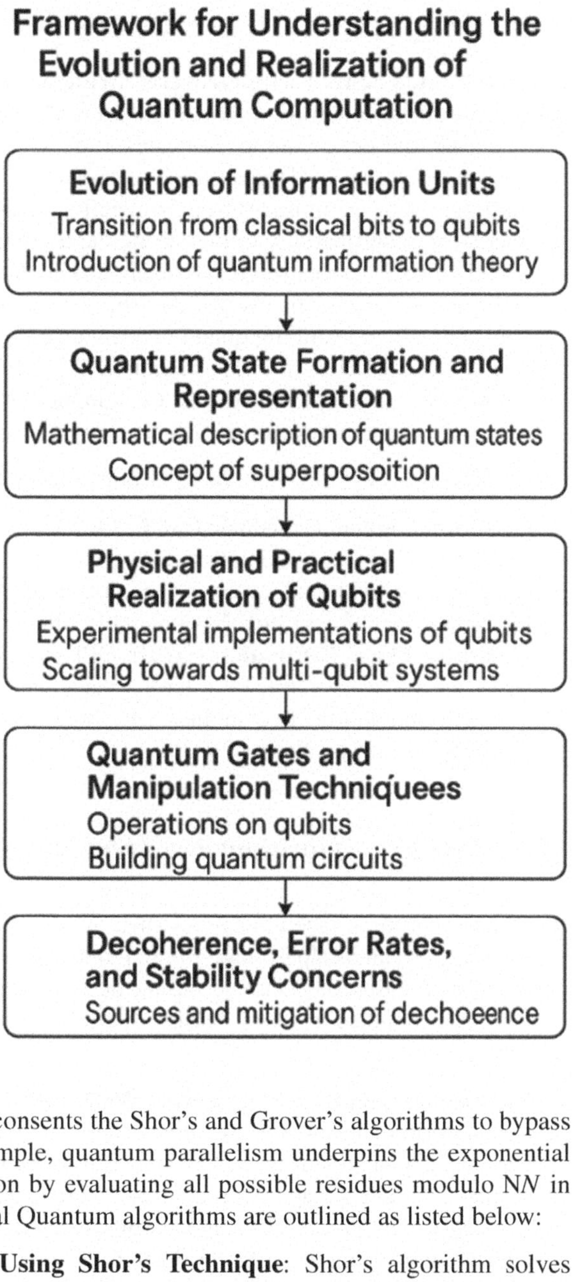

Fig. 4 Quantum computing evaluation

superposition. This property consents the Shor's and Grover's algorithms to bypass classical limitations. For example, quantum parallelism underpins the exponential speedup in Shor's factorization by evaluating all possible residues modulo N*N* in parallel [29]. The foundational Quantum algorithms are outlined as listed below:

i. **Prime Decomposition Using Shor's Technique**: Shor's algorithm solves integer factorization in polynomial time, threatening classical cryptographic systems like RSA. Its core insight reduces factorization to find the period of a modular exponential function, leveraging quantum based Fourier transform (QFT) [28, 29]. The algorithm computes $f(x) = a^x \bmod N$ to identify period r, where $a^r = 1 \bmod N$. Period-finding via QFT allows efficient determination of factors [27].

ii. **Accelerated Search with Grover's Strategy**: Grover's algorithm comes with a quadratic speedup for an unstructured lookup, with applications in database querying and optimization. It locates a marked item in $O(\sqrt{N})$ steps, outperforming classical O(N) brute-force methods [30, 31]. By iteratively amplifying the target state's amplitude through reflections over the mean, Grover's algorithm increases the probability of measuring the accurate solution [31, 32].

iii. **Deterministic Advantage: Deutsch–Jozsa Approach**: The Deutsch-Jozsa algorithm deterministically distinguishes constant from balanced functions with one query, demonstrating quantum advantage for oracle problems. Classically, this requires $\Omega(2n - 1 + 1)$ evaluations [28]. By exploiting superposition and interference, the algorithm evaluates all inputs simultaneously and destructively interferes incorrect outcomes, leaving only the correct answer [28].

Also the core tools in Quantum algorithmic design are listed below:

i. **Quantum Fourier Transform (QFT)**: QFT maps quantum states into frequency space, enabling phase estimation and periodicity detection. It is central to Shor's and phase estimation algorithms, reducing computational complexity from O(N log N) classically to O $(\log^2 N)$ quantumly [28, 29].

ii. **Quantum Phase Estimation (QPE)**: QPE determines the phase of an eigenvector of a unit operator, forming the backbone of algorithms like Shor's and quantum simulation. It combines QFT with controlled operations to resolve eigenvalues with precision [28, 30].

4.2 NISQ Era Algorithms Based on Hybrid Approaches

NISQ-era algorithms face challenges such as noise susceptibility, barren plateaus in optimization landscapes, and limited qubit connectivity. To mitigate these, recent work focuses on adaptive ansatz designs (e.g., ADAPT-VQE) and error-mitigation techniques reviews hardware-efficient analyses tailored for specific quantum processors [28].

i. **Variational Quantum Eigensolver (VQE)**: It is a hybrid based quantum–classical algorithm developed to estimate energy consumption of the ground state of systems at a molecular level, a task critical for quantum chemistry and materials science. By leveraging quantum circuits which are parameterized and loops of classical optimization, VQE circumvents the coherence-time limitations of NISQ devices. For example, VQE has been applied to simulate small molecules like H_2, LiH, and BeH_2, where it computes the lowest energy eigenstate of the molecular Hamiltonian [29]. The algorithm encodes molecular orbitals into qubits via mappings such as Jordan-Wigner or Bravyi–Kitaev transformations, which are explicitly detailed in works discussing quantum chemistry applications [33, 34].

ii. **Quantum Approximate Optimization Algorithm (QAOA)**: It addresses problems of combinatorial optimization as Max-Cut or traveling salesman problems, by preparing a parameterized quantum state that approximates the optimal solution. The algorithm iteratively adjusts parameters using classical optimization to lower a cost function encoded as a quantum observable. QAOA's performance has been benchmarked on graph-based problems, where shallow quantum circuits provide advantages over classical heuristics for specific instances [28, 29]. Notably, QAOA's scalability depends on problem structure and parameter optimization strategies, as highlighted in analyses of hybrid algorithm limitations [28].

Figure 5 illustrates a categorization of quantum application development tools, organized to highlight key frameworks and platforms in the QC ecosystem. The tools are grouped to reflect their affiliations, technical focus, and accessibility. The structure emphasizes both proprietary and open-source solutions which facilitate researchers to design, simulate, and test certain quantum algorithms.

The diversity of tools available for quantum application development are:

i. Strawberry Fields: A photonic QC platform, specializing in continuous variable quantum systems. Enables simulation and optimization of photonic circuits, often used for QML and algorithm prototyping.
ii. PennyLane (Xanadu): A cross-platform library for QML and hybrid quantum–classical workflows. Integrates with photonic quantum hardware (e.g., Xanadu's devices) and simulators like Strawberry Fields.

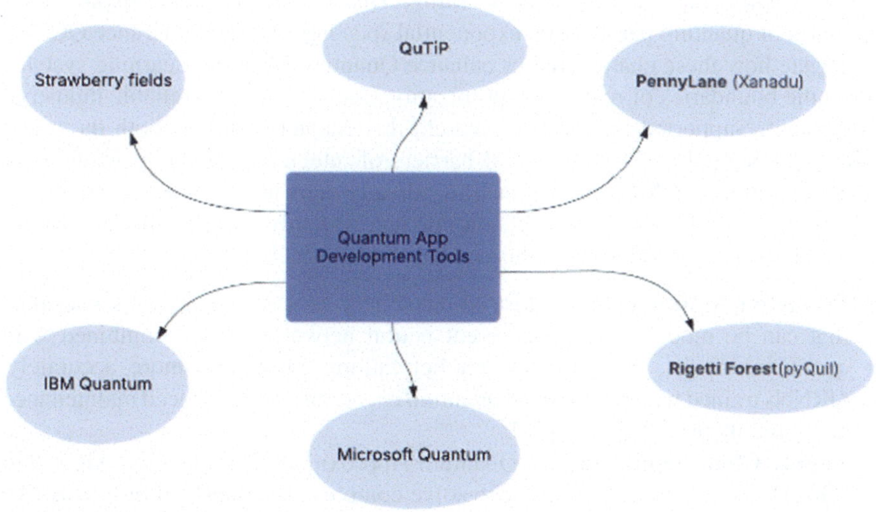

Fig. 5 Free quantum app development tools

iii. IBM Quantum: Provides cloud-based access to superconducting quantum processors via IBM's quantum hardware. Supported by the Qiskit SDK, which allows circuit design, simulation, and execution on real devices.
iv. Microsoft Quantum: A development ecosystem centered on the Q# programming language. Focuses on scalable quantum algorithm development and integration with classical software stacks.
v. Rigetti Forest (pyQuil): A toolkit for programming Rigetti's superconducting quantum computers. Uses the Quil instruction set and pyQuil (Python library) for hybrid quantum–classical programming.

The above tools caters to different hardware architectures (superconducting, photonic) and use cases (machine learning, algorithm design). It highlights the collaborative nature of the quantum ecosystem, where open-source frameworks (e.g., PennyLane, Qiskit) coexist with proprietary platforms (e.g., IBM Quantum) to accelerate innovation.

5 Parallelism, Exponential Speedup, and Fault-Tolerant Quantum Computing

5.1 Introduction

Quantum computing presents revolutionary capabilities that significantly impact industrial automation and machine learning. This particular section explores the principles of quantum parallelism, exponential speedup, and fault tolerance QC. We investigate how these characteristics enhance Quantum Machine Learning systems within the boundaries of real-time optimization, scalable AI, and reliable industrial deployment. Supported by current research, this chapter identifies both the transformative potential and technological barriers of integrating quantum methods in industrial settings. QML applications are already showing promise in predictive maintenance, optimization, and decision-making. As hardware and software mature, QC will become integral to the industrial automation stack.

i. **Predictive Maintenance**: Industrial machinery generates temporal sensor data that can be modeled using recurrent neural networks. When combined with quantum circuits, these models predict failures faster and more accurately. QRNNs trained with parallelized quantum encoding enable reduced maintenance costs and increased uptime [36].
ii. **Supply Chain Optimization**: Quantum Approximate Optimization Algorithm (QAOA) proves to be promising to solve complex combinatorial problems like vehicle routing, delivery scheduling, and warehouse placement. Integrated fault-tolerant architectures ensure continuity in these mission-critical tasks, even with noisy hardware [41].

Modern industries face an explosion of data generated from interconnected sensors, machines, and enterprise platforms. These environments demand intelligent, high-throughput computing systems. Quantum computing, grounded in quantum mechanics, offers three disruptive features: parallelism, exponential speedup, and fault tolerance. These innovations hold the potential to redefine machine learning for industrial applications [35, 35].

5.2 Harmony of Computational Feasibility in Quantum Computing

Quantum computing introduces a triad of capabilities: parallelism, exponential speedup, and fault tolerance that reimagine what is computationally feasible in industrial AI systems.

i. **Parallelism**: Quantum parallelism arises from the property of superposition of the qubits, which enables them to simultaneously represent multiple values. A quantum processor can perform operations on all basis states at once, resulting in massive data throughput compared to classical sequential logic. This property is particularly beneficial in QML. For example, variational quantum circuits (VQCs) allow simultaneous exploration of multiple configurations of a neural network [36]. This reduces training time and enhances optimization in industrial workflows. Additionally, quantum model parallelism is being explored to split tasks across processors, advancing edge computing in distributed factory systems [40].

ii. **Exponential Speedup**: Quantum algorithms provide speedups by leveraging entanglement and amplitude amplification. For instance, Grover's algorithm tends to offer quadratic acceleration for search problems which are unstructured, while Shor's algorithm achieves exponential improvement in integer factorization. In industrial automation, such speedups are crucial for real-time analytics, logistics, and resource planning. QML techniques using quantum kernel methods and quantum support vector machines (QSVMs) accelerate classification and regression tasks [38]. Applications include predictive maintenance, robotic navigation, and smart inventory management. These systems are far superior to the classical systems in terms of convergence time and cost of computation [41].

iii. **Fault-Tolerant Quantum Computing**: Quantum systems are by default, error prone due to decoherence, gate noise, and qubit instability. Fault tolerance is essential to ensure reliable, scalable quantum computation. Modern quantum based error correction(QEC) techniques like surface codes, enable continuous monitoring and correction of errors without collapsing quantum states [34]. Real-time decoders, often assisted by classical AI or reinforcement learning, are being used to reduce QEC latency in critical industrial applications like power grid control and aerospace safety systems [39].

While the potential is vast, current hardware is limited in qubit count and coherence time. Full-scale fault-tolerant quantum systems remain a research focus [40]. Hybrid quantum–classical pathways are emerging as a transitional solution, leveraging the strengths of both worlds. Error correction also presents an overhead challenge, requires often 1000's of such physical qubits for a unit logical qubit. However, software-level improvements in compiler optimization and algorithmic resilience are rapidly progressing [37].

6 Revolutionized Quantum-Based AI Techniques: Data Preprocessing, Prediction, and Optimization

6.1 Introduction

The synergy of Quantum computing and Artificial Intelligence (AI) has transformed how data is processed, interpreted, and optimized in industrial settings. This chapter discusses recent advancements in quantum-based AI techniques, emphasizing their role in transforming data preprocessing, prediction accuracy, and optimization capabilities in industrial automation. Drawing on state-of-the-art literature and real world case studies, it evaluates the potential and hurdles of integrating QML into industrial pipelines, setting the stage for Industry 5.0.

As industries embrace the transition to Industry 4.0 and beyond, the demand for intelligent automation systems has surged. Traditional AI has provided a foundation, but the exponential growth of data complexity requires computational tools that transcend classical limits. QML which is the combination of QC and AI, offers a paradigm shift by exploiting the principles of QC to process data in a much more effective way and learn from fewer examples [42, 45].

i. **Quantum Data Preprocessing**: Industrial data is often noisy, unstructured, and voluminous. Before any AI algorithm can yield insight, this raw data must undergo preprocessing steps like normalization, dimensionality reduction, and feature selection. Quantum approaches to preprocessing, such as quantum principal component analysis (qPCA), reduce the computational cost of these operations significantly [47]. Quantum encoding techniques, including amplitude and basis encoding, represent large datasets in compact quantum states. These encodings can capture complex data patterns in fewer qubits, offering superior scalability compared to classical systems [48].
ii. **Quantum Prediction Techniques**: Prediction in industrial AI involves anticipating failures, demand, resource needs, or system outputs. Quantum neural networks (QNNs) and variational quantum circuits (VQCs) enhance this by enabling more expressive model representations and faster convergence during training. For instance, Siemens leveraged AI-powered sensors integrated with quantum-inspired analytics to reduce machine failure rates [49, 50]. Similarly, quantum support vector machines (QSVMs) exploit the high-dimensional

Hilbert space to achieve better classification margins, especially in sparse or high-noise environments [42].

iii. **Quantum Optimization in Industrial Systems**: Optimization problems are central to industrial automation ranging from production line scheduling to route planning. Classical solvers struggle with NP-hard tasks, especially at scale. Algorithms of QC such as the Quantum Approximate Optimization Algorithm (QAOA) and Grover's algorithm have been successfully applied to such domains [44]. In agriculture and energy management, QAOA has been used to tune IoT-driven control parameters, maximizing efficiency while maintaining environmental thresholds [47]. Additionally, hybrid classical-quantum optimization frameworks allow industries to start integrating quantum solutions without needing full quantum infrastructure [46].

6.2 Case Studies

Despite rapid advancement, integrating QML into industrial automation faces its share of hurdles:

- **Hardware maturity**: Quantum processors are susceptible to decoherence and noise.
- **Data representation**: Efficient and scalable quantum data encoding remains under research.
- **Integration complexity**: Seamless operation between classical and quantum systems requires specialized software and hybrid architectures.

However, the research landscape is rapidly evolving. As shown the developments in quantum error correction and cloud-accessible quantum hardware are making QML increasingly viable [46].

i. **Predictive Maintenance**: A quantum–classical hybrid approach was implemented in CNC machines to predict failures. By using variational circuits and quantum-enhanced pattern recognition, the system achieved a 42% improvement in failure prediction accuracy over classical methods [45].

ii. **Logistics and Supply Chain**: A supply chain optimization use case employed Grover's algorithm and QAOA to reduce route inefficiencies and improve delivery predictability. Quantum optimization models reduced computational overhead and enabled real-time adjustments based on demand fluctuations [43].

Quantum-based AI is unlocking new possibilities for automation, optimization, and predictive analytics in industrial contexts. From more efficient data preprocessing to faster learning and superior optimization, QML offers a transformative upgrade to classical approaches. As the technology matures, its industrial applications will expand, setting a foundation for intelligent systems in the Industry 5.0 era.

7 Applications and Challenges of Quantum Computing

7.1 Applications of Quantum Computing

Quantum computers easily compute massive computations in a parallel manner. As each qubit can store 2 inputs simultaneously, by means of a quantum superposition state, $2n$ inputs can be stored in n qubits. Using universal quantum gates an arbitrary quantum circuit can be designed and the n qubits can be used as the input, that performs an arbitrary computation [2]. In unsorted data search, Grover's algorithm searches a data item in sqrt(N) time in contrast to N in classical computer. Shor's algorithm is another application of QC in which quantum computer has been used for large integer factorization with in a polynomial time [53]. Security agencies are attracted by this algorithm since, with the development of quantum computer the most widely used RSA encryption algorithm which is the public-key encryption based algorithm can be cracked.

The research relevant to QC is however in the initial stage and it has the high potential for revolutionary developments in the following fields such as communication, risk analysis, ML, financial modelling, medical drug design, and cryptography to name a few that speed up innovation in science and global market [51]. Quantum key distribution (QKD) is another real-life example of QC that enables unbreakable encryption by using the base principles of quantum physics. The "no-cloning theorem" stops the replication of an arbitrary quantum state which could be unknown, and hence provides security during data transmission in the quantum channel.

7.2 Challenges of Quantum Computing

In the past few decades, various approaches have been adopted to realize the quantum computers that demonstrates quantum operations. Beyond the fast computational efficiency, QC has its own challenges and limitations to address. An arbitrary quantum circuit can be designed if hardware with universal gates is developed. However, the main technical challenge is to build the gate-based quantum computation level hardware which can have an enormous number of qubits [2]. Every qubit is to be isolated from its respective environment including the neighbouring qubits for better performance. For example, a qubit must be shown as a steady two-level system, irrespective of the material used and must possess a more extensive decoherence time than the operations on the gates [2]. Decoherence refers to the coupling between the qubit and its corresponding environment. In the case of insufficient decoherence time for a gate operation, erroneous outputs are exhibited by the gates. Thus, isolation of qubits from its neighbouring qubits and its environment is important [2]. Another challenge is in case of using Quantum error correction mechanisms, lots of excess qubits are to be managed.

Another challenge in QC is using Quantum Key Distribution(QKD) technique, as this technology has not been disseminated extensively through the information security practitioner's community. At present QKD systems that are commercially available and are in use are generally developed for point-to-point dedicated connections between networks [2]. This makes it impossible to use QKD system in the long distances and still bounded to metropolitan area networks. Another significant issue with a QKD is that there is no quantum repeater that is available commercially. In general, repeaters are used for extending the distance and to data fan out across a network. In QC environment, a photon is prone to interfere with its environment. For a QKD scheme, this type of interference is not a massive problem for data transmission over a short distance through a quantum channel, as the data transmitted is random bits and can be discarded upon identifying errors. In communication channels, for a long-distance, repeaters are needs as signal amplification is needed due to its high signal to noise ratio (SNR). However, it is quite difficult to develop a quantum repeater, since duplication of a transmitted unknown photon is not possible according to the no-cloning theorem [2].

Security is another challenge in using QKD schemes, as the origine of the message cannot be guaranteed and may suffer from man-in-the middle attacks in multipoint networks through eavesdropping while compromising both quantum and classical channels. Hence, to safeguard the QKD schemes, proper authentication functionalities to be incorporated [2]. Another challenging area is achieving tolerance to faults and providing reliability over quantum computations as practically implementing corrections over quantum error is still a problem [52]. Another challenging task is verification of quantum states, as quantum states are delicate in nature and require low temperatures and accurate fabrication to operate on quantum bits. Hence, verification of errors is challenging in quantum computation due to existence of significant errors compared to classical computations.

In terms of QC technological design, fragility is another challenge to be addressed. The main reasons for fragility are because of small coherence times of these qubits and error rates which are large. Hence it is difficult to build a quantum computer with relatively low error rates. The first reason short coherence time happens as the superconducting qubits always forget their information within nanoseconds of time and the second reason being the large error rates quantum operations are unreliable [52]. Apart from the technological design issue, environmental instabilities and small material faults also contribute to the errors in qubits and loose their quantum data. Hence, measures to be taken to protect the qubits information from environmental and other kind of instabilities.

8 Conclusion with Future Directions in Research

8.1 Future Directions in Quantum Computing Research

Quantum computing (QC) holds the capacity to be the next profound technology, with a broad range of real-life applications for Industry and various organizations. QC employs various principles of quantum mechanics, such as superposition and entanglement for speeding up the computations exponentially. Therefore, for scientific advancement a highly relevant study is required to illuminate the evolution of its research. The authors in [11] conducted extensive literature survey and categorised research areas in QC into five domains as follows:

i. Quantum Communication.
ii. Quantum algorithm and simulation.
iii. Quantum enhanced analysis methods.
iv. Quantum cryptography and information security.
v. Quantum Computational Complexity.

Each of these domains answer the specific queries and distinguish them from other research domains in QC. Quantum Communication (QCo) domain concentrates on answering the queries relevant to quantum communication and quantum entanglement. Quantum Algorithm and simulation (QAS) domain specifies research relevant to quantum algorithms development and quantum simulation environments. Quantum Enhanced Analysis Method (QEAM) domain represents quantum artificial intelligence(QAI), quantum image processing(QIP), QML and quantum neural networks(QNN) and so on. Quantum Cryptography and Information Security (QCIS) domain specifies the research on quantum cryptography, quantum information security and quantum key distribution technologies. Quantum Computation Complexity (QCC) domain concentrates on computation complexity i.e. power of quantum-based algorithms on quantum computers [11]. In QCC, the quantum complexity classes, such as EQP, BQP, and NQP which are analogous to quantum P, NP classes specifies the quantum Turing machine.

The authors in [52] identified various ongoing research areas where a lot of innovation is possible and further research to undertaken. Post quantum cryptography is one such area which is at its pinnacle while majority of research has been conducted on complex quantum procedures and their simulations. Other research areas include robotics, distributed quantum computing, energy management, quantum-based portfolio-risk optimization, fraud-detection, and fault-tolerant system and so on.

Another research area is Energy management which is another significant challenge to be addressed, as quantum computers are to be scaled up from an approximate of 50 qubits to nearly 10,000 qubits for solving real-world problems related to computational biology etc., which leads to more energy requirements for computation and cooling systems. Quantum processors work at shallow temperatures using less energy leading to low cost and carbon footprints. Hence there is a necessity to develop energy efficient quantum data centres for better management of energy utilization [52]. Further hybrid applications can be built with high-energy and low-energy portions, where high-energy requirement portion can be processed using quantum computers while low-energy portion of application can be processed by classical computers with the help of cloud infrastructure. Future research in this direction leads to sustainable future for computational requirement.

Post quantum cryptography refers to the cryptographic algorithms that maintain integrity and confidentiality, while preventing the various attacks and safeguards the information. It is another area where a lot of concentration in research is needed in the techniques such as symmetric key quantum resistance, code-based cryptography, hash based-cryptography, multi variate cryptography and so on. Hence, we are required to build agile quantum systems that can accurately predict possible attacks and fix the faults dynamically at runtime [52].

8.2 Conclusion

In this chapter, we provided a basic understanding of Quantum Computing through its core principles, qubits, quantum algorithms. We also discussed the quantum–mechanical phenomena such as qubit superposition, qubit entanglement and their role in solving complex real-life problems. Moving forward we discussed revolutionized quantum-based AI techniques for data processing by leveraging the amazing capabilities of quantum systems. By combining the principles of QC, with AI, ML and DL paved a way for Quantum based Machine Learning (QML), Quantum based Reinforcement Learning (QRL), Quantum based Neural Networks (QNN), and Quantum based Deep Learning (QDL) etc. Classical AI techniques face challenges with high-dimensional data, complex optimization problems, and computationally intensive tasks which can be addressed through QC providing a pathway to solve such problems. Post-quantum cryptography techniques are discussed for secure transmission in quantum channel. We have discussed various open hurdles such as quantum error correction are discussed and future research directions are proposed.

As part of the discussion, five future research domains such as Quantum Communication, Quantum algorithm and simulation, Quantum enhanced analysis methods, Quantum cryptography and information security, Quantum Computational Complexity were some of the few topics proposed. The design strategies to develop a quantum computer which can perform simultaneous operations, and the challenges to be addressed in building a repeater device to amplify the signal. This paper suggests a framework that can be utilized to differentiate various QC techniques that could be applied on the existing classical computing infrastructure. In addition to these challenges energy efficiency, fault-tolerant, quantum internet, QKD etc., are the areas where more research to be carried out. In conclusion, quantum computing being in its infancy stage, is acquiring importance and having applications in solving many real-life complex problems.

References

1. Bhatt, H., Gautam, S.: Quantum computing: a new era of computer science. In: 2019 6th International Conference on Computing for Sustainable Global Development (INDIACom), pp. 558–561. IEEE, USA (2019)
2. Kanamori, Y., Yoo, S.M.: Quantum computing: principles and applications. J. Int. Technol. Inf. Manag. **29**(2), 43–71 (2020)
3. Abdelgaber, N., Nikolopoulos, C.: Overview on quantum computing and its applications in artificial intelligence. In: 2020 IEEE third international conference on artificial intelligence and knowledge engineering (AIKE), pp. 198–199. IEEE, USA (2020)
4. Riedel, M., Cavallaro, G., Benediktsson, J.A.: Practice and experience in using parallel and scalable machine learning in remote sensing from HPC over cloud to quantum computing. In: 2021 IEEE International Geoscience and Remote Sensing Symposium IGARSS, pp. 1571–1574. IEEE, USA (2021)
5. Mafu, M., Senekane, M.: Design and implementation of efficient quantum support vector machine. In: 2021 International conference on electrical, computer and energy technologies (ICECET), pp. 1–4. IEEE, USA (2021)
6. Doan, A.D., Sasdelli, M., Suter, D., Chin, T.J.: A hybrid quantum-classical algorithm for robust fitting. In: Proceedings of the IEEE/CVF conference on computer vision and pattern recognition, pp. 417–427 (2022)
7. Bhat, H.A., Khanday, F.A., Kaushik, B.K., Bashir, F., Shah, K.A.: Quantum computing: fundamentals, implementations and applications. IEEE Open J. Nanotechnol. **3**, 61–77 (2022)
8. Larasati, H. T., Kim, H.: Trends of quantum computing applications to computer vision. In: 2022 International Conference on Platform Technology and Service (PlatCon), pp. 7–12. IEEE, USA (2022)
9. Ullah, U., Maheshwari, D., Gloyna, H.H., Garcia-Zapirain, B.: Severity classification of COVID-19 patients data using quantum machine learning approaches. In: 2022 International Conference on Electrical, Computer, Communications and Mechatronics Engineering (ICECCME), pp. 1–6. IEEE, USA (2022)
10. Manjunath, T.N., Hiremath, A., Niranjanamurthy, M., Peng, S.L., Shrihari, M.R., Pushpa, S.K.: A survey on machine learning techniques using quantum computing. In: 2022 Fourth International Conference on Emerging Research in Electronics, Computer Science and Technology (ICERECT), pp. 1–6. IEEE, USA (2022)
11. Sood, S.K., Pooja: Quantum computing review: a decade of research. IEEE Trans. Eng. Manag. **71**, 6662–6676 (2024). https://doi.org/10.1109/TEM.2023.3284689
12. Yang, Z., Zolanvari, M., Jain, R.: A survey of important issues in quantum computing and communications. IEEE Commun. Surv. Tutorials **25**(2), 1059–1094 (2023)
13. Tripathi, S.M., Upadhyay, H., Soni, J.: Quantum neural network classification-based cyber threat detection in virtual environment. In: 2023 International Conference on Computational Science and Computational Intelligence (CSCI), pp. 391–396. IEEE, USA (2023)
14. Chen, S.Y.C.: An introduction to quantum reinforcement learning (QRL). In: 2024 15th International Conference on Information and Communication Technology Convergence (ICTC), pp. 1139–1144. IEEE, USA (2024)
15. Shafique, M.A., Munir, A., Latif, I.: Quantum computing: circuits, algorithms, and applications. IEEE Access **12**, 22296–22314 (2024)
16. Xiong, H., Tang, Y., Ye, X., Yan, J.: Circuit design and efficient simulation of quantum inner product and empirical studies of its effect on near-term hybrid quantum-classic machine learning. In: Proceedings of the IEEE/CVF Conference on Computer Vision and Pattern Recognition, pp. 26162–26170 (2024)
17. Solanki, A., Kang, S.S., Singla, S.: Comprehensive analysis on image generation using quantum generative adversarial network. In: 2024 First International Conference on Technological Innovations and Advance Computing (TIACOMP), pp. 302–307. IEEE, USA (2024)

18. Dai, X., Wei, T.C., Yoo, S., Chen, S.Y.C.: Quantum machine learning architecture search via deep reinforcement learning. In: 2024 IEEE International Conference on Quantum Computing and Engineering (QCE), vol. 1, pp. 1525–1534. IEEE, USA (2024)
19. Tejaswi, G., Prabhakar, K., Kunar, S.D.: Quantum machine learning for rotating machinery prognostics and health management. In: 2024 International Conference on Sustainable Communication Networks and Application (ICSCNA), pp. 1005–1009. IEEE, USA (2024)
20. Singh, V.K., Singh, A.K.: Harnessing quantum computing for next-generation intelligent systems. In: 2025 International Conference on Intelligent Control, Computing and Communications (IC3), pp. 976–977. IEEE, USA (2025)
21. Vasantrao, K.S., Saxena, A.: Bits to qubits: an overview of quantum computing. In: 2025 International Conference on Intelligent Control, Computing and Communications (IC3), pp. 120–124. IEEE, USA (2025)
22. Sebastian, E., Poonia, R.C.: Compendium of qubit technologies in quantum computing. In: International Conference on Communication and Intelligent Systems, pp. 91–100. Springer, Singapore (2022)
23. Liu, C., Wang, M., Stein, S.A., Ding, Y., Li, A.: Quantum memory: a missing piece in quantum computing units. arXiv preprint arXiv:2309.14432 (2023)
24. Ardila-Perez, L.E., Fuchs, M., Gartmann, R., Scheller, L., Schlösser, M., Heil, R., Roth, C., Bekman, I., Jerger, M., Barends, R., van Waasen, S., Sander, O.: The quantum interface controller: a full-stack, modular, and scalable system for qubit readout and manipulation. In: 2024 IEEE International Conference on Quantum Computing and Engineering (QCE), vol. 2, pp. 466–467. IEEE, USA (2024)
25. Chen, K.C.: Short-depth circuits and error mitigation for large-scale GHZ-state preparation, and benchmarking on IBM'S 127-qubit system. In: 2023 IEEE International Conference on Quantum Computing and Engineering (QCE), vol. 2, pp. 207–210. IEEE, USA (2023)
26. Omar, A.: The Analogy between Magnetic Resonance Imaging and Qubits in Quantum Computing (2025)
27. Ugwuishiwu, C.H., Orji, U.E., Ugwu, C.I., Asogwa, C.N.: An overview of quantum cryptography and Shor's algorithm. Int. J. Adv. Trends Comput. Sci. Eng. **9**(5) (2020)
28. Adedoyin, A., Ambrosiano, J., Anisimov, P., Casper, W., Chennupati, G., Coffrin, C., Coffrin, C., Djidjev, H., Gunter, D., Karra, S., Lemons, N., Lin, S., Lokhov, A.Y.: Quantum algorithm implementations for beginners. arXiv preprint arXiv:1804.03719 (2018)
29. Martyn, J.M., Rossi, Z.M., Tan, A.K., Chuang, I.L.: Grand unification of quantum algorithms. PRX Quantum **2**(4), 040203 (2021)
30. Sahoo, S., Mandal, A.K., Samanta, P.K., Basu, I., Roy, P.: A critical overview on quantum computing. J. Quantum Comput. **2**, 181–192 (2020)
31. Weigold, M., Barzen, J., Leymann, F., Salm, M.: Expanding data encoding patterns for quantum algorithms. In: 2021 IEEE 18th International Conference on Software Architecture Companion (ICSA-C), pp. 95–101. IEEE, USA (2021)
32. Mosca, M.: Quantum algorithms. arXiv preprint arXiv:0808.0369 (2008)
33. Weigold, M., Barzen, J., Leymann, F., Salm, M.: Encoding patterns for quantum algorithms. IET Quantum Commun. **2**(4), 141–152 (2021)
34. Battistel, F., Chamberland, C., Johar, K., Overwater, R.W.J., Sebastiano, F., Skoric, L., Ueno, Y., Usman, M.: Real-time decoding for fault-tolerant quantum computing: progress, challenges and outlook. Nano Futures **7**(3), 032003 (2023). https://doi.org/10.1088/2399-1984/aceba6
35. Dai, F., Hossain, M.A., Wang, Y.: State of the art in parallel and distributed systems: emerging trends and challenges. Electronics **14**(4), 677 (2025). https://doi.org/10.3390/electronics14040677
36. Khurana, S., Nene, M.: Quantum Machine Learning: Unraveling a New Paradigm in Computational Intelligence. Authorea (Authorea) (2024). https://doi.org/10.22541/au.171037552.27623320/v1
37. Njoku, T.: Quantum software engineering: algorithm design, error mitigation, and compiler optimization for fault-tolerant quantum computing. Int. J. Comput. Appl. Technol. Res. (2025). https://doi.org/10.7753/ijcatr1404.1003

38. Perijoc, S.N.: Quantum AI for Application in Industry : A Workflow Model. University of Twente Student Theses (2024). http://essay.utwente.nl/100208/
39. Venkatesha, S., Parthasarathi, R.: Survey on redundancy based-fault tolerance methods for processors and hardware accelerators—trends in quantum computing, heterogeneous systems and reliability. ACM Comput. Surv. **56**(11), 1–76 (2024). https://doi.org/10.1145/3663672
40. Wang, Y., Liu, J.: A comprehensive review of quantum machine learning: from NISQ to fault tolerance. Rep. Prog. Phys. **87**(11), 116402 (2024). https://doi.org/10.1088/1361-6633/ad7f69
41. Zaman, K., Marchisio, A., Hanif, M.A., Shafique, M.: A survey on quantum machine learning: current trends, challenges, opportunities, and the road ahead. arXiv (Cornell University) (2023). https://doi.org/10.48550/arxiv.2310.10315
42. Ahmadi, A.: Quantum computing and AI: the synergy of two revolutionary technologies. Asian J. Manag. Sci. **12**(2), 15–27 (2023). https://doi.org/10.51983/ajes-2023.12.2.4118 [1]
43. Agarwal, P., Alam, M.: Exploring quantum computing to revolutionize big data analytics for various industrial sectors pp. 113–130. Auerbach Publications eBooks (2021). https://doi.org/10.1201/9781003175711-7
44. Andreou, A., Mavromoustakis, C.X., Markakis, E., Bourdena, A., Mastorakis, G.: Sustainable AI with quantum-inspired optimization: enabling end-to-end automation in cloud-edge computing. IEEE Access **1** (2025). https://doi.org/10.1109/access.2025.3554024
45. Bala, I., Ahuja, K., Mijwil, M.M.: Quantum machine learning for Industry 4.0. In: Quantum Computing and Artificial Intelligence: The Industry Use Cases, pp. 415–433 (2025). https://doi.org/10.1002/9781394242399.ch16
46. Chakravarthi, M.K., Kumar, Y.V.P., Reddy, G.P.: Potential technological advancements in the future of process control and automation. In: 2024 IEEE Open Conference of Electrical, Electronic and Information Sciences (eStream), pp. 1–6 (2024). https://doi.org/10.1109/estream61684.2024.10542581
47. Gharib, J., Gahi, Y.: Quantum computing and AI applications in Industry 5.0 use cases. In: Quantum Computing and Artificial Intelligence: The Industry Use Cases, pp. 435–464 (2025). https://doi.org/10.1002/9781394242399.ch17
48. Rane, J., Mallick, S.K., Kaya, Ö., Rane, N.L.: Future research opportunities for artificial intelligence in industry 4.0 and 5.0. Deep Science Publishing (2024)
49. Sonavane, A.S., Aylani, A.: Integration of IoT and quantum computing. In: Advances in computer and electrical engineering book series, pp. 331–358 (2024). https://doi.org/10.4018/979-8-3693-3940-4.ch016
50. Garewal, I.K., Mahamuni, C.V., Jha, S.: Emerging applications and challenges in quantum computing: a literature survey. In: 2024 International Conference on Artificial Intelligence, Big Data, Computing and Data Communication Systems (icABCD), Port Louis, Mauritius, pp. 1–12 2024. https://doi.org/10.1109/icABCD62167.2024.10645271
51. Bova, F., Goldfarb, A., Melko, R.G.: Commercial applications of quantum computing. EPJ Quantum Technol. **8**(1), 2–15 (2021)
52. Gill, S.S., Kumar, A., Singh, H., et al.: Quantum computing: a taxonomy, systematic review and future directions. Softw. Pract. Exper. **52**(1), 66–114 (2022). https://doi.org/10.1002/spe.3039
53. Shor, P.W.: Algorithms for quantum computation: discrete logarithms and factoring. In: Proceedings 35th Annual Symposium on Foundations of Computer Science, pp. 124–134. IEEE, Santa Fe, NM (1994). https://doi.org/10.1109/SFCS.1994.365700

The manufacturer's authorised representative in the EU is Springer Nature Customer Service Centre GmbH, Europaplatz 3, 69115 Heidelberg, Germany. If you have any concerns regarding our products, please contact ProductSafety@springernature.com

Printed and bound by CPI Group (UK) Ltd, Croydon, CR0 4YY
30/09/2026
02225584-0016